Lecture Notes in Computational Science and Engineering 97

Editors:

Timothy J. Barth
Michael Griebel
David E. Keyes
Risto M. Nieminen
Dirk Roose
Tamar Schlick

For further volumes:
http://www.springer.com/series/3527

Jochen Garcke • Dirk Pflüger
Editors

Sparse Grids and Applications - Munich 2012

Springer

Editors
Jochen Garcke
Institut für Numerische Simulation
Universität Bonn
Bonn
Germany

Dirk Pflüger
Institute for Parallel and Distributed Systems
Universität Stuttgart
Stuttgart
Germany

ISSN 1439-7358 ISSN 2197-7100 (electronic)
ISBN 978-3-319-04536-8 ISBN 978-3-319-04537-5 (eBook)
DOI 10.1007/978-3-319-04537-5
Springer Cham Heidelberg New York Dordrecht London

Library of Congress Control Number: 2014936993

Mathematics Subject Classification (2010): 65D99, 65M12, 65N99, 65Y20, 65N12, 62H99

Printed on acid-free paper

Springer is part of Springer Science+Business Media (www.springer.com)

Preface

Sparse grids have gained increasing interest in recent years for the numerical treatment of high-dimensional problems. While classical numerical discretization schemes fail in more than three or four dimensions, sparse grids make it possible to overcome the "curse of dimensionality" to some degree—extending the number of dimensions that can be dealt with.

The second Workshop on Sparse Grids and Applications (SGA2012), which took place from July 2 to 6 in 2012, demonstrated once again the importance of this numerical discretization scheme. Organized by Hans-Joachim Bungartz, Jochen Garcke, Michael Griebel, Markus Hegland and Dirk Pflüger, more than 40 researchers from 11 different countries have presented and discussed the current state of the art of sparse grids and their applications. Thirty-three talks covered their numerical analysis as well as efficient data structures, and the range of applications extended to uncertainty quantification settings and clustering, to name but a few examples. This volume of LNCSE collects selected contributions from attendees of the workshop.

More than 20 years after the term "sparse grids" was coined by Christoph Zenger in Munich, the SGA was hosted by his former institution, the Department of Informatics of the Technische Universität München, together with the new Institute for Advanced Study (IAS). Financial support of the IAS is kindly acknowledged. We especially thank Christoph Kowitz and Valeriy Khakhutskyy for their effort and enthusiasm in the local organization of the workshop, and the IAS staff, especially Stefanie Hofmann and Sigrid Wagner, for their assistance.

Bonn, Germany — Jochen Garcke
Stuttgart, Germany — Dirk Pflüger
December 2013

Contents

Efficient Pseudorecursive Evaluation Schemes for Non-adaptive Sparse Grids

Gerrit Buse, Dirk Pflüger, and Riko Jacob

Abstract In this work we propose novel algorithms for storing and evaluating sparse grid functions, operating on regular (not spatially adaptive), yet potentially dimensionally adaptive grid types. Besides regular sparse grids our approach includes truncated grids, both with and without boundary grid points. Similar to the implicit data structures proposed in Feuersänger (Dünngitterverfahren für hochdimensionale elliptische partielle Differntialgleichungen. Diploma Thesis, Institut für Numerische Simulation, Universität Bonn, 2005) and Murarasu et al. (Proceedings of the 16th ACM Symposium on Principles and Practice of Parallel Programming. Cambridge University Press, New York, 2011, pp. 25–34) we also define a bijective mapping from the multi-dimensional space of grid points to a contiguous index, such that the grid data can be stored in a simple array without overhead. Our approach is especially well-suited to exploit all levels of current commodity hardware, including cache-levels and vector extensions. Furthermore, this kind of data structure is extremely attractive for today's real-time applications, as it gives direct access to the hierarchical structure of the grids, while outperforming other common sparse grid structures (hash maps, etc.) which do not match with modern compute platforms that well. For dimensionality $d \leq 10$ we achieve good speedups on a 12 core Intel Westmere-EP NUMA platform compared to the results

G. Buse (✉)
TU München, München, Germany
e-mail: buse@in.tum.de

D. Pflüger
Institute for Parallel and Distributed Systems, Universität Stuttgart, Stuttgart, Germany
e-mail: dirk.pflueger@ipvs.informatik.uni-stuttgart.de

R. Jacob
ETH Zurich, Zurich, Switzerland
e-mail: rjacob@inf.ethz.ch

J. Garcke and D. Pflüger (eds.), *Sparse Grids and Applications - Munich 2012*, Lecture Notes in Computational Science and Engineering 97, DOI 10.1007/978-3-319-04537-5_1, presented in Murarasu et al. (Proceedings of the International Conference on Computational Science—ICCS 2012. Procedia Computer Science, 2012). As we show, this also holds for the results obtained on Nvidia Fermi GPUs, for which we observe speedups over our own CPU implementation of up to 4.5 when dealing with moderate dimensionality. In high-dimensional settings, in the order of tens to hundreds of dimensions, our sparse grid evaluation kernels on the CPU outperform any other known implementation.

1 Introduction

Sparse grids, as introduced in [19] for the solution of high-dimensional partial differential equations, are widely used to tackle problems that are hard or impossible to solve with conventional discretizations because of the *curse of dimensionality*. In contrast to full isotropic grids that require $\mathcal{O}(N^d)$ sampling points to discretize a d-dimensional domain with mesh-width $1/N$, the number of sparse grid points has a much weaker dependency on the dimensionality d. For sufficiently smooth functions only $\mathcal{O}(N \cdot (\log N)^{d-1})$ grid points are necessary to obtain a similar approximation quality as for full grids. The trade-off are complicated hierarchical data structures and algorithms which are difficult to match with modern compute platforms, and since the gap between existing implementations' requirements and future hardware features seems to widen, there is constant need for improvement. Where real-time requirements meet the need for incremental algorithms in high-dimensional settings, efficient and hardware-aware algorithms and data structures become crucial. Just consider applications in the context of computational steering or other parameter-dependent simulation tasks that require to store whole vector fields (e.g., a 3-D pressure field) for each grid point in the parameter space [4].

The settings that triggered our work were such real-time applications with the need for incremental algorithms and the minimization of storage space. The corresponding core algorithms (evaluation, hierarchization, dehierarchization, etc.) have been developed and optimized for the most general case of spatial adaptivity [18]. As a consequence, the underlying data structures are designed for flexibility and performance in arbitrary settings, but make no use of any regularity in the grids. In contrast, the closely related combination technique as introduced in [10] is widely used to tackle similar problems, though with a completely different algorithmic approach. It does not require complex data structures in its direct form, but where large vector fields have to be stored for each grid point either a large memory overhead has to be invested, or one is back to less efficient data structures. Besides the loss of the inherently incremental representation this is another reason to restrict ourselves to the use of direct sparse grids.

A comparison of the characteristics of both approaches in our setting yields, that often spatial adaptivity is not crucial, and dimensional adaptivity as used with the combination technique [7] leads to good results. We therefore focus on so-called *truncated* sparse grids [1, 15], a special form of dimensionally adaptive grids, for which storage efficient data structures exist. We propose a slim, array-based data structure that preserves the intrinsic recursive property of sparse grids, and we show how *evaluation*, one of the sparse grid core algorithms, can perfectly take advantage of this. Our goal is to achieve maximum performance for our specifically designed evaluation kernels on modern multicore architectures as well as Nvidia Fermi GPUs. Comparisons with similar approaches are presented that demonstrate how a *pseudorecursive* implementation is superior to both a sophisticated spatially adaptive and thus recursive implementation and the most optimized implementation known to us that has been designed for our setting.

An overview of related work (Sect. 2) and a brief introduction of the notation we use (Sect. 3) are followed by the description of our array-based data structure in Sect. 4. Section 5 motivates the term *pseudorecursive*, and it is shown why these algorithmic variants are superior to current state-of-the-art implementations. Results for all sparse grid bases introduced in Sect. 3 are presented in Sect. 6. Finally, the paper is summarized and concluded in Sect. 7.

2 Related Work

Examples of applications that depend on the performance of the function evaluation can be found, e.g., in interpolation, classification or regression settings. In [11], classification and regression are performed with spatially adaptive sparse grids with tens of thousands of grid points in moderate dimensionalities but for large data sets. There, evaluation is the hot spot, and in order to avoid recursive algorithms and data structures, the authors rely on a brute force approach instead of investing in algorithmic optimizations and complex data structures. Performance is drawn from the massive parallelism of a hybrid compute environment, consisting of several CPUs and accelerators. The authors observe, that it does not hurt to carry out unnecessary computations as long as the code is tailored to the specific platform and a high degree of parallelism is maintained at all times.

The authors of [12] try to achieve a performance boost for evaluation in adaptive tensor approximation spaces through the use of multi-dimensional segment trees. As will become clear when discussing our compact data structure in Sect. 4, such memory intensive support constructs are not required here, and thus bear no advantage for our setting.

In other applications, spatial adaptivity is not generally required. The interpolation problem of the astrophysics scenario described in [6] could be solved with competitive quality of results for both regular and spatially adaptive sparse grids. In [9], dimensionally adaptive sparse grids had been employed for numerical quadrature as a more generalized form of regular sparse grids with the combination technique. The combination technique (see [10]), which is bound to work with grids of a regular nature, has successfully been employed on the same kind of grid structure in classification and machine learning [7].

From these settings and applications we draw our motivation to use *truncated sparse grids* as described in [1, 15]. For example, truncation can be very helpful in

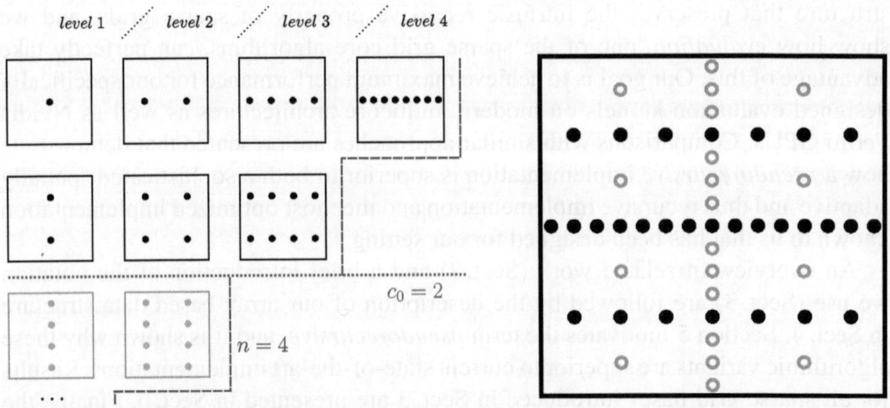

Fig. 1 Illustration of a truncated sparse grid in 2-D for level $n = 4$ and truncation vector $\mathbf{c} = [2, 4]^T$. *Left*: the subspace table and indicated constraints n and $\mathbf{c}$. *Right*: the resulting truncated sparse grid with omitted grid points grayed out. Note that dimension 1 is not truncated since $c_1 = 4 = n$

settings where high-dimensional singularities violate the smoothness assumptions such as for option pricing in finance. These grids also allow for anisotropic grid resolution; they are, however, more restrictive than general dimensionally adaptive grids. Figure 1 gives an idea of a truncated sparse grid for $d = 2$, for which a truncation vector $\mathbf{c} = [2, 4]^T$ was specified (for more details see Sect. 3). From an algorithmic point of view these grids are extremely attractive for us. While they provide a higher degree of flexibility than regular sparse grids, the a priori known structure enables one to optimize algorithms and data structures with respect to certain data access patterns. *Truncated sparse grids* also resemble *anisotropic sparse grids* as introduced in [8], for which the sparse-grid-typical hyperplane-cut $|\mathbf{l}|_1 = n + d - 1$ at the diagonal of the subspace tableau can be tilted to allow for arbitrary cutting planes. With truncated sparse grids on the other hand, up to d axis-aligned hyperplanes can be introduced in addition to the diagonal cut, such that grid refinement can be truncated in each dimension separately.

A fully implicit data structure for regular sparse grids is based on a linear array and an enumeration of the grid points and was first suggested in [5]. The idea of such a no-overhead data structure had been adapted in similar, hardware-aware implementations in [3,14,15], where the sparse grid core algorithms *hierarchization* and *evaluation* were optimized for commodity hardware.

The idea we follow here builds upon these two approaches. We also design a slim sparse grid data structure requiring only a linear coefficient array and a small lookup table to compute the unique location of each grid point's coefficient in the data array. We additionally optimize for data locality, the use of caches, and the vectorization capability of the algorithms and data structure, which is urgently necessary on modern commodity computers. Sections 4 and 5 provide a thorough comparison of state-of-the-art implementations and our approach. Note that in the simultaneous publication [13], the author shows that a variation of the implicit data structure presented in this paper is also suitable for an efficient implementation of the *hierarchization* algorithm.

3 Notation and Sparse Grid Bases

In the following, we assume that the concept of sparse grids and hierarchical bases is known and refer for further reading to [2]. Therefore, this section does not contain an extensive introduction to sparse grids, but merely a brief summary of the notation we adopt from [2, 18]. We will consider three different hierarchical bases commonly used with sparse grids, all of which were implemented and benchmarked for this work and are briefly described in this section. More information on the whys and hows of these bases and the results of the experiments are found in Sect. 6.

3.1 The Piecewise Linear Hat Function Basis

Our domain is the d-dimensional hypercube without and with boundaries, and it is denoted by

$$\Omega = [0; 1]^d. \tag{1}$$

A common choice of one-dimensional basis functions is the linear hat function basis. It builds upon the "mother function of all hat functions"

$$\phi(x) = \begin{cases} x+1 & \text{for } -1 \leq x < 0\ , \\ 1-x & \text{for } 0 \leq x < 1\ , \\ 0 & \text{else}\ . \end{cases} \tag{2}$$

The hierarchical basis functions $\phi_{l,i}(x)$ on levels $l \in \mathbb{N}$ with supports $[(i-1)/2^l, (i+1)/2^l]$ and grid points located at $x_{l,i} = i/2^l$ are then obtained by scaling and translation according to

$$\phi_{l,i}(x) = \phi(2^l \cdot x - i)\ ,\ i \in \mathcal{I}_l\ , \tag{3}$$

where $\mathcal{I}_l$ is the hierarchical index set given as

$$\mathcal{I}_l = \{i \in \mathbb{N} : 1 \leq i \leq 2^l - 1, i \text{ odd}\}. \tag{4}$$

This leads to a *multi-level subspace splitting* of the space of piecewise linear functions V_n on a full grid with mesh width $h_n := 2^{-n}$

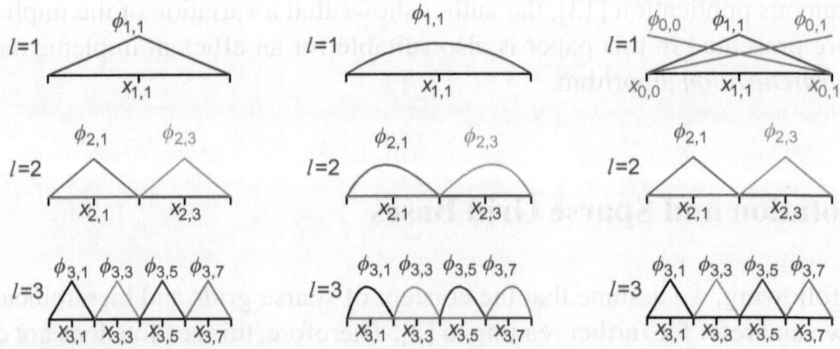

Fig. 2 *Left*: the piecewise linear hat function basis in 1-D. *Center*: the piecewise polynomial basis in 1-D. *Right*: the extended hat function basis with boundary grid points

$$V_n = \bigoplus_{l \leq n} W_l, \tag{5}$$

with *hierarchical increments* or *subspaces* W_l

$$W_l = span\{\phi_{l,i}(x) : i \in \mathcal{I}_l\}. \tag{6}$$

The first three levels of the one-dimensional hat function basis can be seen in the left part of Fig. 2.

For d dimensions the level-index notation l, i is extended to multi-variate indices $\mathbf{l} = (l_0, \dots, l_{d-1})$, $\mathbf{i} = (i_0, \dots, i_{d-1})$ and index sets $\mathcal{I}_\mathbf{l} = \{\mathbf{i} \mid i_j \in \mathcal{I}_{l_j},\ j = 0, \dots, d-1\}$. The d-dimensional basis functions are defined via a tensor product construction as

$$\phi_{\mathbf{l},\mathbf{i}}(\mathbf{x}) = \prod_{j=0}^{d-1} \phi_{l_j,i_j}(x_j) \tag{7}$$

and accordingly span the d-dimensional subspaces $W_\mathbf{l} = span\{\ \phi_{\mathbf{l},\mathbf{i}} \mid \mathbf{i} \in \mathcal{I}_\mathbf{l}\ \}$.

The approximation space of a regular sparse grid of level n is then obtained via

$$V_n = \bigoplus_{|\mathbf{l}|_1 \leq n+d-1} W_\mathbf{l}, \tag{8}$$

and in the case of a truncated sparse grid with truncation vector $\mathbf{c} \in \{1, \dots, n\}^d$ by

$$V_{n,\mathbf{c}} = \bigoplus_{|\mathbf{l}|_1 \leq n+d-1, \mathbf{l} \leq \mathbf{c}} W_\mathbf{l}, \tag{9}$$

with $\mathbf{l}' \leq \mathbf{l} \Leftrightarrow l'_j \leq l_j,\ 0 \leq j < d$.

3.2 The Piecewise Polynomial Basis

The piecewise polynomial basis functions are introduced in [2] as a generalization of the piecewise linear hat functions. Following the tensor product approach, the d-dimensional basis functions $\phi_{\mathbf{l},\mathbf{i}}^{(\mathbf{p})}(\mathbf{x})$ are defined as

$$\phi_{\mathbf{l},\mathbf{i}}^{(\mathbf{p})}(\mathbf{x}) = \prod_{j=0}^{d-1} \phi_{l_j,i_j}^{(p_j)}(x_j)\,, \tag{10}$$

where $\mathbf{p} = (p_0, \ldots, p_{d-1})$ denotes the respective polynomial degree of the one-dimensional functions $\phi_{l_j,i_j}^{(p_j)}$. As before, they have supports $[(i_j-1)/2^{l_j},\ (i_j+1)/2^{l_j}]$ and grid points at $x_{l_j,i_j} = i_j/2^{l_j}$.

Each one-dimensional $\phi_{l,i}^{(p)}$ is a hierarchical Lagrangian basis polynomial uniquely constructed with roots at the locations of its ancestors' grid points. One typically starts with a quadratic function on level one (first two roots at $x = 0$ and $x = 1$), and increases the polynomial degree with every level l such that $p = l + 1$ holds. Of course, the maximum degree can be limited. Figure 2 (center) shows the first three levels of the one-dimensional basis functions.

3.3 The Piecewise Linear Hat Function Basis with Boundaries

To account for the case of non-zero boundaries, the two hierarchical bases can be extended by two linear basis functions $\phi_{0,0}$ and $\phi_{0,1}$ that are associated with grid points on the boundaries:

$$\begin{aligned} \phi_{0,0}(x) &= 1 - x\,, \\ \phi_{0,1}(x) &= x\,. \end{aligned} \tag{11}$$

In Fig. 2 (right), these functions are included for the piecewise linear hat basis on the first level $l = 0$.

3.4 Sparse Grid Evaluation

The sparse grid interpolant $u_h : \Omega \to \mathbb{R}$ of level n is defined as

$$u_h(\mathbf{x}) = \sum_{|\mathbf{l}|_1 \le n+d-1} \sum_{\mathbf{i} \in \mathcal{I}_\mathbf{l}} \alpha_{\mathbf{l},\mathbf{i}} \cdot \phi_{\mathbf{l},\mathbf{i}}(\mathbf{x}). \tag{12}$$

Evaluating u_h at a point $\mathbf{x} \in \Omega$ corresponds to the task of collecting all non-zero contributions of basis functions $\phi_{\mathbf{l},\mathbf{i}}$ at position $\mathbf{x}$. Because of the disjoint supports of basis functions of the same subspace $W_{\mathbf{l}}$, at most one basis function in $W_{\mathbf{l}}$ has a non-zero contribution at position $\mathbf{x}$. This property can be exploited in an efficient implementation of sparse grid evaluation.

4 Data Structures for Regular Sparse Grids

This section provides an overview of common sparse grid data structures and explains in detail the data layout used in our implicit array-based data structure. Special focus is also set on the comparison with the data structure described in [14, 15], as results later are given in terms of speedup over results obtained for that data structure.

4.1 *Common Data Structures for Hierarchical Bases*

In the world of spatially adaptive sparse grids, hash maps are currently the dominating data structure [18]. Level and index vectors can be used as hash keys which allows programmers to stick close to the mathematical notation when implementing complex algorithms. But one has to pay for this convenience by giving up control. The hash map fully takes care of the data storage. This guarantees good mean performance for random data access but barely leaves space for algorithm- or cache-specific optimizations in the case that data access patterns are known. As a result, even compute-intensive algorithms may happen to be memory latency bound.

In [14, 15], the authors employ an implicit data structure for regular and truncated sparse grids, which facilitates the use of a plain array to efficiently store the grid's coefficients. Its storage overhead is negligible, as it merely requires a small lookup table to efficiently calculate a bijective mapping between the grid's points and a range of unique integer indices. The mapping relies on the enumeration of all subspaces in the grid, treating each subspace as a small- to medium-sized full grid, which is why we will refer to it as the *subspace-based data structure* from now on. One can also think of the subspace tableau, for which an efficient iterator exists. The subspace-based data structure shares many properties of our pseudorecursive data structure and was shown to be able to store the same kinds of sparse grids. It was also previously employed in a similar setting like ours, which renders it a perfect object for comparisons. A more detailed discussion of its characteristics follows in the next subsection.

For completeness, pointer-based tree structures need to be mentioned as a good fit for settings requiring spatial adaptivity, even though the hash map variant is nowadays generally preferred over them. The major downsides of the trees are a large memory footprint ($2\,d$ pointers per node typically reflect parent-children

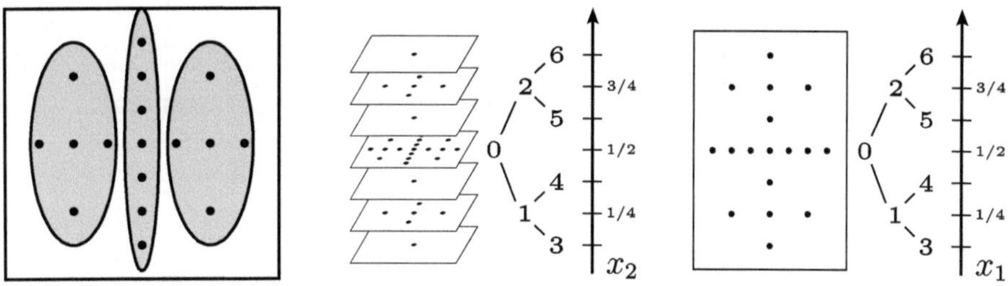

Fig. 3 *Left*: a regular sparse grid in 2-D is composed of a 1-D sparse grid of the same level, and two 2-D sparse grids of lower level (which are in turn composed of three smaller grids of lower level resp. dimensionality). *Middle* and *right*: the recursive application of the decomposition rule is indicated for sparse grids in 3-D and 2-D, eventually leading to a sequence of 1-D grids (depicted as *rows* of points in the image on the *right*)

relationships) and lack of support for random data access. Furthermore, the experiments on pointer-based variants of sparse grids done in [14] have already shown their inferiority to the implicit data structures in the context of regular grids.

4.2 A Recursive Data Layout

The recurrent patterns and formulas that are inherent to sparse grids have already been addressed in [2]. A d-dimensional regular sparse grid of level n can be decomposed into one $(d-1)$-dimensional sparse grid of the same level and two d-dimensional sparse grids of level $n-1$. For the 2-D case this decomposition is illustrated in the left part of Fig. 3. The center and the right hand side of Fig. 3 show how a 3-D grid (set of planes in the center image) can thus be decomposed into a sequence of 1-D grids (rows of points in the image to the right). Using a level-wise enumeration scheme like indicated in the binary trees in Fig. 3 rather than recursive in-place expansion of subgrids leads to a sequence of one-dimensional grids, which reminds of a multi-dimensional *breadth first traversal* of the resulting tree-like structure. Figure 4 depicts this for the case of $d = 4$ dimensions and sparse grid level $n = 3$. Transforming the 4-D grid into a sequence of 1-D grids leads to a serialized representation in which each grid point can be assigned a unique index from a set of contiguous integer indices. Note how the four subgrids on the right are listed *after* the two "parent" grids in the center, following the level-wise approach. It is also exactly these four subgrids that would be cut off from the 4-D grid, if a truncation vector $\mathbf{c} = [3, 3, 3, 2]^T$ were defined. This is because these subgrids contain grid points that are on level 3 with respect to the 4th dimension. As a general rule, specifying truncation corresponds to omitting subgrids in the recursive representation. Since this changes other subgrids' sizes, the lookup table used to calculate grid point indices needs to be adjusted accordingly.

Finally, we provide yet another graphical representation to illustrate how the order of the stored coefficients aligns with the geometrical order of the grid points.

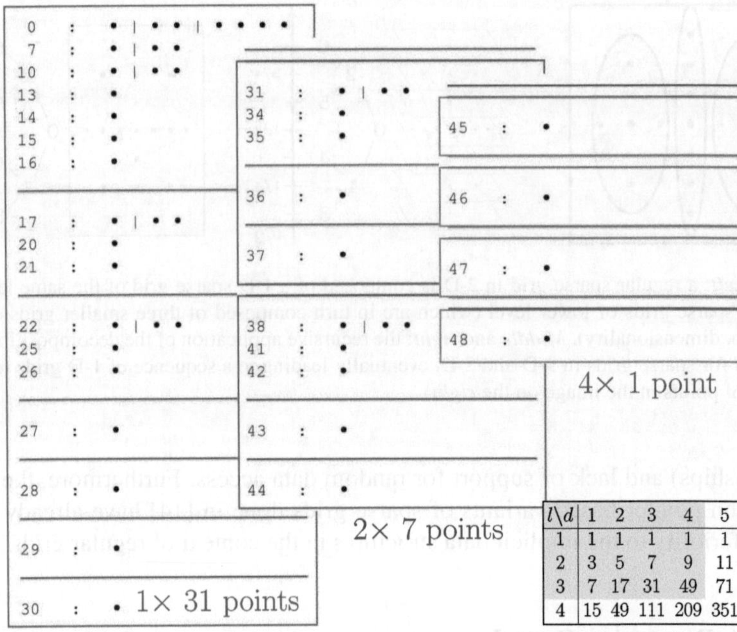

$l \backslash d$	1	2	3	4	5
1	1	1	1	1	1
2	3	5	7	9	11
3	7	17	31	49	71
4	15	49	111	209	351

Fig. 4 A 4-D sparse grid ($n = 3$) of 49 points is recursively decomposed into all of its lower-dimensional constituents: Each row of *dots* corresponds to a 1-D subgrid. *Horizontal separators* denote where 2-D subgrids start and end. *Colored boxes* frame the 3-D subgrids. *Vertical lines* (in rows) further mark the levels of the sparse grid. The offset before each row allows to match sizes of subgrids with the numbers of grid points in the lookup table (*right bottom*). Note how appending two 4-D grids of level 2 and four 4-D grids of level 1 would lead to a 5-D grid of level 3 with 71 points

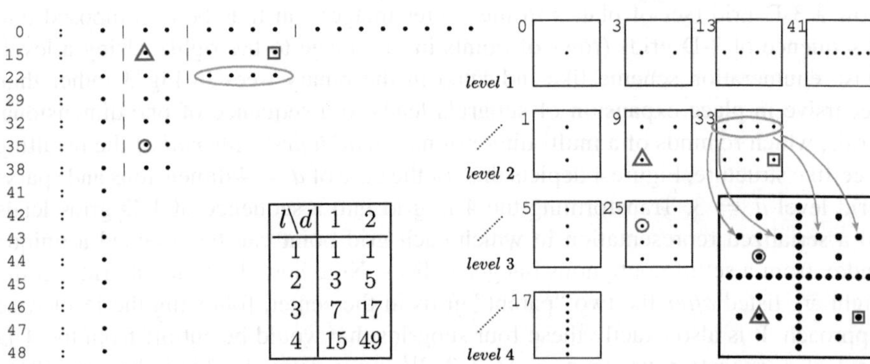

$l \backslash d$	1	2
1	1	1
2	3	5
3	7	17
4	15	49

Fig. 5 The positions of coefficients of a regular sparse grid are compared ($d = 2, n = 4$) for our recursive data layout (*left*) and the subspace-based data layout (*right*)

Figure 5 compares the new recursive layout and a classical data layout motivated by the subspace-wise treatment of coefficients. The latter one is the subspace-based layout proposed in [14]. It is a level-wise enumeration of all subspaces in the grid, each of them stored as a full grid. The numbers in front of the rows (left image) and next to the subspaces (right image) denote the offset of the respective sequence of coefficients in its corresponding data structure.

The example of the four marked grid points (or coefficients, respectively) shows how in both layouts sequences of these coefficients are aligned with respect to the x_0 (first) dimension. This property proves advantageous when implementing fast transformations between both layouts. We have already accomplished this in an efficient parallel implementation, but the discussion of the resulting permutation is not part of this work.

5 Recursive and Pseudorecursive Algorithms and Traversals

This section revolves around one of the classical prototypes of sparse grid algorithms: the evaluation of a sparse grid function u_h (12), which is implemented based on our implicit pseudorecursive data structure. First, a closer look at the properties of our data structure points out its main advantages. This is the basis for the novel algorithms. We then sketch recursive and pseudorecursive algorithms for efficient grid traversal and explain why the pseudorecursive approach leads to better results, especially in high-dimensional settings. A separate discussion of algorithms for truncated grids is not contained in this section, as the presented algorithms for regular sparse grids only need to be modified in a single line to support these grids as well.

5.1 Data Access Patterns

The red-colored squares in Fig. 6 correspond to the trace of our algorithm for sparse grid evaluation at a certain point. We call coefficients that correspond to basis functions which are affected by the evaluation *relevant* or *affected coefficients*. We refer to a basis function as being affected, if the evaluation point either lies within the interior of its support or on its left boundary.

The underlying grid in Fig. 6 is a regular three-dimensional sparse grid of level $n = 4$. While most of the rows do not contain red squares at all and thus need not to be visited, those rows which do often contain more than one square. This is due to the hierarchical structure of the sparse grid basis: Each row contains a one-dimensional subgrid in level-wise representation. If the root node of such a grid is affected (potentially non-zero) by the evaluation, so will be exactly one child on each level below it. This leads to a clustering of affected coefficients, which is very desirable from the perspective of computer caches and will be advantageous for the pseudorecursive evaluation algorithm. Consequently, large chunks of coefficients can be skipped between such clusters, as there are possibly large sub-structures of the sparse grid that do not contribute to the result.

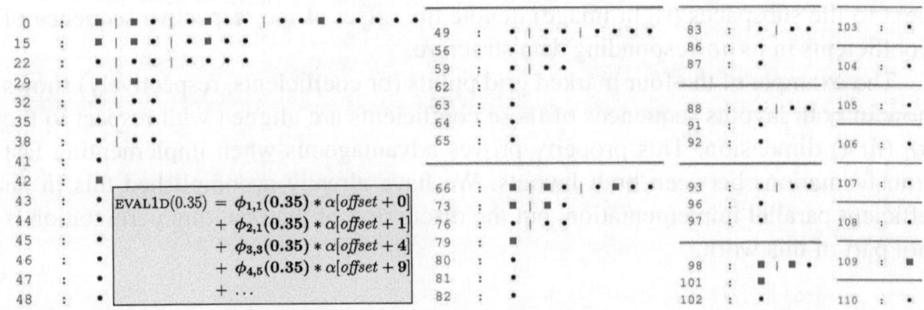

Fig. 6 The relevant coefficients of a regular sparse grid ($d = 3, n = 4$) in the recursive data layout are marked as *red squares* for a run of *evaluation* at point $\mathbf{x} = [0.35, 0.15, 0.8]^T$

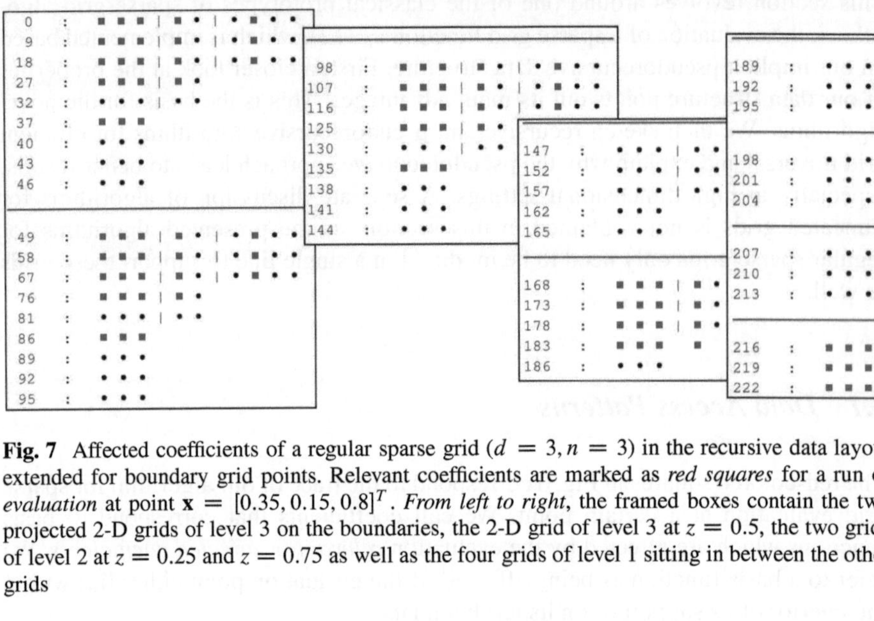

Fig. 7 Affected coefficients of a regular sparse grid ($d = 3, n = 3$) in the recursive data layout extended for boundary grid points. Relevant coefficients are marked as *red squares* for a run of *evaluation* at point $\mathbf{x} = [0.35, 0.15, 0.8]^T$. *From left to right*, the framed boxes contain the two projected 2-D grids of level 3 on the boundaries, the 2-D grid of level 3 at $z = 0.5$, the two grids of level 2 at $z = 0.25$ and $z = 0.75$ as well as the four grids of level 1 sitting in between the other grids

Another observation is that the pattern of relevant coefficients within each affected row is the same. This is because the recursive descent in each of the one-dimensional subgrids depends on the same coordinate x_0. As a very effective optimization, the offsets per row as well as the evaluated one-dimensional basis functions $\phi_{l,i}(x_0)$ (bold in the formula in Fig. 6) can be precomputed and stored in stencil form. A function EVAL1D can then be defined to compute this simple sum, and it can be applied to each of the affected rows efficiently. How these partial results are in turn combined to the global result in an efficient manner is the topic of Sect. 5.4.

Figure 7 shows how the layout presented in Sect. 4 is extended in a consistent way to support boundary grid points. Just as in Fig. 6, the trace of a run of *evaluation*

Algorithm 1 Recursive sparse grid traversal for evaluation. Result computations are stripped from the algorithm for readability and generality. The output is a sequence of offsets of visited rows in the grid, e.g., "0, 15, 29, 42, 66, 73, 79, 98, 101, 109" for the setup in Fig. 6

```
function RECDIM(offset, d, l, ref, i)
    if d = 0 then
        print(offset)                                      } arrived at affected row,
        return                                             } do work (e.g., EVAL1D)
    end if
    subgrid ← gridSize(d, l)
    recDim(offset + i * subgrid, d − 1, l, 0, 0)           } recursive descent into dimension
    if l > 1 then
        if liesLeftOfGridPoint(x_d) then                   }
            î ← 2 * i                                      }
        else                                               } "refine" in this dimension,
            î ← 2 * i + 1                                  } turn left or right
        end if                                             }
        recDim(offset + 2^ref * subgrid, d, l − 1, ref + 1, î) }
    end if
end function
...
recDim(0, dim − 1, level, 0, 0)                            } top level function call
```

is again highlighted, clearly showing that affected coefficients form even larger clusters when compared to the case of no boundaries. Data locality is thus even better. Note that what we see here are what we call "trapezoidal grids" (boundary grid points are only introduced at the positions of projections of inner grid points, see [18]), for which we just need to extend the recursive description by a prepended level 0 consisting of two boundary functions.

5.2 A Recursive Algorithm

Algorithm 1 uses the classical recursive scheme for sparse grid traversal that couples a recursive descent in the dimension with a recursive traversal of the sparse grid's levels. We will refer to the latter part as *refining*. The algorithm reflects the recursive grid construction (see Sect. 4.2), but due to the evaluation only one of the $(d-1)$-dimensional subgrids is affected. The respective sizes of all lower-dimensional subgrids can be precomputed and stored in a small lookup table (cf. Fig. 4). In the given algorithm this table is accessed via the function *gridSize*.

Note that in order to reduce code complexity as much as possible, the code only covers the case of no-boundary grids (extension to grids with boundaries is however straightforward). Furthermore, all statements specific to the accumulation of the result have been stripped from the code, and instead a sequence of row offsets is printed out, belonging to those rows relevant to the evaluation.

The formal parameter d of the recursive function RECDIM marks the currently focused dimension. It defines the context of parameter *ref*, which counts the steps of recursive refinement in dimension d, and parameter i, which encodes the path taken during refining. While parameter *offset* intuitively logs the current index position in the grid structure, parameter l globally tracks how many higher sparse grid levels still exist for the current position in the grid.

The algorithm is already quite efficient, but it still bears the downside that the recursive call stack needs to be almost fully discarded for changes in the slow dimensions d (e.g., $d = dim - 1$), which then requires an immediate rebuilt. The pseudorecursive solution presented in the following section manages to circumvent this problem.

5.3 A Pseudorecursive Formulation

Algorithm 2 can be seen as an iterative version of Algorithm 1. The formal parameters of function RECDIM that used to lie on the recursive call stack are mapped almost one-to-one to variables on an actual stack s. Traversal order and output of both algorithms are exactly the same, however, there is a significant difference in the runtime complexity.

Obviously, the number of visited rows is identical when both algorithms are executed on the same grid. But the cost of a jump from one affected row to the next differs. The pseudorecursive variant in Algorithm 2 manages to compute this jump in constant time. For the variant relying on actual recursion the cost of the jump is defined by the depth of the call stack of function RECDIM. In the worst case this depth is limited by the number of dimensions d, and so the complexity of this jump is $\mathcal{O}(d)$. Since the amortized cost is lower than that, there is barely a notable difference in the runtime for smaller grids. This changes for higher dimensionality d, and the lower runtime complexity of the pseudorecursive variant leads to a considerable speedup over the recursive variant, as the empirical results will show.

5.4 Efficient Calculation of the Tensor Product

The clustering of affected coefficients is only one beneficial property of the recursive data layout. Additionally, algorithms descending recursively into the dimension offer the opportunity to successively combine the d one-dimensional basis functions ϕ_{l_j,i_j}, $0 \leq j < d$ to the final tensor product (7).

Figure 8 demonstrates a function evaluation of our implementation (left) and of the subspace-based implementation (right) at the example of a small three-dimensional sparse grid. Our procedure resembles Horner's method, where the computational effort is kept to a minimum. In the given example, only 9 instead of 20 multiplications need to be computed. But more importantly, we achieve a

Algorithm 2 Pseudorecursive sparse grid traversal for evaluation. Result computations are stripped from the algorithm for readability and generality. The output is a sequence of offsets of visited rows in the grid, e.g., "0, 15, 29, 42, 66, 73, 79, 98, 101, 109" for the setup in Fig. 6

```
function PSEUDOREC(dim, level)
    s.push(offset = 0, d = dim − 1, l = level, i = 0)
    finished ← false
    while ! finished do
        offset_new ← s.offset + s.i * gridSize(s.d, s.l)
        print(offset_new)                                  } arrived at affected row,
        d_new ← 0                                          } do work (e.g., EVAL1D)
        if s.l = 1 then
            d_new ← s.d + 1
            s.pop()                                        } return from "refining"
            offset_new ← s.offset + s.i * gridSize(s.d, s.l)
        end if
        if s.d ≠ d_new then
            s.push(offset = offset_new, d = d_new, l = s.l, i = 0)   } recursive descent into
        end if                                                       } dimension
        if d_new ≠ dim − 1 then
            if liesLeftOfGridPoint(x_d) then
                î ← 2 * s.i
            else
                î ← 2 * s.i + 1                            } "refine" in this dimen-
            end if                                         } sion, turn left or right
            s.i ← î                                        } (s.l(2) is 2nd
            ref ← s.l(2) − s.l                             } element from top)
            s.offset ← s.offset + 2^ref * gridSize(d_new + 1, s.l)
            s.l ← s.l − 1
        else
            finished ← true                                } finished
        end if
    end while
end function
...
pseudoRec(d, n)                                            } top level function call
```

considerable reduction of the number of 1-D basis function evaluations (19 vs. 30), which can be rather expensive. Numerically this makes sense, too, as partial results of expectedly similar size are combined additively, reducing the risk of cancellation and floating point errors.

An implementation of this approach requires to manage additional stacks for partial products and results, which are eventually merged into a global result. For piecewise linear and polynomial bases it is possible to seamlessly integrate these computations into both traversal algorithms presented in the previous subsections. In the case of these two bases, 2 resp. 3 more stacks are required compared to the algorithm shown above, and the low runtime complexity of the pseudorecursive variant can be preserved.

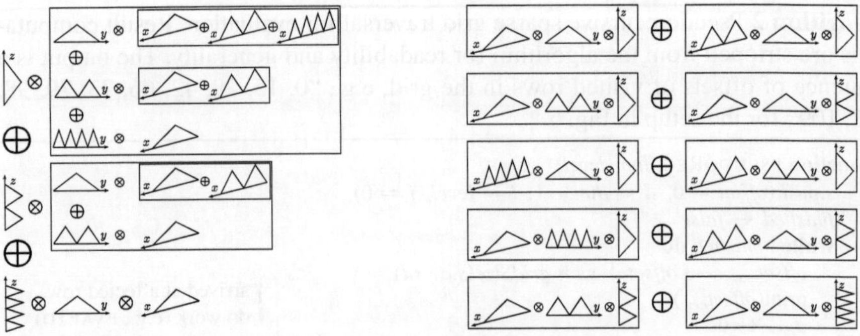

Fig. 8 Scheme of tensor product calculation during the evaluation of a regular sparse grid ($d = 3$, $n = 3$). *Boxes* represent parentheses, $\oplus$ denotes classical addition, $\otimes$ multiplication. *Left*: our successive approach. *Right*: subspace-wise computation

5.5 A Pseudorecursive Variant for GPUs

Evaluation is typically not executed for a single data point, but rather for a large number of sampling points or training data. A high degree of parallelism is therefore desirable which makes GPUs an adequate choice.

The subspace-based approach has led to good results for *evaluation* on Nvidia's Tesla C1070 [14]. While the on-chip resources like registers and fast scratchpad memory were too scarce to run complex traversals like Algorithm 2 on the C1070, devices from Nvidia's Fermi generation ship with an increased amount of fast scratchpad memory (48 or 16 kB, depending on the execution mode) and 32k registers per streaming processor. Targeting "full occupancy" of the Fermi device, i.e. the optimum of 1,536 active threads on each streaming processor, this still leaves merely 21 registers and 32 bytes of scratchpad memory to each thread, which poses a challenge to the pseudorecursive variant of our algorithm. Even for the case of $d = 3$, with an unmodified algorithm we already need to store at least 48 bytes for the evaluation point $\mathbf{x}$ and the d-, l- and i-stacks from Algorithm 2. Note that even though recursion is generally supported on Fermi, explicit control over the recursive stack is still our only chance to achieve good performance, which renders the pseudorecursive variant our only option.

To account for the shifted constraints of an efficient GPU implementation, we devised a Fermi-tailored variant of the pseudorecursive traversal algorithm using Nvidia OpenCL [17]. Note that in the beginning, we used Nvidia's CUDA framework [16], but our decision for OpenCL was then mainly motivated by the JIT compiler's capabilities, which helps saving scarce resources (e.g., registers, scratchpad memory) by allowing source code modifications at application runtime. In our explanations we stick to OpenCL terminology rather than CUDA terminology, as no Nvidia-specific extensions to the OpenCL API are used.

Table 1 The table shows how certain constructs from Algorithm 2 were mapped to more GPU-friendly ones

CPU		Fermi GPU	
Construct	Required space	Required space	Construct
$\mathbf{x}$	$d \cdot sizeof(float)$	$d \cdot sizeof(float)$	$\mathbf{x}$: unchanged
$eval1d.offset$	$n \cdot sizeof(int)$	1 *int* register	$path1d$: binary tree path encoded in bits
$eval1d.phi$	$n \cdot sizeof(float)$	–	–
$s.d$	$min(d,n) \cdot sizeof(int)$	–	–
$s.i$	$min(d,n) \cdot sizeof(int)$	1 *int* register	i_{int}: array entries dynamically encoded in bits
$s.l$	$min(d,n) \cdot sizeof(int)$	$d \cdot sizeof(int)/32$	$\mathbf{l}$: level vector shared across a warp
$s.offset$	$min(d,n) \cdot sizeof(int64)$	1 *int* register	*offset*: simple offset counter
2 stacks for tensor product	$2d \cdot sizeof(float)$	1 *float* register	phi_{d2}: keeps $\prod_{j=2}^{d-1} \phi_{l_j,i_j}(x_j)$

Each thread must not exceed the usage of 21 registers and 32 bytes of shared memory, otherwise full occupancy is not possible and performance breaks down

5.5.1 Optimizing Storage Consumption

Table 1 yields a comparison of variables and arrays used in the original implementation of Algorithm 2 and the GPU-adjusted one, for which optimization of memory consumption was the top design goal. The most important changes are that

1. bitwise encoding is used to get rid of integer arrays,
2. the two stencil arrays for EVAL1D are replaced by one integer, and
3. we ceased to rely on stacks.

The actual OpenCL source code with its explicit thread and memory management is too technical and lengthy to be included here. We therefore focus on giving remarks on the implications of the changes listed in Table 1.

Evaluating a sparse grid function means traversing a multi-dimensional tree on a complex path. On this traversal the current position is typically encoded in a level-index-vector-pair $(\mathbf{l}, \mathbf{i})$ or a similar construct like the set of stacks we use in our pseudorecursive algorithm. For our GPU implementation we go back to actual vectors, because it gives us the opportunity to compress the index vector $\mathbf{i}$ into a single integer i_{int} as described in Fig. 9. The image shows that we can encode $\mathbf{i}$ into a sequence of $\sum_{j=0}^{d-1}(l_j - 1) = n - 1$ bits, and so an integer of 32 bits is big enough for a sparse grid of level $n = 33$. But we can also decrease memory consumption for the level vector $\mathbf{l}$ by taking advantage of the GPU's threading model. On Fermi, a streaming processor has one instruction unit for simultaneous control of 32 scalar processors. In OpenCL, the threads running on these processors are called *work items*, and the 32 work items of a streaming processor make up a *warp*. With the work items of a warp being perfectly synchronized, the level vector $\mathbf{l}$ can be made a shared resource, reducing its memory footprint by a factor of 32.

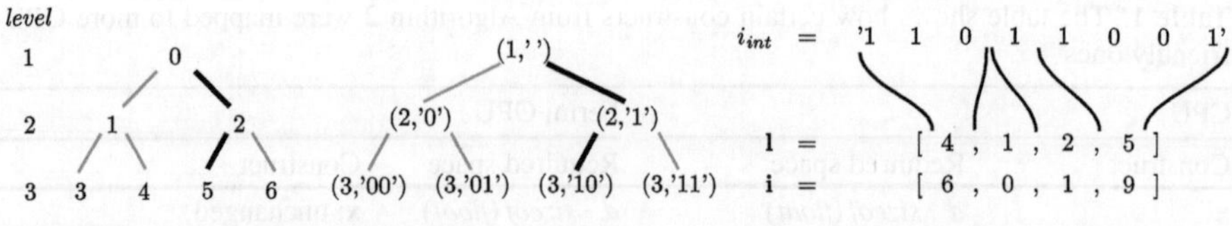

Fig. 9 *Left* and *center*: The previously introduced enumeration of grid points (cf. Fig. 3) can be translated to a level-index-like one. A pair (l, i) describes each point's respective level l and its index i on this level. i is given as bit sequence to show how it encodes the path to a point like in a bitwise trie, giving access to all its parents via subtraction and bit shifts. *Right*: a whole index vector $\mathbf{i}$ is encoded into one integer i_{int}. The level vector $\mathbf{l}$ is needed to decode i_{int} again, as the number of bits used to encode i_j into i_{int} is given by $l_j - 1$, respectively

---

Algorithm 3 GPU implementation of EVAL1D, using integer *path1d* to reconstruct the 1-D evaluation

---

function EVAL1D(*offset, l, path1d, x*)
  $result \leftarrow 0$
  for $1 \leq l' \leq l$ **do**
    $idx \leftarrow (2^{l'-1} - 1) + (path1d >> (l - l'))$
    $i \leftarrow 2 * (path1d >> (l - l')) + 1$
    $\phi \leftarrow max(0, 1 - |2^{l'} \cdot x - i|)$
    $result \leftarrow result + \phi * \alpha[offset + idx]$
  end for
  return *result*
end function

However, replacing the stacks of the original algorithm by vectors brings us back to the complexity of the recursive algorithm 1. It requires adding an inner loop (maximum loop size d) to the main loop. In this loop the vector pair $(\mathbf{l}, \mathbf{i})$ and the offset in the coefficient array need to be updated. In order to avoid full computation of the tensor product, we introduce a variable $phi_{d2} = \prod_{j=2}^{d-1} \phi_{l_j,i_j}(x_j)$ as indicated in Table 1. We treat it as a one-element-stack, and as it only needs to be recomputed when $\phi_{l_j,i_j}(x_j)$ changes for one of the "slow" dimensions with index $j \geq 2$, we save quite a lot of floating point operations.

Finally, we cannot afford to store the two stencil arrays for the EVAL1D-function on the GPU, so we use the technique from Fig. 9 to encode the path of the 1-D evaluation into an integer *path1d*. Algorithm 3 shows how EVAL1D can still be computed efficiently, totally avoiding conditional branches. To evaluate a 1-D grid along the path highlighted in the left part of Fig. 9, we need to call EVAL1D with the parameters $l = 3$ and $path1d = 2$. Obviously, the direct application of the stencils is faster; however, we now have practically no related storage consumption any more.

5.5.2 Coalesced Memory Access

Since memory access is a performance critical factor in a GPU implementation, we choose a particular layout of the work items. Each warp of 32 work items

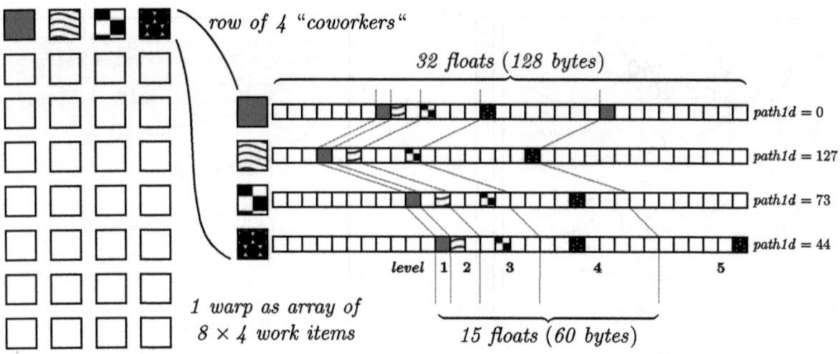

Fig. 10 *Left*: a warp is mapped to an array of 8 × 4 work items, each of which is in charge of one sparse grid evaluation, while groups of four (rows) act as coworkers. *Right*: in qualitatively depicted linear storage, levels 1–4 of a 1-D grid are shown to reside in lines of 15 floats. Regardless of the 1-D evaluation path (note *path1d* values on the *very right*), at most 2 memory transactions are thus necessary to access a 1-D grid's affected coefficients via coalesced load. The access strategy is indicated by matching patterns for work items and coefficients

is structured as 8 groups of 4 coworkers as seen in the left part of Fig. 10. The coworkers of each group simultaneously access the first 4 coefficients of each 1-D grid of level $l \geq 4$, taking advantage of coalesced memory access and reducing the overall number of memory transactions. This is done explicitly, as we do not want to heavily rely on the small L1 cache of the Fermi architecture, which is shared among 1,536 concurrent threads. The group size of 4 is derived from the size of memory transactions, which is 128 bytes on Fermi. The first 4 levels of a 1-D grid comprise 15 coefficients (summing up to 60 bytes), and so chances are high that the first 4 affected coefficients of a 1-D evaluation can be loaded in one memory transaction.

The maximum runtime reduction due to this optimization was 44 %, and it was observed for the case of a regular 7-D sparse grid of level 9. In average and over a whole range of levels and dimensionalities we were still measuring a runtime improvement of around 10 %, even though this included plenty of small grids. And they do not benefit greatly from this optimization, as they do not contain many subgrids of higher level. pseudoRec

6 Results

To demonstrate the advantages of the implementation of our novel data structures, algorithms and optimizations, we present performance comparisons with other highly optimized, state-of-the-art implementations of sparse grid evaluation. The complexity of all algorithms depends on the level n and the dimensionality d, which is why we provide all results in the form of speedups measured over this two-dimensional parameter space. First, our implementation competes with highly optimized algorithms operating on hash maps. These results are followed by an

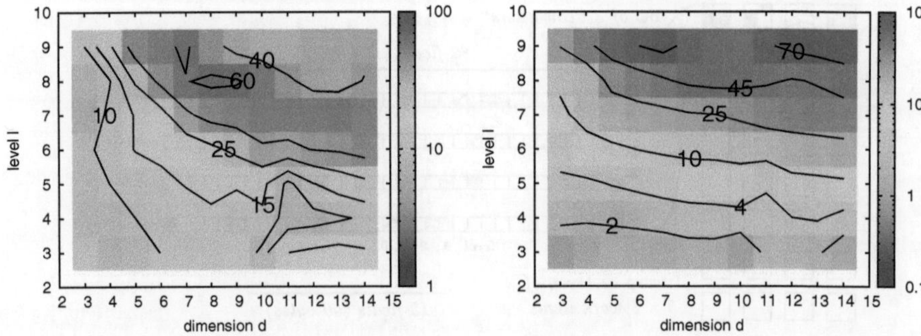

Fig. 11 *Left*: the speedup of our evaluation over the parallelized, recursive SG^{++} implementation. *Right*: the speedup of our evaluation over the parallelized and vectorized brute force evaluation scheme implemented in SG^{++}

extensive comparison of our implementation with the subspace-based one. Finally, the suitability of GPUs for our purpose of fast sparse grid evaluation is shown.

All speedup measurements have been obtained on a dual socket hexacore Intel Xeon X5690 Westmere-EP system clocked at 3.46 GHz (Intel Hyper Threading and Turbo disabled) and equipped with 96 GB DDR3 memory (1333 MHz, no ECC). The operating system is Scientific Linux 6 (kernel 2.6.32–71.29.1.el6.x86_64), and the compiler used is GNU's g++ version 4.4.4. Note that for vectorization we used Intel's SSE, SSE2 and SSSE3 instruction sets instead of relying on the compiler's auto-vectorization capabilities, which are known to be quite limited. We parallelize our algorithms with OpenMP and account for the parallel environments' comparatively long setup times, by ensuring that benchmarked times never drop below about 2 s. This is consistently done for all variations in grid dimensionality and level, which leads to cardinalities from a few thousands to hundreds of thousands of evaluation points. The interpretation of the presented speedup results does not require these numbers, so we do not include them here to avoid unnecessary complexity and confusion. For each run, a new set of random evaluation points $\mathbf{x}_j \in [0;1]^d$ was created, which was then used to time and compare all implementations under identical conditions. Lastly, for each test case a series of ten runs has been executed to account for system-specific performance oscillations.

6.1 Comparison with a Hash-Map-Based Data Structure

The SG^{++} toolbox provides programmers with a broad range of tools and interfaces to use spatially adaptive sparse grids to tackle various high-dimensional tasks (e.g., interpolation, quadrature, classification) [18]. Most algorithms in the toolbox have been optimized and parallelized to achieve best performance on modern multi-core systems. In Fig. 11 we compare our pseudorecursive evaluation kernel with two different kernels provided by SG^{++}, which are both parallelized via OpenMP. The first one (left image) is the default evaluation kernel, relying on an optimized recursive grid traversal to visit only the affected basis functions. The second one (right image) achieves its performance by avoiding conditional branches and complex traversals. In a brute force fashion all basis functions are evaluated (also those not affected), which allows for a perfect static load balancing scheme and makes vectorization straightforward. The approach has been employed very successfully for classification tasks [11], especially when grids do not grow too big. Note that it is targeted at fully, spatially adaptive sparse grids.

It can be seen at first glance, that our implementation outperforms both implementations from SG^{++} in our setting of truncated and regular sparse grids. For smaller grids, the ratio between affected and not affected basis functions is rather large, a fact from which the brute force variant benefits. However, the larger the grids grow, the more important become algorithmic optimizations, and so we see speedups of up to 95 over the brute force variant, but "only" 63 over the recursive variant.

It needs to be mentioned that the primary floating point type in the SG^{++} toolbox is the 64 bit double, for which we have not implemented a vectorized version yet. Using SSE-vectorization we thus might expect an additional speedup for our implementation of about 1.2 or even more.

6.2 Comparison with the Subspace-Based Data Structure

The subspace-based implementation is the only other implementation known to us that is specifically optimized for regular sparse grids. It is also the implementation that so far performs best for evaluation of these grids. We have therefore chosen the three types of basis functions presented in Sect. 3 (linear no-boundary, linear with trapezoidal boundary, polynomial no-boundary), to analyze the performance gain of our algorithms over the subspace-based ones in detail. We start however by investigating the cache behavior of both approaches. Because of the context of related results for the linear hat function basis, we also include the comparison of our recursive and pseudorecursive algorithms in this part.

We only had access to a sequential implementation of the subspace-based approach that works with single precision floats. Therefore, the speedups presented in this subsection have all been measured for non-parallel and non-vectorized implementations of the pseudorecursive and the subspace-based approach.

6.2.1 Memory Access

The heterogeneous structure of our pseudorecursive algorithm makes it difficult to decide whether they are bound by memory latency or computation, or whether the

Table 2 Statistics for L1 and L2 data cache misses (DCL1 and DCL2) for the pseudorecursive and the subspace-based variant

	Regular grid without boundaries				Regular grid with boundaries			
	Pseudorecursive		Subspace-based		Pseudorecursive		Subspace-based	
dim	DCL1	DCL2	DCL1	DCL2	DCL1	DCL2	DCL1	DCL2
2	0.003	0.001	0.09	0.009	0.033	0.012	0.019	0.012
3	0.052	0.003	0.226	0.012	1.727	0.053	2.753	0.043
4	1.164	0.072	2.286	0.029	10.45	3.43	22.44	10.65
5	3.4	0.254	7.57	1.5	47.2	24.3	116.7	76.4
6	7.6	2.7	17.1	9.5	207.9	95	569.7	423.6
7	14.3	5.8	34.4	27.3	881	356	2,666	2,109
8	25.6	10.3	65.2	55.7	3,670	1,283	12,072	9,440

Each number in the table denotes the cache misses in millions, measured for 10,000 sequential evaluations of a level 8 sparse grid of the respective dimensionality

conditional statements in the main loop induce branch mispredictions and pipeline stalls that might hinder performance. In this analysis, we want to demonstrate that our recursive data layout generally improves memory access as compared to the subspace-based data layout. Therefore, we employ PAPI[1] to examine the cache behavior of the respective evaluation algorithms more closely. PAPI gives programmatic access to modern CPUs' performance monitoring units (PMU), and we use the interface to read out the PMU counters for L1 and L2 data cache misses. As an exception, these measurements are not obtained on our Westmere-EP test platform, but on an Intel Ivy Bridge i5-3320M CPU, which supports an extended set of performance counters. We perform our tests for the hat function basis with and without boundary extension. The results are listed in Table 2 and they clearly show two tendencies:

1. The pseudorecursive algorithm and data structure generally lead to a considerably lower absolute number of L1 data cache misses.
2. For the subspace-based variant it is much more likely that an L1 cache miss leads to an L2 cache miss.

Both algorithmic variants have been tested under exactly the same conditions, and for each of the grid types both variants need to load exactly the same number of affected coefficients.

Keeping in mind that even with the array-based data structures evaluating a sparse grid function introduces scattered data access to the coefficient array, these results are very satisfying. We manage to considerably decrease the load on the deeper layers of the memory hierarchy, thus weakening the dependency on the component that is slowest in the pipeline and has the highest power consumption.

[1] http://icl.cs.utk.edu/papi.

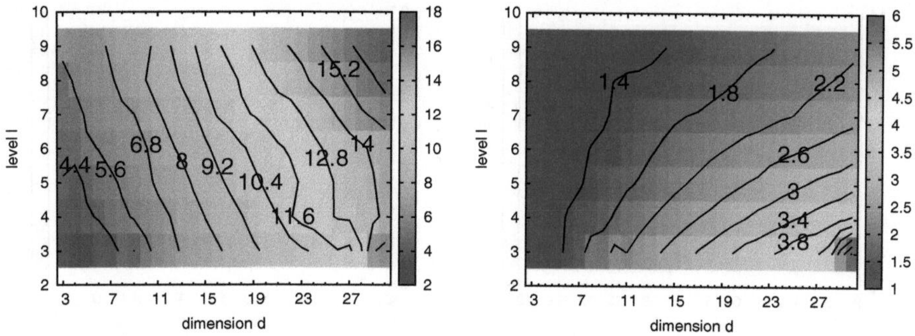

Fig. 12 *Left*: the speedup of our pseudorecursive evaluation scheme over the subspace-based one for the hat function basis. *Right*: the speedup of the pseudorecursive scheme over the actual recursive scheme, also for the hat function basis. Mind the different color scales

6.2.2 The Hat Function Basis

Evaluating these basis functions does generally not require much computational effort. As detailed before, our implementation optimizes the tensor product calculation and causes a lot less cache misses as compared to the subspace-based implementation which is compute bound due to the repeated computation of the tensor product (7).

Figure 12 shows that we achieve good speedups over the subspace-based implementation. They monotonically increase for higher level or dimensionality. For the largest test case ($d = 31, n = 10$) our implementation is about 18 times faster. Note that this behavior continues: tests with $d = 100$ and $n = 6$ even lead to speedups around 70.

For the comparison of our recursive and pseudorecursive algorithms we have chosen the same setting (linear hat functions without boundaries). Both algorithms operate on our recursive data layout, but the speedup plot in the right part of Fig. 12 shows the indicated asymptotic influence of the higher runtime complexity of the recursive variant. While for $d = 3$ it is almost at level with the pseudorecursive variant, we can see that it is two to six times slower for $d = 30$.

6.2.3 The Piecewise Polynomial Function Basis

We picked this basis for the benchmark to have a test case which is computationally more expensive, but structurally identical to that of the hat function basis. For the subspace-based variant this poses a challenge, as the increasing polynomial degree p causes the cost of directly calculating the full tensor product to grow from $\mathcal{O}(d)$ to $\mathcal{O}(d + p)$ for each of the affected basis functions. At the same time, for our pseudorecursive solution the runtime complexity of the algorithm does not change at all.

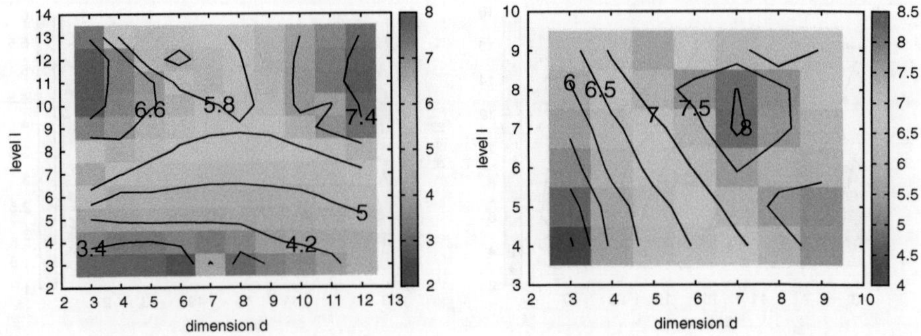

Fig. 13 *Left*: the speedup of our implementation over the subspace-based one for the polynomial basis functions. *Right*: the speedup of our implementation over the subspace-based one for the extended hat function basis with boundary basis functions. Mind the different parameter ranges and color scales

Note the speedups in Fig. 13 (left), which are already greater for moderate dimensionality d as compared to the previous case of the computationally cheaper hat function basis (cf. left part of Fig. 12). Since the locations of the affected basis functions and thus the traversal of the grid do not change when switching from one basis to the other, the difference must stem from a reduced number of arithmetic floating point operations. Besides, the increasing asymptotic dependency on the level n becomes obvious, as it takes a direct influence on the polynomial degree p.

6.2.4 The Extended Linear Basis with Boundary Functions

The special characteristic we want to put to test with this scenario is not related to the complexity of the basis functions. Instead, we will show that our data structures and algorithms are well suited to handle a well-known problem of sparse grids: With the introduction of boundary grid points the total number of grid points explodes already for small d, such that the boundaries become predominant for computations.

As pointed out in the discussion of the recursive data layout in Sect. 5.1, the extension to boundaries even leads to an improved clustering of affected coefficients during the evaluation. Moreover, we can easily extend the stencils used in the EVAL1D-function of our pseudorecursive variant to also cover the two affected points on the boundary of the 1-D subgrids. The subspace-based layout on the other hand does not benefit from the extension. Affected coefficients associated with boundary grid points are scattered evenly all over the coefficient array, each being contained in its own subspace.

And indeed, even for the smallest grids we immediately observe speedups of at least 4 in Fig. 13 (right). The number of arithmetic operations is as low as for the original hat function basis, thus the reason lies in the data locality shown by our extended data structure.

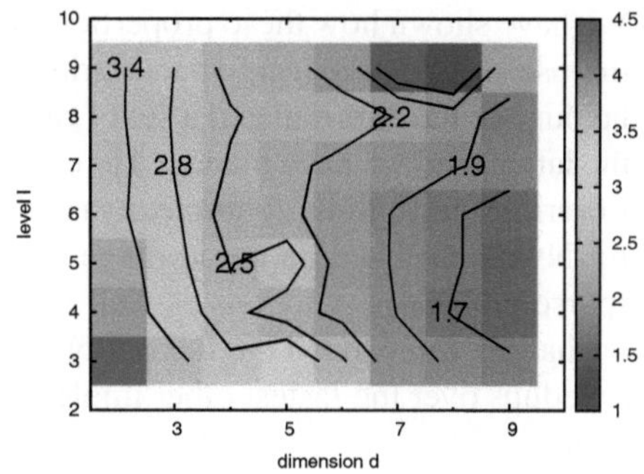

Fig. 14 The speedup of the GPU implementation over the fully parallelized and vectorized CPU implementation

6.3 Performance Comparison for CPU and GPU

In this last analysis, the single precision performance of our OpenCL implementation for the GPU is measured against the single precision performance of our parallelized and vectorized pseudorecursive CPU implementation. On the one side we use an Nvidia GTX 580 with 1.5 GB main memory, on the other side it is again the dual socket Westmere-EP platform, using 12 OpenMP threads and SSE vectorization. We choose the hat function basis without boundary extension as the test scenario.

The CPU implementation uses an optimized function for multiple evaluations, which only allocates the stacks once during initialization. A general limitation of the GPU implementation is given by the hardware constraints mentioned in Sect. 5.5, which make it less flexible. This can be seen in the speedup graph in Fig. 14, where the speedup of the GPU version decreases for increasing d. It has to be mentioned that while working perfectly for lower dimensionality, the implementation is not stable for $d \geq 10$, as OpenCL seems to show incorrect behavior when swapping registers and shared memory contents into the GPUs main memory. Still, the GPU achieves a speedup of 2 and more for grids of reasonable size, which is beneficial, be it only for the financial aspect.

7 Conclusions

The need for fast algorithms on sparse grids arises in various settings. Here, and motivated by our setting, we focused exemplarily on evaluation, achieving the best possible performance for regular and truncated sparse grids, both with and without boundary grid points. It has been shown before, that specialized data structures for these kinds of grids are indispensable when trying to exploit the full potential of modern compute platforms. We therefore proposed a compact, array-based data layout, that preserves some of the intrinsic recursive properties of sparse grids.

We have shown how these properties are beneficial for data access, as they enhance the use of caches during sparse grid evaluation. Motivated by the particular data layout, we have formulated a set of recursive and pseudorecursive algorithms. With the latter kind we have managed to overcome shortcomings of the recursive variant, lowering the runtime (complexity) and offering a suitable, non-recursive solution for Nvidia GPUs. In a broad range of tests, covering different implementations of sparse grid basis functions as well as grids with explicit boundary representation, we have demonstrated the flexibility of our approach and have achieved impressive speedups over the fastest other implementations of sparse grid evaluation we know about.

Our next endeavor is, most of all, the integration of our approach into the actual application that motivated our work. On the technical side, we will refine the vectorized version of our algorithm for Intel's Haswell generation of processors. With the corresponding AVX2 instruction set, a more consistent API for mixed floating point and integer arithmetic will be provided with access to vector registers of the same width for both data types.

Acknowledgements This publication is based on work supported by Award No. UK-C0020, made by King Abdullah University of Science and Technology (KAUST). The second author would like to thank the German Research Foundation (DFG) for financial support of the project within the Cluster of Excellence in Simulation Technology (EXC 310/1) at the University of Stuttgart. Special thanks go to Matthias Fischer, who helped with the implementation of the different sparse grid bases.

References

1. J. Benk, D. Pflüger, Hybrid parallel solutions of the Black-Scholes PDE with the truncated combination technique, in *Proceedings of the High Performance Computing and Simulation (HPCS)* (IEEE, Madrid, 2012), pp. 678–683
2. H.-J. Bungartz, M. Griebel, Sparse grids. Acta Numer. **13**, 147–269 (2004)
3. G. Buse, R. Jacob, D. Pflüger, A. Murarasu, A non-static data layout enhancing parallelism and vectorization in sparse grid algorithms, in *Proceedings of the 11th International Symposium on Parallel and Distributed Computing – ISPDC 2012* (IEEE, Los Alamitos, 2012), pp. 195–202
4. D. Butnaru, G. Buse, D. Pflüger, A parallel and distributed surrogate model implementation for computational steering, in *Proceedings of the 11th International Symposium on Parallel and Distributed Computing – ISPDC 2012* (IEEE, Los Alamitos, 2012), pp. 203–210
5. C. Feuersänger, Dünngitterverfahren für hochdimensionale elliptische partielle Differentialglei-chung. Diploma Thesis, Institut für Numerische Simulation, Universität Bonn, 2005
6. M. Frommert, D. Pflüger, T. Riller, M. Reinecke, H.-J. Bungartz, T. Enßlin, Efficient cosmological parameter sampling using sparse grids. Mon. Not. R. Astron. Soc. **406**, 1177–1189 (2010)
7. J. Garcke, A dimension adaptive sparse grid combination technique for machine learning. ANZIAM J. **48**, C725–C740 (2007)
8. J. Garcke, M. Griebel, Classification with sparse grids using simplicial basis functions. Intell. Data Anal. **6**, 483–502 (2002)
9. T. Gerstner, M. Griebel, Dimension-adaptive tensor-product quadrature. Computing **71**, 65–87 (2003)

10. M. Griebel, The combination technique for the sparse grid solution of PDE's on multiprocessor machines. Parallel Process. Lett. **2**, 61–70 (1992)
11. A. Heinecke, D. Pflüger, Multi- and many-core data mining with adaptive sparse grids, in *Proceedings of the 8th ACM International Conference on Computing Frontiers*, vol. 29 (ACM, New York, 2011), pp. 1–29
12. R. Hiptmair, G. Phillips, G. Sinha, Multiple point evaluation on combined tensor product supports. Numer. Algorithms **63**, 317–337 (2012)
13. R. Jacob, Efficient regular sparse grid hierarchization by a dynamic memory layout, in *Sparse Grids and Applications*, Munich 2012 (Springer, Berlin, 2013)
14. A. Murarasu, J. Weidendorfer, G. Buse, D. Butnaru, D. Pflüger, Compact data structure and scalable algorithms for the sparse grid technique, in *Proceedings of the 16th ACM Symposium on Principles and Practice of Parallel Programming* (Cambridge University Press, New York, 2011), pp. 25–34
15. A. Murarasu, G. Buse, J. Weidendorfer, D. Pflüger, A. Bode, fastsg: a fast routines library for sparse grids, in *Proceedings of the International Conference on Computational Science – ICCS 2012*. Procedia Computer Science (Elsevier, 2012)
16. Nvidia, *CUDA C Programming Guide* (Nvidia, Santa Clara, 2012)
17. Nvidia, *OpenCL Programming Guide for the CUDA Architecture* (Nvidia, Santa Clara, 2012)
18. D. Pflüger, *Spatially Adaptive Sparse Grids for High-Dimensional Problems* (Dr. Hut, München, 2010)
19. C. Zenger, Sparse grids, in *Parallel Algorithms for Partial Differential Equations*. Notes on Numerical Fluid Mechanics, vol. 31 (Vieweg, Braunschweig, 1991), pp. 241–251

Stochastic Collocation for Elliptic PDEs with Random Data: The Lognormal Case

Oliver G. Ernst and Björn Sprungk

Abstract We investigate the stochastic collocation method for parametric, elliptic partial differential equations (PDEs) with lognormally distributed random parameters in mixed formulation. Such problems arise, e.g., in uncertainty quantification studies for flow in porous media with random conductivity. We show the analytic dependence of the solution of the PDE w.r.t. the parameters and use this to show convergence of the sparse grid stochastic collocation method. This work fills some remaining theoretical gaps for the application of stochastic collocation in case of elliptic PDEs where the diffusion coefficient is not strictly bounded away from zero w.r.t. the parameters. We illustrate our results for a simple groundwater flow problem.

1 Introduction

The elliptic boundary value problem

$$-\nabla \cdot (a(\mathbf{x},\omega)\, \nabla p(\mathbf{x},\omega)) = f(\mathbf{x},\omega) \qquad \text{in } D, \qquad \mathbb{P}\text{-a.s.}, \tag{1a}$$

$$p(\mathbf{x},\omega) = g(\mathbf{x}) \qquad \text{on } \partial D, \qquad \mathbb{P}\text{-a.s.}, \tag{1b}$$

with random coefficient a and random source f, resp. its weak form, is of particular interest for studies on uncertainty quantification (UQ) methods. It is a rather simple mathematical model to study and, at the same time, of practical relevance, e.g., in groundwater flow modelling. There, the conductivity coefficient a is typically uncertain and therefore modeled as a random field, in particular, a lognormal random field [11].

O.G. Ernst (✉) • B. Sprungk
Department of Mathematics, TU Chemnitz, Reichenhainer Str. 41, 09126 Chemnitz, Germany
e-mail: oliver.ernst@mathematik.tu-chemnitz.de; bjoern.sprungk@mathematik.tu-chemnitz.de

J. Garcke and D. Pflüger (eds.), *Sparse Grids and Applications - Munich 2012*, Lecture Notes in Computational Science and Engineering 97,
DOI 10.1007/978-3-319-04537-5_2,

Recent methods for solving random PDEs such as stochastic collocation or Galerkin methods use a truncated Karhunen-Loève expansion of the random fields a and f in order to separate the deterministic and random parts of the problem (1) as well as reduce the randomness to a finite or countable number of random variables. This truncation leads to high-dimensional parametric problems, and approximation methods which are suited for higher dimensions, such as sparse grid collocation, have been successfully applied to this problem [2, 3, 18, 19]. In these works one often finds the assumption that the coefficient a is uniformly bounded away from zero, i.e., there exists a constant $c > 0$ such that $a(\mathbf{x}, \omega) \geq c$ $\mathbb{P}$-a.s. for all $\mathbf{x} \in D$. While this assumption simplifies the analysis, it fails to cover the important case where a has a (multivariate) lognormal distribution. For instance, in [2, 18, 19] the authors ensure uniform positivity by taking a to be the sum of a lognormal field and a positive constant $a_{\min}$. In [6] the analysis of full tensor-product collocation given in [2] is extended to the case of non-uniformly bounded coefficients a, but for deterministic sources f and homogeneous Dirichlet boundary conditions. Moreover, many works consider only the primal form (1) of the diffusion equation, but for many applications the numerical simulation of system (1) in mixed form

$$a^{-1}(\mathbf{x}, \omega)\, \mathbf{u}(\mathbf{x}, \omega) - \nabla p(\mathbf{x}, \omega) = 0 \quad \text{in } D, \tag{2a}$$

$$\nabla \cdot \mathbf{u}(\mathbf{x}, \omega) = -f(\mathbf{x}, \omega) \quad \text{in } D, \tag{2b}$$

$$p(\mathbf{x}, \omega) = g(\mathbf{x}) \quad \text{on } \partial D, \tag{2c}$$

$\mathbb{P}$-almost surely, is more appropriate. This is the case, for instance, if the flux $\mathbf{u}$ is of particular interest, see [12] for numerical examples. In [4] a first study of stochastic Galerkin methods for mixed problems was given, but again the assumptions on a made there do not apply to lognormal or non-uniformly bounded random fields.

In this work, we fill the remaining gaps and present a convergence analysis of sparse grid collocation for (2) without assuming the existence of a deterministic $a_{\min} > 0$ such that $a(\mathbf{x}, \omega) \geq a_{\min}$. Therefore, we introduce in Sect. 2 the parametric variational problem under consideration and prove in Sect. 3 a regularity result for its solution. In Sect. 4 we then conduct the proof of convergence of sparse grid stochastic collocation in unbounded parameter domains for approximating smooth functions. Section 5 illustrates the theoretical results for a simple elliptic boundary value problem in mixed form and Sect. 6 closes with concluding remarks.

2 The Parametric Variational Problem

In this section we briefly recall how the elliptic boundary value problem (BVP) (1) with random diffusion coefficient $a(\mathbf{x}, \omega)$ is transformed into a BVP containing a high-dimensional parameter. We shall restrict our considerations to the mixed formulation (2).

2.1 *Finite-Dimensional Noise via Karhunen-Loève Expansion*

Given a probability space $(\Omega, \mathcal{A}, \mathbb{P})$ we denote by $L^2(\Omega, \mathcal{A}, \mathbb{P}; \mathbb{R})$ the space of second-order real-valued random variables. We make the *finite-dimensional noise* assumption whereby the randomness in the coefficient $a(\mathbf{x}, \omega)$ and right hand side $f(\mathbf{x}, \omega)$ can be completely described by a finite set of M Gaussian random variables $\xi_1, \ldots, \xi_M \in L^2(\Omega, \mathcal{A}, \mathbb{P}; \mathbb{R})$.

Assumption 1. *There exist measurable functions* $\tilde{a} : \mathbb{R}^M \to L^\infty(D)$ *and* $\tilde{f} : \mathbb{R}^M \to L^2(D)$ *and* M *independent Gaussian random variables* $\xi_1, \ldots, \xi_M \in L^2(\Omega, \mathcal{A}, \mathbb{P}; \mathbb{R})$, *such that*

$$a(\mathbf{x}, \omega) = \tilde{a}(\mathbf{x}, \xi_1(\omega), \ldots, \xi_M(\omega)) \quad \textit{and} \quad f(\mathbf{x}, \omega) = \tilde{f}(\mathbf{x}, \xi_1(\omega), \ldots, \xi_M(\omega))$$

hold $\mathbb{P}$*-almost surely almost everywhere in* D.

We shall identify a with $\tilde{a}$ and f with $\tilde{f}$ in the following. Such finite-dimensional noise arises, e.g., when a random field is approximated by its truncated Karhunen-Loève expansion (KLE) [13].

Example 1 (KLE for lognormal random field). For a lognormal random field a, it is convenient to truncate the KLE of its logarithm $\log a$, yielding

$$a(\mathbf{x}, \omega) \approx a_M(\mathbf{x}, \omega) := \exp\left(\psi_0(\mathbf{x}) + \sum_{m=1}^{M} \sqrt{\lambda_m}\psi_m(\mathbf{x})\xi_m(\omega)\right), \tag{3}$$

where $\psi_0(\mathbf{x}) := \mathbb{E}[\log a(\mathbf{x}, \cdot)]$ and $\{(\lambda_m, \psi_m)\}_{m\geq 0}$ denotes the sequence of eigenpairs of the covariance operator C associated with $\log a$,

$$(C\psi)(\mathbf{x}) = \int_D c(\mathbf{x}, \mathbf{y})\psi(\mathbf{y})\,\mathrm{d}\mathbf{y}, \quad \text{where } c(\mathbf{x}, \mathbf{y}) = \operatorname{Cov}(\log a(\mathbf{x}, \cdot), \log a(\mathbf{y}, \cdot)), \tag{4}$$

and where the $\{\xi_m\}_{m\geq 0}$ are i.i.d. standard normally distributed random variables. For a discussion on approximating a directly by a (generalized) truncated polynomial chaos expansion see [10]. For an analysis of the effect of truncating the KLE see [6]. We neglect any truncation error in the following and identify a_M with a resp. $\tilde{a}$.

2.2 *The Parametric Elliptic Problem in Mixed Variational Form*

We set $\boldsymbol{\xi} := (\xi_1, \ldots, \xi_M)$ and denote by $\rho(\boldsymbol{\xi}) = \prod_{m=1}^{M} \frac{\exp(-\xi_m^2/2)}{\sqrt{2\pi}}$ the joint probability density function (pdf) of the i.i.d standard normally distributed $\xi_1, \ldots, \xi_M$.

We rewrite the random mixed elliptic problem (2) as the parametric mixed elliptic problem

$$a^{-1}(\mathbf{x},\boldsymbol{\xi})\,\mathbf{u}(\mathbf{x},\boldsymbol{\xi}) - \nabla p(\mathbf{x},\boldsymbol{\xi}) = 0 \qquad \text{in } D, \tag{5a}$$

$$\nabla\cdot\mathbf{u}(\mathbf{x},\boldsymbol{\xi}) = -f(\mathbf{x},\boldsymbol{\xi}) \qquad \text{in } D, \tag{5b}$$

$$p(\mathbf{x},\boldsymbol{\xi}) = g(\mathbf{x}) \qquad \text{on } \partial D, \tag{5c}$$

where the equations are taken to hold $\rho d\boldsymbol{\xi}$-almost everywhere.

To state the weak mixed form of (5), we assume $g \in H^{1/2}(\partial D)$ and introduce the space

$$H(\mathrm{div}; D) = \{\mathbf{v} \in L^2(D) : \nabla\cdot\mathbf{v} \in L^2(D)\} \tag{6}$$

with norm $\|\mathbf{v}\|^2_{H(\mathrm{div};D)} = \|\mathbf{v}\|^2_{L^2(D)} + \|\nabla\cdot\mathbf{v}\|^2_{L^2(D)}$ as well as the bilinear and linear forms

$$A_{\boldsymbol{\xi}}(\mathbf{u},\mathbf{v}) = \int_D a^{-1}(\mathbf{x},\boldsymbol{\xi})\mathbf{u}(\mathbf{x})\cdot\mathbf{v}(\mathbf{x})\,d\mathbf{x}, \tag{7}$$

$$B(\mathbf{v},q) = -\int_D q(\mathbf{x})\ \nabla\cdot\mathbf{v}(\mathbf{x})\,d\mathbf{x}, \tag{8}$$

$$h_{\boldsymbol{\xi}}(q) = -\int_D f(\mathbf{x},\boldsymbol{\xi})\,q(\mathbf{x})\,d\mathbf{x}, \tag{9}$$

$$\ell(\mathbf{v}) = -\int_{\partial D} g(\mathbf{x})\,\mathbf{v}(\mathbf{x})\cdot\mathbf{n}(\mathbf{x})\,d\mathbf{x}, \tag{10}$$

for $\mathbf{u},\mathbf{v} \in H(\mathrm{div}; D)$ and $q \in L^2(D)$, where in the last line $\mathbf{n}$ denotes the unit outward normal vector along the boundary ∂D and the integral is understood as a linear functional on $H^{1/2}(\partial D)$, see [9, Appendix B.3]. The weak form of (5) then reads

$$A_{\boldsymbol{\xi}}(\mathbf{u}(\cdot,\boldsymbol{\xi}),\mathbf{v}) + B(\mathbf{v}, p(\cdot,\boldsymbol{\xi})) = \ell(\mathbf{v}) \qquad \forall\mathbf{v} \in H(\mathrm{div}; D), \tag{11a}$$

$$B(\mathbf{u}(\cdot,\boldsymbol{\xi}), q) = h_{\boldsymbol{\xi}}(q) \qquad \forall q \in L^2(D), \tag{11b}$$

$\rho d\boldsymbol{\xi}$-almost everywhere. Hence, setting $\mathcal{S} := L^2(\mathbb{R}^M, \mathcal{B}(\mathbb{R}^M), \rho d\boldsymbol{\xi})$, where $\mathcal{B}$ denotes the σ-algebra of Borel sets, and $\mathcal{V} := L^2(D) \times H(\mathrm{div}; D)$, we are thus seeking a solution $(p,\mathbf{u}) \in \mathcal{S}\otimes\mathcal{V}$ which satisfies (11) $\rho d\boldsymbol{\xi}$-a.e. That such a solution exists and is unique will be shown in Sect. 3.

Remark 1. Note that, due to Assumption 1 and the continuous (hence measurable) dependence of the solution $(p,\mathbf{u})$ of a variational problem such as (11) on the coefficient a and the source term f, we can deduce by means of the Doob-Dynkin lemma [17, Lemma 1.13, p. 7] that the solution of (11) may be identified with the weak solution of (2) by way of

$$(p(\omega), \mathbf{u}(\omega)) = (p(\boldsymbol{\xi}), \mathbf{u}(\boldsymbol{\xi})), \qquad \boldsymbol{\xi} = \boldsymbol{\xi}(\omega),$$

$\mathbb{P}$-almost surely as functions in $L^2(D)$ and $H(\mathrm{div}; D)$, respectively.

3 Analytic Dependence on the Parameter

Subsequently, we denote by $(\cdot, \cdot)$ the inner product in $L^2(D)$, where for vector-valued functions we set $(\mathbf{u}, \mathbf{v}) := \int_D \mathbf{u}(\mathbf{x}) \cdot \mathbf{v}(\mathbf{x})\, \mathrm{d}\mathbf{x}$.

In this section we prove existence and analytic dependence of the solution $(p(\boldsymbol{\xi}), \mathbf{u}(\boldsymbol{\xi}))$ of the mixed problem (11) on the parameter $\boldsymbol{\xi}$. In particular, we will prove analyticity of $(p(\cdot), \mathbf{u}(\cdot))$ in a subdomain of $\mathbb{C}^M$. To this end, we consider problem (11) with the parameter vector $\boldsymbol{\xi}$ extended to *complex* values $\boldsymbol{\zeta} = \boldsymbol{\xi} + i\boldsymbol{\eta} \in \mathbb{C}^M$, $\boldsymbol{\xi}, \boldsymbol{\eta} \in \mathbb{R}^M$ along with suitable extensions of the functions $a(\mathbf{x}, \cdot)$ and $f(\mathbf{x}, \cdot)$. To ensure well-posedness of (11) for this complex extension

$$A_{\boldsymbol{\zeta}}(\mathbf{u}, \mathbf{v}) = \int_D a^{-1}(\mathbf{x}, \boldsymbol{\zeta}) \nabla \mathbf{u}(\mathbf{x}) \cdot \overline{\nabla \mathbf{v}(\mathbf{x})}\, \mathrm{d}\mathbf{x},$$

of $A_{\boldsymbol{\xi}}$, and $h_{\boldsymbol{\zeta}}(q) = -\int_D f(\mathbf{x}, \boldsymbol{\zeta})\, \overline{q(\mathbf{x})}\, \mathrm{d}\mathbf{x}$, we restrict the complex parameter $\boldsymbol{\zeta}$ to the domain

$$\Sigma := \left\{ \boldsymbol{\zeta} \in \mathbb{C}^M : a_{\min}(\boldsymbol{\zeta}) > 0 \quad \text{and} \quad a_{\max}(\boldsymbol{\zeta}) < +\infty \right\},$$

where

$$a_{\max}(\boldsymbol{\zeta}) := \operatorname{ess\,sup}_{\mathbf{x} \in D} \operatorname{Re} a(\mathbf{x}, \boldsymbol{\zeta}), \quad a_{\min}(\boldsymbol{\zeta}) := \operatorname{ess\,inf}_{\mathbf{x} \in D} \operatorname{Re} a(\mathbf{x}, \boldsymbol{\zeta}).$$

For a general Banach space W, we denote by $L^q_\rho(\mathbb{R}^M; W)$ the Bochner space $L^q(\mathbb{R}^M, \mathcal{B}(\mathbb{R}^M), \rho d\boldsymbol{\xi}; W)$ of W-valued functions of $\boldsymbol{\xi}$ and make the following assumptions for proving the existence of a solution to (11) for *real-valued* parameters $\boldsymbol{\xi} \in \mathbb{R}^M$:

Assumption 2. *The data a, f and g defining problem* (11) *satisfy*

(1) $g \in H^{1/2}(\partial D)$,
(2) $a \in L^q_\rho(\mathbb{R}^M; L^\infty(D))$ *for all* $q \in [1, \infty)$,
(3) $a_{\min}(\boldsymbol{\xi}) > 0$ *for all* $\boldsymbol{\xi} \in \mathbb{R}^M$ *and* $1/a_{\min} \in L^q_\rho(\mathbb{R}^M; \mathbb{R}_+)$ *for all* $q \in [1, \infty)$,
(4) $f \in L^{q^*}_\rho(\mathbb{R}^M; L^2(D))$ *for some* $q^* > 2$.

Note that, under Assumption 2, we have $\mathbb{R}^M \subset \Sigma$. We can now state

Lemma 1 (cf. [4, Lemma 2.3]). *Let Assumption 2 be satisfied. Then there exists a unique solution* $(p, \mathbf{u}) \in \mathcal{S} \otimes \mathcal{V}$ *of* (11).

Lemma 1 will be proven together with the following existence and continuity result for the solution to (11) for *complex* parameters $\boldsymbol{\zeta} \in \mathbb{C}^M$. In order to state this result, we introduce the spaces

$$C_\sigma(\Sigma; W) := \{v : \Sigma \to W \text{ continuous, strongly measurable and}$$
$$\|v\|_{C_\sigma} = \max_{\boldsymbol{\zeta}\in\Sigma} \sigma(\operatorname{Re}\boldsymbol{\zeta})\|v(\boldsymbol{\zeta})\|_W < \infty\},$$

where $\sigma : \mathbb{R}^M \to \mathbb{R}_+$ is an arbitrary nonnegative weight function and W a Banach space.

Assumption 3. *For* $\sigma : \mathbb{R}^M \to \mathbb{R}_+$ *there holds*

(5) $f \in C_\sigma(\Sigma; L^2(D))$ *and* $a \in C_\sigma(\Sigma; L^\infty(D))$,
(6) $a_{\max} \in C_\sigma(\Sigma; \mathbb{R})$ *and* $1/a_{\min} \in C_\sigma(\Sigma; \mathbb{R})$.

Lemma 2. *Let Assumptions 2 and 3 be satisfied. Then for each* $\boldsymbol{\zeta} \in \Sigma$ *there exists a unique* $(p(\boldsymbol{\zeta}), \mathbf{u}(\boldsymbol{\zeta}))$ *which solves* (11) *with* $(p, \mathbf{u}) \in C_{\sigma^4}(\Sigma; \mathcal{V})$.

Proof (of Lemmas 1 and 2). We first observe that, for $\mathbf{u}, \mathbf{v} \in H(\text{div}; D)$ and $q \in L^2(D)$, we obtain

$$\begin{aligned}
|A_{\boldsymbol{\zeta}}(\mathbf{u}, \mathbf{v})| &= \left|\left(a^{-1}(\boldsymbol{\zeta})\mathbf{u}, \mathbf{v}\right)\right| \le \frac{1}{a_{\min}(\boldsymbol{\zeta})} \|\mathbf{u}\|_{H(\text{div};D)} \|\mathbf{v}\|_{H(\text{div};D)},\\
|B(\mathbf{v}, q)| &= |(q, \nabla\cdot\mathbf{v})| \le \|q\|_{L^2(D)} \|\mathbf{v}\|_{H(\text{div};D)},\\
|\ell(\mathbf{v})| &\le \|\mathbf{v}\|_{H(\text{div};D)} \|g\|_{H^{1/2}(\partial D)},\\
|h_{\boldsymbol{\zeta}}(q)| &= |(f(\boldsymbol{\zeta}), q)| \le \|f(\boldsymbol{\zeta})\|_{L^2(D)} \|q\|_{L^2(D)}.
\end{aligned}$$

Moreover, $A_{\boldsymbol{\zeta}}$ is coercive on

$$\begin{aligned}
V &= \{\mathbf{v} \in H(\text{div}; D) : B(\mathbf{v}, q) = -(q, \nabla\cdot\mathbf{v}) = 0 \quad \forall q \in L^2(D)\}\\
&= \{\mathbf{v} \in H(\text{div}; D) : \|\nabla\cdot\mathbf{v}\|_{L^2(D)} = 0\},
\end{aligned}$$

since for $\mathbf{v} \in V$ there holds

$$\operatorname{Re} A_{\boldsymbol{\zeta}}(\mathbf{v}, \mathbf{v}) = \operatorname{Re}\left(a^{-1}(\boldsymbol{\zeta})\mathbf{v}, \mathbf{v}\right) \ge \operatorname{ess\,inf}_{\mathbf{x}\in D} \operatorname{Re}\left(a^{-1}(\mathbf{x}, \boldsymbol{\zeta})\right) \|\mathbf{v}\|^2_{L^2(D)} \ge \frac{\|\mathbf{v}\|^2_{H(\text{div};D)}}{a_{\max}(\boldsymbol{\zeta})}.$$

According to [5, p. 136], for any $q \in L^2(D)$ there exists $\mathbf{v}_q \in V$ such that

$$\|\nabla\cdot\mathbf{v}_q - q\|_{L^2(D)} = 0 \quad \text{and} \quad \|\mathbf{v}_q\|_{H(\text{div};D)} \le C_D \|q\|_{L^2(D)},$$

with a constant C_D depending only on the domain D. Thus, the inf-sup-condition follows since, for any $q \in L^2(D)$,

$$\sup_{\mathbf{v}\in H(\mathrm{div};D)} \frac{B(\mathbf{v},q)}{\|\mathbf{v}\|_{H(\mathrm{div};D)}} \geq \frac{(q,\,\nabla\cdot\mathbf{v}_q)}{\|\mathbf{v}_q\|_{H(\mathrm{div};D)}} = \frac{\|q\|^2_{L^2(D)}}{\|\mathbf{v}_q\|_{H(\mathrm{div};D)}} \geq \frac{\|q\|_{L^2(D)}}{C_D}.$$

Therefore, by applying [5, Theorem II.1.1], resp. its generalization to variational problems in complex Hilbert spaces (hereby applying the complex version of the Lax-Milgram-lemma), we obtain for each $\boldsymbol{\zeta} \in \Sigma$ a unique solution $(p(\boldsymbol{\zeta}), \mathbf{u}(\boldsymbol{\zeta}))$ to the associated deterministic variational problem. Moreover, there holds

$$\|\mathbf{u}(\boldsymbol{\zeta})\|_{H(\mathrm{div};D)} \leq \|g\|_{H^{1/2}(\partial D)}\, a_{\max}(\boldsymbol{\zeta}) + 2C_D \frac{a_{\max}(\boldsymbol{\zeta})}{a_{\min}(\boldsymbol{\zeta})} \|f(\boldsymbol{\zeta})\|_{L^2(D)},$$

$$\|p(\boldsymbol{\zeta})\|_{L^2(D)} \leq 2\, C_D\, \|g\|_{H^{1/2}(\partial D)} \frac{a_{\max}(\boldsymbol{\zeta})}{a_{\min}(\boldsymbol{\zeta})} + 2C_D \frac{a_{\max}(\boldsymbol{\zeta})}{a^2_{\min}(\boldsymbol{\zeta})} \|f(\boldsymbol{\zeta})\|_{L^2(D)}.$$

Further, we observe that $p : \mathbb{R}^M \to L^2(D)$ and $\mathbf{u} : \mathbb{R}^M \to H(\mathrm{div}; D)$ are measurable, since they are continuous functions of a, f and g.

By applying the Hölder inequality for the exponents $r = q^*/2$ and $q > 0$ (such that $1/r + 1/q = 1$) and by taking into account the above estimate and the assumptions, we easily obtain that $p \in \mathcal{S} \otimes L^2(D)$ and $\mathbf{u} \in \mathcal{S} \otimes H(\mathrm{div}; D)$, which yields $(p, \mathbf{u}) \in \mathcal{S} \otimes \mathcal{V}$. Uniqueness follows immediately. The continuity of $(p(\boldsymbol{\zeta}), \mathbf{u}(\boldsymbol{\zeta}))$ w.r.t. $\boldsymbol{\zeta} \in \Sigma$ follows from our assumptions on the continuity of a, f w.r.t. $\boldsymbol{\zeta}$. Finally, $p \in C_{\sigma^4}(\Sigma; L^2(D))$ and $\mathbf{u} \in C_{\sigma^4}(\Sigma; H(\mathrm{div}; D))$ follow again directly from the estimates above and the assumptions. This completes the proof. □

In an analogous way to [8, Lemma 2.2] we can show the analyticity of the parameter-to-solution map $\boldsymbol{\zeta} \mapsto (p(\boldsymbol{\zeta}), \mathbf{u}(\boldsymbol{\zeta}))$.

Lemma 3. *Let Assumptions 2 and 3 be satisfied and let the functions* $a^{-1} : \Sigma \to L^\infty(D)$ *and* $f : \Sigma \to L^2(D)$ *be analytic. Then also the mapping* $\boldsymbol{\zeta} \mapsto (p(\boldsymbol{\zeta}), \mathbf{u}(\boldsymbol{\zeta}))$ *is analytic in* Σ.

Proof. We prove the statement by showing the existence of each partial complex derivative $\partial_m(p(\boldsymbol{\zeta}), \mathbf{u}(\boldsymbol{\zeta}))$, $m = 1, \ldots, M$. A deep theorem by Hartogs [14] then yields analyticity as a function of all M complex variables. Therefore, we fix $m \in \{1, \ldots, M\}$, denote by e_m the m-th coordinate in $\mathbb{R}^M$ and set for $z \in \mathbb{C}\backslash\{0\}$

$$(q_z, \mathbf{v}_z)(\boldsymbol{\zeta}) := \frac{(p, \mathbf{u})(\boldsymbol{\zeta} + z\mathbf{e}_m) - (p, \mathbf{u})(\boldsymbol{\zeta})}{z}.$$

Note, that Σ is an open set due to the continuity of $a_{\max}$ and $a_{\min}$. Therefore, for each $\boldsymbol{\zeta} \in \Sigma$, there exists ϵ_{ζ} such that, for $|z| \leq \epsilon_{\zeta}$, the solution $(p, \mathbf{u})(\boldsymbol{\zeta} + z\mathbf{e}_m)$ and thus also the quotient above are well defined.

To simplify the presentation, we rewrite the variational problem (11) as a coupled linear system in the corresponding dual spaces, denoting again by $A_{\zeta} : H(\mathrm{div}; D) \to H(\mathrm{div}; D)^*$ the linear mapping $[A_{\zeta}\mathbf{u}](\mathbf{v}) := (a^{-1}(\boldsymbol{\zeta})\mathbf{u}, \mathbf{v})$, by

$B : L^2(D) \to H(\mathrm{div}; D)^*$ the linear map $[Bp](\mathbf{v}) := (p, \nabla \cdot \mathbf{v})$ and by $B^\top : H(\mathrm{div}; D) \to L^2(D)^*$ the map $[B^\top \mathbf{u}](q) := (q, \nabla \cdot \mathbf{u})$. Moreover, by ℓ and h_ξ we denote the linear functionals corresponding to the right hand side of (11). Thus, the variational problem (11) reads

$$\begin{pmatrix} A_\xi \mathbf{u} + Bp \\ B^\top \mathbf{u} \end{pmatrix} = \begin{pmatrix} \ell \\ h_\xi \end{pmatrix}. \tag{12}$$

Hence, by denoting $\boldsymbol{\zeta}_z = \boldsymbol{\zeta} + z\mathbf{e}_m$ we have

$$\begin{pmatrix} A_\xi \mathbf{v}_z + Bq_z \\ B^\top \mathbf{v}_z \end{pmatrix} - \begin{pmatrix} \frac{A_\xi - A_{\xi_z}}{z} \mathbf{u}(\boldsymbol{\zeta}_z) \\ 0 \end{pmatrix} = \begin{pmatrix} 0 \\ \frac{h_{\xi_z} - h_\xi}{z} \end{pmatrix},$$

i.e., the pair $(q_z, \mathbf{v}_z)$ solves the linear system (12) for the right hand side

$$L_z := \frac{1}{z} \begin{pmatrix} -(A_{\xi_z} - A_\xi)\mathbf{u}(\boldsymbol{\zeta}_z) \\ h_{\xi_z} - h_\xi \end{pmatrix}.$$

We now show that

$$L_z \to L_0 := \begin{pmatrix} -\partial_m A_\xi \mathbf{u}(\boldsymbol{\zeta}) \\ \partial_m h_\xi \end{pmatrix} \qquad \text{as } z \to 0,$$

where $[\partial_m A_\xi \mathbf{u}](\mathbf{v}) := (\partial_m a^{-1}(\boldsymbol{\zeta})\mathbf{u}, \mathbf{v})$ and $\partial_m h_\xi(q) := (\partial_m f(\boldsymbol{\zeta}), q)$. Note first that there holds

$$\lim_{h \to 0} \left\| \frac{h_{\xi + z\mathbf{e}_m} - h_\xi}{z} - \partial_m h_\xi \right\|_{L^2(D)^*} = 0,$$

which can be easily seen by applying the Cauchy-Schwarz inequality and the assumption about the analyticity of f. Moreover, we have

$$\begin{aligned} \left\| \frac{A_{\xi_z} - A_\xi}{z} \mathbf{u}(\boldsymbol{\zeta}_z) - \partial_m A_\xi \mathbf{u}(\boldsymbol{\zeta}) \right\|_{H(\mathrm{div};D)^*} &\le \left\| \frac{A_{\xi_z} - A_\xi}{z} \right\| \|\mathbf{u}(\boldsymbol{\zeta}_z) - \mathbf{u}(\boldsymbol{\zeta})\|_{H(\mathrm{div};D)} \\ &+ \left\| \frac{A_{\xi_z} - A_\xi}{z} - \partial_m A_\xi \right\| \|\mathbf{u}(\boldsymbol{\zeta})\|_{H(\mathrm{div};D)}. \end{aligned}$$

There holds

$$\lim_{z \to 0} \|\mathbf{u}(\boldsymbol{\zeta} + z\mathbf{e}_m) - \mathbf{u}(\boldsymbol{\zeta})\| = 0,$$

since $\mathbf{u}(\boldsymbol{\zeta})$ depends continuously on $\boldsymbol{\zeta}$ as shown before. Furthermore, there holds

$$\frac{\left|\int_D \frac{a^{-1}(\zeta_z)-a^{-1}(\zeta)}{z}\mathbf{u}\cdot\mathbf{v}\,\mathrm{d}\mathbf{x}\right|}{\|\mathbf{u}\|_{H(\mathrm{div};D)}\,\|\mathbf{v}\|_{H(\mathrm{div};D)}} \le \left\|\frac{a^{-1}(\zeta_z)-a^{-1}(\zeta)}{z}\right\|_{L^\infty(D)} \to \|\partial_m a^{-1}(\zeta)\|_{L^\infty(D)}$$

as $z \to 0$ due to the analyticity of a^{-1}. Thus, we have

$$\|(A_{\zeta_z} - A_\zeta)/z\| \le \|(a^{-1}(\zeta_z) - a^{-1}(\zeta))/z\|_{L^\infty(D)} \to \|\partial_m a^{-1}(\zeta)\|_{L^\infty(D)}$$

as $z \to 0$. By linearity we obtain with the same argument

$$\left\|\frac{A_{\zeta_z} - A_\zeta}{z} - \partial_m A_\zeta\right\| \le \left\|\frac{a^{-1}(\zeta_z) - a^{-1}(\zeta)}{z} - \partial_m a^{-1}(\zeta)\right\|_{L^\infty(D)} \to 0$$

as $z \to 0$, which finally yields $L_z \to L_0$ as $z \to 0$. Again, by the continuous dependence of the solution of (12) on the right hand side, we conclude

$$(q_z, \mathbf{v}_z) \to (\partial_m p(\zeta), \partial_m \mathbf{u}(\zeta)) \quad \text{as } z \to 0,$$

where $(\partial_m p(\zeta), \partial_m \mathbf{u}(\zeta))$ solves (12) for the right hand side L_0. We have thus established that $(p(\zeta), \mathbf{u}(\zeta))$ possesses the partial derivative $(\partial_m p(\zeta), \partial_m \mathbf{u}(\zeta))$ in the m-th (complex) direction, which completes the proof. □

Example 2 (lognormal diffusion coefficient). We consider a coefficient

$$a(\mathbf{x}, \zeta) = \exp\left(\phi_0(\mathbf{x}) + \sum_{m=1}^{M} \phi_m(\mathbf{x})\zeta_m\right)$$

with (real-valued) $\phi_m \in L^\infty(D)$ for $m = 0, \dots, M$. Let ρ be the M-dimensional standard normal probability density function. Setting $b_m := \|\phi_m\|_{L^\infty(D)}$ for $m = 0, \dots, M$, then for all $\zeta \in \Sigma$ with

$$\Sigma = \left\{\zeta \in \mathbb{C}^M : \sum_{m=1}^{M} b_m\,|\mathrm{Im}\,\zeta_m| < \frac{\pi}{2}\right\} \tag{13}$$

there holds

$$a_{\min}(\zeta) \ge \exp\Big(-b_0 - \sum_{m=1}^{M} b_m|\xi_m|\Big)\,\cos\Big(\sum_{m=1}^{M} b_m|\eta_m|\Big) > 0,$$

$$a_{\max}(\zeta) \le \exp\Big(b_0 + \sum_{m=1}^{M} b_m|\xi_m|\Big),$$

where $\boldsymbol{\zeta} = \boldsymbol{\xi} + i\boldsymbol{\eta}$. Furthermore, a then satisfies the assumptions of Lemma 3 for Σ as given in (13) and the weighting function $\sigma(\boldsymbol{\xi}) = \sigma_1(\xi_1)\cdots\sigma_M(\xi_M)$, where $\sigma_m(\xi_m) = \exp\big(-b_m|\xi_m|\big)$, $m = 1,\ldots,M$.

Remark 2. Note that if, in Example 2, the expansion functions $\{\phi_m\}_{m=1}^M$ in addition have disjoint supports, then a satisfies the assumptions of Lemma 3 for the larger domain

$$\Sigma = \{\boldsymbol{\zeta} \in \mathbb{C}^M : b_m \,|\,\mathrm{Im}\,\zeta_m| < \pi/2\},$$

since then

$$\begin{aligned}\mathrm{Re}\, a(\mathbf{x},\boldsymbol{\zeta}) &= \exp\Big(\phi_0(\mathbf{x}) + \sum_{m=1}^{M}\phi_m(\mathbf{x})\xi_m\Big)\cos\Big(\sum_{m=1}^{M}\phi_m(\mathbf{x})\eta_m\Big)\\ &\geq \exp\Big(-b_0 - \Big(\max_m b_m|\xi_m|\Big)\Big)\cos\Big(\max_m b_m|\eta_m|\Big).\end{aligned}$$

4 Sparse Grid Collocation

Stochastic collocation in the context of UQ or parametric problems can be described roughly as a method for approximating a function $u : \mathbb{R}^M \to W$ with values in, say, a separable Banach space W from the span of n linearly independent functions $\{u_j : \mathbb{R}^M \to W\}_{j=1}^n$ given only the values of u at certain distinct points in the parameter domain $\mathbb{R}^M$. Suitable finite-dimensional function spaces are determined by the smoothness of u as a function of the parameter. Since the solution of (11) depends smoothly on $\boldsymbol{\xi}$, as was shown in the previous section, we consider approximations by global interpolating polynomials as done in, e.g., [2,3,6,18,19,23].

Therefore, let $\chi_k = \{\xi_{k,1},\ldots,\xi_{k,n_k}\}$, $k = 1,2,\ldots$, be a given sequence of node sets in $\mathbb{R}$ and

$$(\mathcal{I}_k v)(\xi) := \sum_{j=1}^{n_k} v\left(\xi_{k,j}\right)\ell_{k,j}(\xi)$$

be the associated Lagrange interpolation operator with the Lagrange basis polynomials $\ell_{k,j}$. We further define the difference operators $\Delta_k = \mathcal{I}_k - \mathcal{I}_{k-1}$ for $k \geq 1$, where $\mathcal{I}_0 := 0$. Then the (Smolyak) sparse grid stochastic collocation operator is defined as

$$\mathcal{A}_{q,M} := \sum_{|\mathbf{i}|\leq q+M}\Delta_{i_1}\otimes\cdots\otimes\Delta_{i_M} = \sum_{q+1\leq|\mathbf{i}|\leq q+M} c_{q,M}(\mathbf{i})\,\mathcal{I}_{i_1}\otimes\cdots\otimes\mathcal{I}_{i_M}, \quad (14)$$

where $|\mathbf{i}| := i_1 + \ldots + i_M$ and

$$c_{q,M}(\mathbf{i}) = (-1)^{q+M-|\mathbf{i}|} \binom{M-1}{q+M-|\mathbf{i}|},$$

cf. [22]. The sparse grid associated with $\mathcal{A}_{q,M}$ consists of the points

$$\mathcal{H}_{q,M} := \bigcup_{q+1 \le |\mathbf{i}| \le q+M} \chi_{i_1} \times \cdots \times \chi_{i_M} \subset \mathbb{R}^M. \tag{15}$$

One may choose $\mathcal{X}_k$ to be the roots of the n_k-th Hermite polynomial (w.r.t. to the weight $\rho_m(\xi) = e^{-\xi^2/2}/\sqrt{2\pi}$), since these nodal points yield maximally convergent interpolations (cf. [21]) and this choice also simplifies the computation of moments of $\mathcal{A}_{q,M}u$ w.r.t. the weight $\rho = \prod_m \rho_m$.

For bounded parameter domains and constant density $\rho_m \equiv$ const, popular sequences of nodal sets are Gauss-Legendre and Clenshaw-Curtis nodes. For sparse grid collocation based on these sequences a convergence analysis is given in [19], where it is indicated that a similar analysis applies to Gauss-Hermite nodes. We carry out this analysis in the following.

Assumption 4. *There exist constants $c > 0$ and $\varepsilon_m > 0$, $m = 1, \ldots, M$, such that*

$$\rho_m(\xi_m) = \frac{\exp(-\xi_m^2/2)}{\sqrt{2\pi}} \le c \exp(-\varepsilon_m \xi_m^2)\sigma_m^2(\xi_m), \qquad m = 1, \ldots, M, \tag{16}$$

and the weighting function has the product structure $\sigma(\boldsymbol{\xi}) = \prod_{m=1}^M \sigma_m(\xi_m)$.

Note that Assumption 4 implies that $C_\sigma(\mathbb{R}^M; W)$ is continuously embedded in $L^2_\rho(\mathbb{R}^M; W)$, since for $v \in C_\sigma(\mathbb{R}^M; W)$ there holds

$$\begin{aligned}\int_{\mathbb{R}^M} \|v(\boldsymbol{\xi})\|_W^2 \, \rho(\boldsymbol{\xi}) \, \mathrm{d}\boldsymbol{\xi} &\le \|v\|^2_{C_\sigma(\mathbb{R}^M;W)} \int_{\mathbb{R}^M} \frac{\rho(\boldsymbol{\xi})}{\sigma^2(\boldsymbol{\xi})} \, \mathrm{d}\boldsymbol{\xi} \\ &\le c\|v\|^2_{C_\sigma(\mathbb{R}^M;W)} \prod_{m=1}^M \int_{\mathbb{R}^M{}_m} \exp(-\varepsilon_m \xi_m^2) \, \mathrm{d}\boldsymbol{\xi} < \infty.\end{aligned}$$

The same is true of the restrictions of functions in $C_\sigma(\Sigma; W)$, since $\mathbb{R}^M \subset \Sigma \subset \mathbb{C}^M$ due to Assumption 2.

Theorem 1 (cf. [19, Theorem 3.18]). *Let W be a separable Banach space, let $u : \mathbb{R}^M \to W$ admit an analytic extension to the domain*

$$\Sigma_\tau = \{\boldsymbol{\zeta} \in \mathbb{C}^M : |\operatorname{Im} \zeta_m| \le \tau_m, \, m = 1, \ldots, M\}, \qquad \boldsymbol{\tau} = (\tau_1, \ldots, \tau_M),$$

and, in addition, $u \in C_\sigma(\Sigma_\tau; W)$, *i.e.,*

$$\max_{\zeta \in \Sigma_\tau} \sigma(\operatorname{Re} \zeta) \, \|u(\zeta)\|_W < +\infty,$$

where $\rho(\xi) = \prod_m \rho_m(\xi_m)$ *and* $\sigma(\xi) = \prod_m \sigma_m(\xi_m)$ *satisfy Assumption 4. Then the error of the sparse grid collocation approximation* $\mathcal{A}_{q,M} u$ *based on Gauss-Hermite nodes* χ_k *where*

$$|\chi_k| = n_k = \begin{cases} 1, & k = 1, \\ 2^{k-1} + 1, & k > 1, \end{cases}$$

can be bounded by

$$\|u - \mathcal{A}_{q,M} u\|_{L^2_\rho} \le \frac{C_r^{M+1} - C_r}{C_r - 1} \begin{cases} \exp\left(-q \frac{\log 2}{2} \left(R \frac{e}{\sqrt{2}} - 1\right)\right), & \text{if } 0 \le q \le \frac{2M}{\log 2}, \\ \exp\left(-R \frac{M}{\sqrt{2}} \sqrt{2^{q/M}} + \frac{q}{2} \log 2\right), & \text{otherwise,} \end{cases} \tag{17}$$

where $C_r = C(2 + \sqrt{8\pi/r}/\log 2)$ *and*

$$r := \min_{m=1,\dots,M} \tau_m, \qquad R := \sqrt[M]{\tau_1 \cdots \tau_M}.$$

In particular, for $0 \le q \le \frac{2M}{\log 2}$ *there holds*

$$\|u - \mathcal{A}_{q,M} u\|_{L^2_\rho} \le \tilde{C}(r, R, M) \, N^{-\nu_1}, \qquad \nu_1 = \frac{\log 2}{2(2.1 + \log M)} \left(\frac{eR}{\sqrt{2}} - 1\right), \tag{18}$$

where $N = |\mathcal{H}_{q,M}|$ *and* $\tilde{C}(r, R, M) = C(r) \frac{1 - C(r)^M}{1 - C(r)} \sqrt{2^{eR/\sqrt{2} - 1}}$.
Conversely, for $q > \frac{2M}{\log 2}$ *there holds*

$$\|u - \mathcal{A}_{q,M} u\|_{L^2_\rho} \le \frac{C_r^{M+1} - C_r}{C_r - 1} \frac{N^2}{M^2} e^{-\frac{R}{\sqrt{2}} M N^{\nu_2}}, \qquad \nu_2 = \frac{\log 2}{2M(2.1 + \log M)}. \tag{19}$$

Proof. The proof follows closely the procedure for showing convergence of $\mathcal{A}_{q,M}$ w.r.t. Clenshaw-Curtis nodes given in [19]. Since only certain steps need to be modified we only mention these here and refer to [19] for further details.

Step 1: Show $\|u - \mathcal{A}_{q,M}u\|_{L^2_\rho} \le \sum_{k=1}^{M} R(q,k)$.
According to the proof of [19, Lemma 3.4], there holds

$$I - \mathcal{A}_{q,M} = \sum_{k=2}^{M}\left[\tilde{R}(q,k) \bigotimes_{m=k+1}^{M} I\right] + \left(I - \mathcal{A}_{q,1}\right)\bigotimes_{m=2}^{M} I,$$

where

$$\tilde{R}(q,k) = \sum_{\substack{\mathbf{i}\in\mathbb{N}^{k-1}\\ |\mathbf{i}|\le q+k-1}} \bigotimes_{m=1}^{k-1} \Delta_{i_m} \otimes \left(I - \mathcal{I}_{\hat{i}_k}\right)$$

and $\hat{i}_k = 1 + q - \sum_{m=1}^{k-1}(i_m - 1)$. Further, the term $\tilde{R}(q,k)$ can bounded using the results given in the Appendix:

$$\begin{aligned}
\|\tilde{R}(q,k)u\|_{L^2_\rho} &\le \sum_{\substack{\mathbf{i}\in\mathbb{N}^{k-1}\\ |\mathbf{i}|\le q+k-1}} \left\| \bigotimes_{m=1}^{k-1} \Delta_{i_m} \otimes \left(I - \mathcal{I}_{\hat{i}_k}\right) u \right\|_{L^2_\rho} \\
&\le \sum_{\substack{\mathbf{i}\in\mathbb{N}^{k-1}\\ |\mathbf{i}|\le q+k-1}} C^k \left(\sqrt{2^{\hat{i}_k}}\right) e^{-\frac{1}{2}\left(\sum_{m=1}^{k-1}\tau_m\sqrt{2^{i_m}} + \tau_k\sqrt{2^{\hat{i}_k+1}}\right)} \prod_{m=1}^{k-1}\left(\sqrt{2^{i_m}} + 1\right) \\
&= C^k \sum_{\substack{\mathbf{i}\in\mathbb{N}^{k}\\ |\mathbf{i}|= q+k}} \exp\left(-\frac{1}{2}h(\mathbf{i},k)\right),
\end{aligned}$$

where $h(\mathbf{i},k) = \sum_{m=1}^{k} \tau_m\sqrt{2^{i_m}} - (\log 2) i_m$. Moreover, we obtain by applying results from [2, Sect. 4]

$$\begin{aligned}
\|\left(I - \mathcal{A}_{q,1}\right)u\|_{L^2_\rho} = \|(I - \mathcal{I}_{n_{q+1}})u\|_{L^2_\rho} &\le C\left(\sqrt{2^{q+1}}\right)\exp\left(-\frac{\tau_1}{\sqrt{2}}\sqrt{2^{q+1}}\right) \\
&= \sum_{\substack{i\in\mathbb{N}^{1}\\ |i|= q+1}} C\left(\sqrt{2^{i}}\right)\exp\left(-\frac{\tau_1}{\sqrt{2}}\sqrt{2^{i}}\right).
\end{aligned}$$

Therefore, setting

$$R(q,k) := C^k \sum_{\substack{\mathbf{i}\in\mathbb{N}^{k}\\ |\mathbf{i}|= q+k}} \exp\left(-\frac{1}{2}h(\mathbf{i},k)\right),$$

we arrive at the bound $\left\|\left(I - \mathcal{A}_{q,M}\right)u\right\|_{L^2_\rho} \le \sum_{k=1}^{M} R(q,k)$.

Step 2: Estimate $R(q,k)$.
Computing the minimum of $h(\cdot,k)$ on the set $\{\mathbf{x} \in \mathbb{R}^k : x_1 + \cdots + x_k = q + k\}$ yields the optimal point $\mathbf{i}^* = (i_1^*, \ldots, i_k^*)$ with

$$i_m^* = 1 + q/k + \frac{2}{k} \sum_{n=1}^{k} \log_2(\tau_n/\tau_m), \qquad m = 1, \ldots, k.$$

Moreover, expanding $h(\cdot,k)$ at $\mathbf{i}^*$ up to second order yields for any $\mathbf{i} \in \mathbb{N}^k$ with $|\mathbf{i}| = q + k$

$$\begin{aligned} h(\mathbf{i},k) &= h(\mathbf{i}^*,k) + \underbrace{\nabla h(\mathbf{i}^*,k) \cdot (\mathbf{i} - \mathbf{i}^*)^T}_{=0} + \frac{1}{2} (\mathbf{i} - \mathbf{i}^*) \cdot \nabla^2 h(\iota,k) \cdot (\mathbf{i} - \mathbf{i}^*)^T \\ &= k 2^{(q+k)/(2k)} \prod_{m=1}^{k} \sqrt[k]{\tau_m} - (\log 2)(q+k) + \frac{1}{2} \sum_{m=1}^{k} \tau_m \frac{(\log 2)^2}{4} 2^{\iota_m/2} (i_m - i_m^*)^2 \\ &\geq k 2^{(q+k)/(2k)} \prod_{m=1}^{k} \sqrt[k]{\tau_m} - (\log 2)(q+k) + r \frac{(\log 2)^2}{8} \sum_{m=1}^{k} (i_m - i_m^*)^2, \end{aligned}$$

where $\iota_m \in [\min(i_m, i_m^*), \max(i_m, i_m^*)]$ for $m = 1, \ldots, M$.

Without loss of generality we may assume that $\tau_1 \geq \tau_2 \geq \ldots \geq \tau_M$. Thus, we have for any $k = 1, \ldots, M$

$$\prod_{m=1}^{k} \sqrt[k]{\tau_m} \geq \prod_{m=1}^{M} \sqrt[M]{\tau_m} =: \sqrt[M]{\boldsymbol{\tau}}$$

and there holds furthermore

$$\begin{aligned} R(q,k) &\leq C^k \exp\left(\frac{q}{2} \log 2 - k \frac{\sqrt[M]{\boldsymbol{\tau}}}{2} 2^{(q+k)/(2k)} \right) \sum_{\substack{\mathbf{i} \in \mathbb{N}^k \\ |\mathbf{i}| = q+k}} \prod_{m=1}^{k} e^{r \log^2 2/8 (i_m - i_m^*)^2} \\ &\leq C^k \exp\left(\frac{q}{2} \log 2 - k \frac{\sqrt[M]{\boldsymbol{\tau}}}{2} 2^{(q+k)/(2k)} \right) \prod_{m=1}^{k} \sum_{i=1}^{q+1} e^{r \log^2 2/8 (i - i_m^*)^2} \\ &\leq C^k \exp\left(\frac{q}{2} \log 2 - k \frac{\sqrt[M]{\boldsymbol{\tau}}}{2} 2^{(q+k)/(2k)} \right) \left(2 + \sqrt{\frac{8\pi}{r \log^2 2}} \right)^k \\ &= C_r^k \exp\left(\frac{q}{2} \log 2 - k \frac{\sqrt[M]{\boldsymbol{\tau}}}{2} 2^{(q+k)/(2k)} \right), \end{aligned}$$

where we have used $\{\mathbf{i} \in \mathbb{N}^k : |\mathbf{i}| = q + k\} \subset \{\mathbf{i} \in \mathbb{N}^k : |\mathbf{i}| \leq q + k\}$ in the second and [19, Lemma A.1] in the next-to-last last line.

Step 3: Combine Previous Steps
The remaining steps are analogous to the proof of [19, Theorem 3.7] and [19, Theorem 3.10], respectively, using the bound for $N = |\mathcal{H}_{q,M}|$ from [19, Lemma 3.17]

$$\frac{\log N}{2.1 + \log M} - 1 \leq q \leq \log_2(N/M - 1).$$

□

Remark 3. Note that Theorem 1 states algebraic convergence of $\mathcal{A}_{q,M}u$ w.r.t. the number of collocation nodes N in the regime $q \leq 2M/\log 2$ and subexponential convergence in the regime $q > 2M/\log 2$. Typically, in applications with $M \geq 3$ a level $q > 2M/\log 2$ is seldom feasible.

Remark 4. Note, that our proof takes into account different widths τ_m of the strips of analyticity for different dimensions ξ_m in contrast to the corresponding proofs of [19, Lemmas 3.4 and 3.16]. Moreover, we would like to mention that the proofs in [19], in particular the estimates of the term $\|\tilde{R}(q,k)u\|_{L^2_\rho}$ given there, require also that u possesses an analytic continuation to a product subdomain $\prod_{m=1}^M \Sigma_0$ of $\mathbb{C}^M$, $\Sigma_0 \subset \mathbb{C}$. This condition is, however, never explicitly assumed or shown to hold in [19]. Rather, the authors only state one-dimensional regularity results, i.e., results on the domain of analytic continuation of u w.r.t. each ζ_m, $m = 1, \ldots, M$, separately, with the remaining coordinates $\xi_n \in \mathbb{R}$, $n \neq m$, kept fixed and *real*. However, this type of one-dimensional regularity is not sufficient for concluding analyticity of u in a product domain in $\mathbb{C}^M$. As we have seen in the proof of Lemma 3, the results on the one-dimensional complex domain of analytic continuation of u w.r.t. ζ_m, $m = 1, \ldots, M$, need to hold for all fixed, *complex* coordinates $\zeta_n \in \Sigma_0$, $n \neq m$.

Combining the result above with our investigations of the previous section, we conclude

Corollary 1 (Convergence in Case of Lognormal Diffusion). *Let a problem* (11) *with a diffusion coefficient of the form*

$$a(\mathbf{x}, \boldsymbol{\xi}) = \exp\left(\phi_0(\mathbf{x}) + \sum_{m=1}^{M} \phi_m(\mathbf{x})\xi_m\right)$$

be given and let the assumption of Lemma 3 be satisfied. Then there holds for $0 \leq q \leq \frac{2M}{\log 2}$ *and* $N = |\mathcal{H}_{q,M}|$

$$\|u - \mathcal{A}_{q,M}u\|_{L^2_\rho} \leq \tilde{C}(r, R, M)\, N^{-\frac{\log 2}{2(2.1+\log M)}\left(\frac{eR}{\sqrt{2}}-1\right)}$$

where $\tilde{C}(r, R, M)$ *is according to Theorem 1 and where*

$$R \geq \frac{\pi - \varepsilon}{2M\left(\|\phi_1\|_{L^\infty(D)} \cdots \|\phi_M\|_{L^\infty(D)}\right)^{1/M}}$$

for any $\varepsilon > 0$*.*

Proof. Given the statement of Theorem 1 and the observations made in Example 2, we simply maximize $\sqrt[M]{\tau_1 \cdots \tau_M}$ under the constraint

$$\sum_{m=1}^{M} \tau_m \|\phi_m\|_{L^\infty(D)} = \frac{\pi - \varepsilon}{2}$$

for an arbitrary $\varepsilon > 0$. This yields the optimal point

$$\tau_m^* = \frac{\pi - \varepsilon}{2M\,\|\phi_m\|_{L^\infty(D)}}, \qquad m = 1, \ldots, M,$$

and, furthermore, $\sqrt[M]{\tau_1^* \cdots \tau_M^*} = (\pi - \varepsilon)/\left(2M(\|\phi_1\|_{L^\infty(D)} \cdots \|\phi_M\|_{L^\infty(D)})^{1/M}\right)$. □

5 Numerical Example

We illustrate the theoretical results of the previous sections for a simple elliptic boundary value problem in mixed form as arises, e.g., in groundwater flow modeled by Darcy's law. In addition, we examine how the convergence of stochastic collocation is affected by properties of the lognormal diffusion coefficient a. Although the results given in the previous section are valid for a general random field with a representation of the form given in (3), we wish to relate common properties of Gaussian random fields such as their mean, variance etc. to the convergence of the stochastic collocation method.

The Gaussian random field $\log a$ is uniquely determined by its mean and covariance functions. The mean ϕ_0 of $\log a$ does not affect the convergence of the stochastic collocation approximation as Corollary 1 shows, but the covariance plays a more important role, since it determines the representation (3). Generally speaking, covariance functions are characterized by a variance parameter, correlation length and its degree of smoothness. The latter may also be expressed in terms of a parameter, as is the case for the Matérn family of covariance functions (see, e.g., [7]). However, since the smoothness of the covariance function controls the asymptotic decay of the eigenvalues of the associated covariance operator and the correlation length determines the length of a preasymptotic plateau preceding the asymptotic decay, both will affect the length M of a truncated Karhunen-Loève expansion with sufficiently small truncation error. Hence, by relating the smoothness and the correlation length to M, we will illustrate the effect of increasing M and increasing σ on the convergence of the stochastic collocation approximations in the following.

A Simple Groundwater Flow Model

The PDE under consideration is of the form (1) with source term $f \equiv 0$, boundary data $g(x_1, x_2) = 3\left(x_1^2 + (1 - x_2)^2\right)^{1/2}$, and lognormal coefficient a on the unit square $D = [0,1]^2$ in $\mathbb{R}^2$. In particular, we assume for the Gaussian random field $\log a$ a mean $\phi_0(\mathbf{x}) \equiv 1$ and a stationary and isotropic two-point covariance function given by

$$\operatorname{Cov}(\log a(\mathbf{x}), \log a(\mathbf{y})) = \sigma^2 \exp\left(-\|\mathbf{x} - \mathbf{y}\|^2\right).$$

Thus, the approximation $\log a_M(\mathbf{x}, \boldsymbol{\xi})$ is the truncated KLE of this Gaussian random field, i.e.,

$$a_M(\mathbf{x}, \boldsymbol{\xi}) = \exp\left(1 + \sum_{m=1}^{M} \phi_m(\mathbf{x})\, \xi_m\right), \tag{20}$$

where $\phi_m(\mathbf{x}) = \sigma^2 \sqrt{\lambda_m} \psi_m(\mathbf{x})$ and $\{(\lambda_m, \psi_m)\}_{m=1,\dots,M}$ are the first M eigenpairs (in order of decreasing eigenvalues) of

$$\lambda \psi(\mathbf{x}) = \int_{[0,1]^2} \exp\left(-\|\mathbf{x} - \mathbf{y}\|^2\right) \psi(\mathbf{y})\, \mathrm{d}\mathbf{y}.$$

Figure 1 displays the exponential decay of the eigenvalues λ_m and the norms $\|\phi_m\|_{L^\infty(D)}$ of the corresponding lognormal diffusion coefficient.

Remark 5. Note that, in geophysical applications such as hydrogeology, one usually encounters "rougher" random fields with a covariance function, e.g., of Matérn type, see [7]. However, the above model for $\log a$ is sufficient for our purpose of illustrating how the convergence of stochastic collocation depends on M and σ as explained above.

Note that, as the variance parameter σ of the random field $\log a$ increases, so does the L^∞-norm of the expansion functions ϕ_m, and therefore the rate of convergence for the stochastic collocation should decrease according to Corollary 1. We will demonstrate this in the following.

For the spatial discretization we use Raviart-Thomas finite elements of lowest order [5] for the flux and piecewise constants for the head variable. Thus, $p(\cdot, \boldsymbol{\xi})$ is approximated as a piecewise constant and $\mathbf{u}(\cdot, \boldsymbol{\xi})$ as a piecewise linear function.

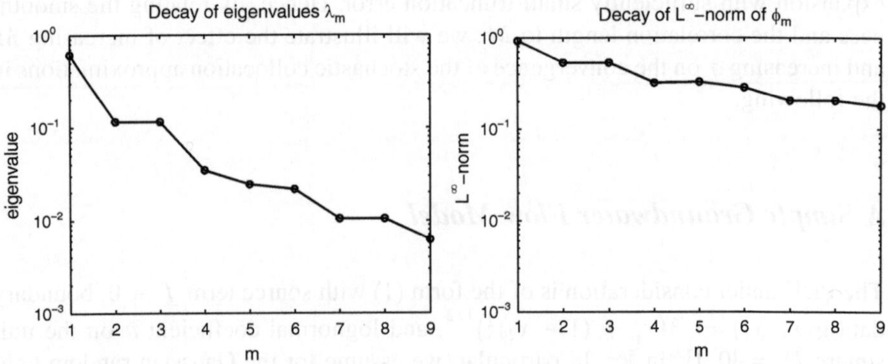

Fig. 1 *Left*: decay of the eigenvalues of the covariance operators associated with $\log a$. *Right*: decay of the $L^\infty(D)$-norm of the ϕ_m in (20)

Moreover, the domain D is decomposed into 4,206 triangles resulting in 10,595 spatial degrees of freedom. Hence, the space $\mathcal{V} = L^2(D) \times H(\text{div}; D)$ is replaced by the cartesian product $\mathcal{V}_h \subset \mathcal{V}$ of the finite dimensional approximation spaces and the continuous solution pair $(p, \mathbf{u})$ by the semidiscrete pair $(p_h, \mathbf{u}_h)$. Note that this does not influence the analysis of the previous sections, we merely apply the statements of Lemma 3 and Theorem 1 to the finite-dimensional subspaces. The full—i.e., collocation and finite element approximation—error can be obtained by appealing to standard finite element approximation theory, (cf. e.g., [1,20]).

Solution and Convergence

In Fig. 2 we show for illustration the computational domain D, the triangular mesh and the mean head (left) and streamlines of the mean flux (right) of the solution obtained by sparse grid collocation with level $q = 5$ for a truncated KLE containing $M = 9$ terms.

We observe (at least algebraic) convergence of the stochastic collocation approximation for head and flux in the left plot in Fig. 3. Here and in the following we estimate the $L^2(\mathbb{R}^M; W)$-error of the stochastic collocation approximations by a sparse quadrature method applied to the error $\|p_h(\boldsymbol{\xi}) - \mathcal{A}_{q,M}\, p_h(\boldsymbol{\xi})\|_{L^2(D)}$ and $\|\mathbf{u}_h(\boldsymbol{\xi}) - \mathcal{A}_{q,M}\mathbf{u}_h(\boldsymbol{\xi})\|_{H(\text{div};D)}$, respectively. We have chosen the sparse Smolyak quadrature operator corresponding to a stochastic collocation approximation of a high level q^*, i.e.,

$$\mathbb{E}\left[\|p_h - \mathcal{A}_{q,M}\, p_h\|^2_{L^2(D)}\right] \approx \sum_{\boldsymbol{\xi}_j \in \mathcal{H}_{q^*,M}} w_j \, \|p_h(\boldsymbol{\xi}_j) - \mathcal{A}_{q,M}\, p_h(\boldsymbol{\xi}_j)\|^2_{L^2(D)},$$

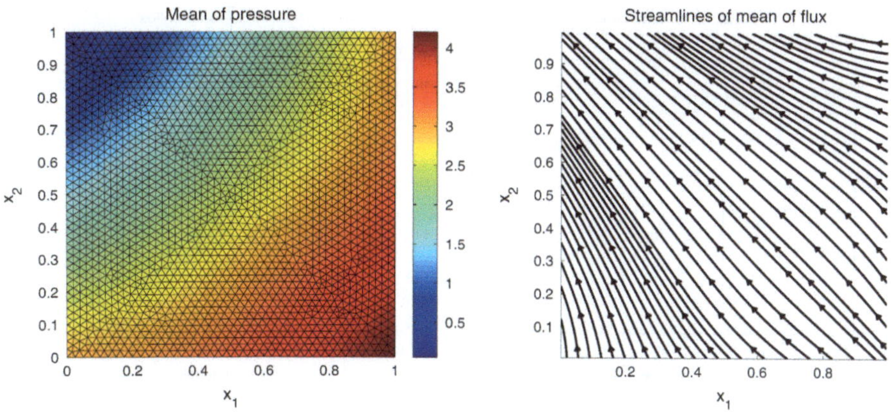

Fig. 2 *Left*: the mean of the pressure head approximation $\mathcal{A}_{5,9}p_h$; *right*: streamlines of the mean of the flux approximation $\mathcal{A}_{5,9}\mathbf{u}_h$

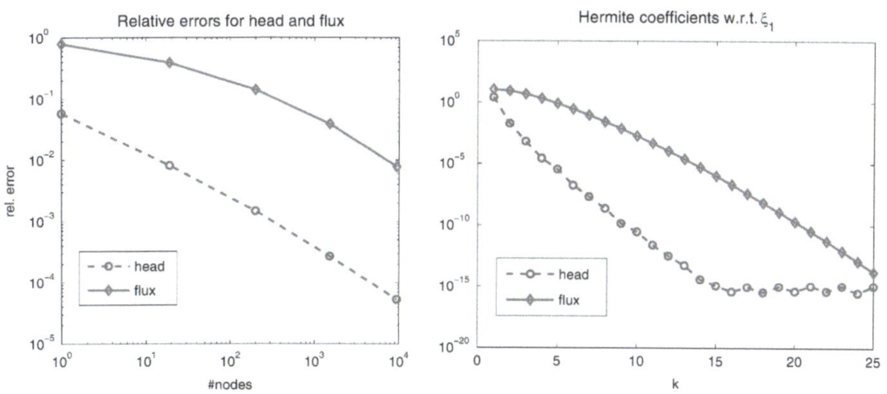

Fig. 3 *Left*: relative errors for head and flux for $M = 9$ and $q = 0, \ldots, 4$. *Right*: estimated Hermite coefficients of p_h and $\mathbf{u}_h$ for the first 25 Hermite polynomials w.r.t. ξ_1

where $\{w_j : \boldsymbol{\xi}_j \in \mathcal{H}_{q^*,M}\}$ are the weights of the sparse quadrature operator associated with $\mathcal{A}_{q^*,M}$. For the results shown in Fig. 3 we used $q^* = 5$. Note that we have also applied a Monte Carlo integration for the error estimation above for comparison which showed no substantial difference to the quadrature procedure above. The error estimation for $\mathcal{A}_{q,M}\mathbf{u}_h$ was obtained in the same way. We observe that the relative error for the flux does not immediately decay at the asymptotic rate. This is due to a preasymptotic phase of slower decay of the Hermite coefficients of $\mathbf{u}_h$. We display the Hermite coefficients for the first 25 Hermite polynomials in ξ_1 for p_h and $\mathbf{u}_h$ on the right hand side of Fig. 3. The preasymptotic slow decay of the coefficients in case of the flux is clearly visible. However, both errors apparently decay at a much greater rate than the estimate in Corollary 1 would suggest.

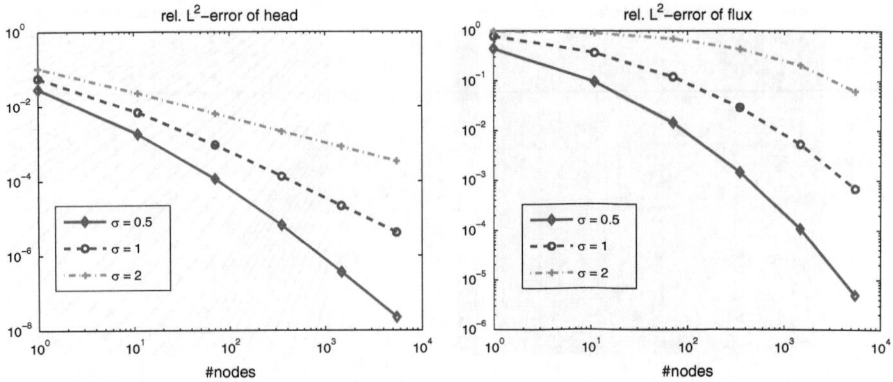

Fig. 4 Estimated relative $L^2(\mathbb{R}^M; W)$ errors of the sparse grid stochastic collocation approximations for pressure head p_h and flux $\mathbf{u}_h$ for $M = 5$ but different values of σ. The level q of $\mathcal{A}_{q,5}$ varies from 0 to 5

Influence of the Input Variance σ

We fix $M = 5$ and vary the variance parameter $\sigma \in \{1/2, 1, 2\}$. For all three values of σ we choose a quadrature level of $q^* = 6$ for the error estimation. The results are shown in Fig. 4. We observe the expected behaviour that for increased σ the convergence rate is reduced.

Influence of the Parameter Dimension M

We set $\sigma = 1$ and let $M \in \{3, 6, 9\}$. As quadrature levels for the error estimation we choose $q^* = 8$ for $M = 3$, $q^* = 7$ for $M = 6$, and $q^* = 6$ for $M = 9$. The results are shown in Fig. 5. Again, the results are according to the conjecture that for increased dimension M the convergence rate decreases.

Remark 6. In view of the decelerating effect of large variance σ and roughness of a random field a (requiring large M for small truncation error) on the convergence rate of stochastic collocation, certain advanced Monte Carlo methods (such as quasi- or multilevel Monte Carlo) might be preferable for certain applications in subsurface physics where such rough random fields of high variance are common. We refer to the results in [7] for a comparison of the Monte Carlo and stochastic collocation method in case of a real-world subsurface flow problem. However, while efficient for estimating moments, probabilities or other statistical properties (so-called *quantities of interest*), Monte Carlo methods do not yield an approximate solution function of the PDE problem with random data as does stochastic collocation, which may serve as a cheap, sufficiently accurate surrogate model in many situations.

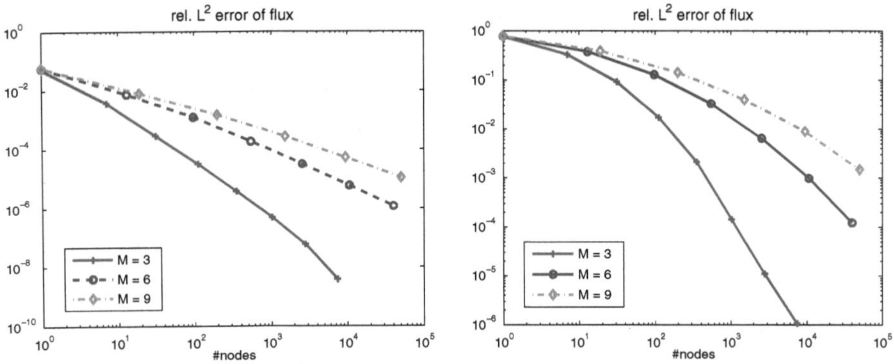

Fig. 5 Estimated relative $L^2(\mathbb{R}^M; W)$ errors of the sparse grid stochastic collocation approximations for pressure head p_h and flux $\mathbf{u}_h$ for $\sigma = 1$ but different values of M. The level q of $\mathcal{A}_{q,M}$ varies from 0 to 7 for $M = 3$, from 0 to 6 for $M = 6$ and from 0 to 5 for $M = 9$

6 Conclusions

In this paper we have filled some remaining theoretical gaps for the application of sparse grid stochastic collocation to diffusion equations with a random, lognormally distributed diffusion coefficient. In particular, we have shown the smooth dependence of the solution of the associated parametric variational problems on the parameter under natural assumptions. This extends previous work [4] on random mixed elliptic problems to a broader and practically relevant class of diffusion coefficients. In addition, we have given a complete convergence proof for sparse grid stochastic collocation using basic random variables with unbounded supports, which was previously only hinted at in the literature as a remark [19]. Both results combine to form the theoretical foundation for applying stochastic collocation to interesting real world problems [7, 12] which we have illustrated for a simple groundwater flow model. The qualitative behavior of the approximation bounds indicate the limitations of stochastic collocation when applied to problems with diffusion coefficients displaying roughness or short correlation length.

Acknowledgements The authors gratefully acknowledge financial support from the DFG Priority Programme 1324 as well as helpful discussions with Lorenzo Tamellini, Fabio Nobile and Alexei Bespalov.

Appendix

We will here prove some auxiliary results used in the previous sections. In particular, we want to generalize some results from [2, Sect. 4] to multi-dimensional interpolation. The first concerns the uniform boundedness of the operator

$$\mathcal{I}_{k_1} \otimes \cdots \otimes \mathcal{I}_{k_M} : C_\sigma(\mathbb{R}^M; W) \to L^2_\rho(\mathbb{R}^M; W),$$

where σ and ρ are according to (16). It can be shown by an obvious generalization of [2, Lemma 4.2] that

$$\|\mathcal{I}_{k_1} \otimes \cdots \otimes \mathcal{I}_{k_M} v\|_{L^2_\rho(\mathbb{R}^M;W)} \le C(\rho,\sigma)\, \|v\|_{C_\sigma(\mathbb{R}^M;W)},$$

where the constant $C(\rho,\sigma)$ is independent of $\mathbf{k} = (k_1,\dots,k_M)$. In the following, let $\mathcal{P}_n$ denote the space of all univariate polynomials up to degree n. We state

Lemma 4 (cf. [2, Lemma 4.3]). *For every function $v \in C_\sigma(\mathbb{R}^M; W)$ there holds*

$$\|v - \Delta_{\mathbf{k}} v\|_{L^2_\rho(\mathbb{R}^M;W)} \le C^M \inf_{w \in \mathcal{P}_{\mathbf{n}_{\mathbf{k}-1}} \otimes W} \|v - w\|_{C_\sigma(\mathbb{R}^M;W)}$$

where $\Delta_{\mathbf{k}} = \Delta_{k_1} \otimes \cdots \otimes \Delta_{k_M}$ and $\mathcal{P}_{\mathbf{n}_{\mathbf{k}-1}} = \mathcal{P}_{n_{k_1-1}} \otimes \cdots \otimes \mathcal{P}_{n_{k_M-1}}$.
In particular, there holds

$$\left\| v - \bigotimes_{m=1}^{M-1} \Delta_{k_m} \otimes (I - \mathcal{I}_{k_M}) v \right\|_{L^2_\rho(\mathbb{R}^M;W)} \le C^M \inf_{w \in \mathcal{P}_{\mathbf{n}_{\mathbf{k}-1}} \otimes W} \|v - w\|_{C_\sigma(\mathbb{R}^M;W)}.$$

Proof. We consider a separable function $v(\boldsymbol{\xi}) = v_1(\xi_1)\cdots v_M(\xi_M) \in C_\sigma(\mathbb{R}^M; W)$. Note that the set of separable functions is dense in $C_\sigma(\mathbb{R}^M; W)$. Further, let $w \in \mathcal{P}_{\mathbf{n}_{\mathbf{k}-1}} \otimes W$ be arbitrary. There holds $\mathcal{I}_{\mathbf{k}-1} w = w$ and

$$\begin{aligned}
\|v - \Delta_{\mathbf{k}} v\|^2_{L^2_\rho(\mathbb{R}^M;W)} &= \prod_{m=1}^M \|v_m - \Delta_{k_m} v_m\|^2_{L^2_{\rho_m}(\mathbb{R};W)} \\
&\le \prod_{m=1}^M 2\Big(\|v_m - \mathcal{I}_{k_m} v_m\|^2_{L^2_{\rho_m}(\mathbb{R};W)} + \|v_m - \mathcal{I}_{k_m-1} v_m\|^2_{L^2_{\rho_m}(\mathbb{R};W)}\Big) \\
&\le \prod_{m=1}^M 4\Big(\|v_m - w_m\|^2_{L^2_{\rho_m}(\mathbb{R};W)} + \|\mathcal{I}_{k_m}(v_m - w_m)\|^2_{L^2_{\rho_m}(\mathbb{R};W)} \\
&\qquad + \|v_m - w_m\|^2_{L^2_{\rho_m}(\mathbb{R};W)} + \|\mathcal{I}_{k_m-1}(v_m - w_m)\|^2_{L^2_{\rho_m}(\mathbb{R};W)}\Big) \\
&\le 4^M \prod_{m=1}^M C^2 \|v_m - w_m\|^2_{C_{\sigma_m}(\mathbb{R};W)} \\
&= C^M \|v - w\|^2_{C_\sigma(\mathbb{R}^M;W)}.
\end{aligned}$$

The statement follows by density. □

Lemma 5 ([16]). *Let $v(\zeta)$ be an analytic function in the strip $\Sigma_\tau = \{\zeta \in \mathbb{C} : |\operatorname{Im}\zeta| < \tau + \epsilon\}$, $\epsilon > 0$. A necessary and sufficient condition that the Fourier-Hermite series*

$$v(\zeta) = \sum_{n=0}^{\infty} v_n h_n(\zeta), \quad v_n = \int_{\mathbb{R}} v(\xi) h_n(\xi)\, \mathrm{d}\xi,$$

where $h_n(\xi) = e^{-\xi^2/2} H_n(\xi)$ and $H_n(\xi) = \frac{(-1)^n}{\sqrt{\pi^{1/2} 2^n n!}} e^{\xi^2} \partial^n (e^{-\xi^2})$, converge, is that for every $\beta \in [0, \tau + \epsilon)$ there exists $C(\beta)$ such that

$$|v(\xi + i\eta)| \le C(\beta) e^{-|\xi|\sqrt{\beta^2 - \eta^2}}, \qquad y \in \mathbb{R},\ |\eta| \le \beta.$$

In this case the Fourier coefficients satisfy

$$v_n \le C e^{-\tau\sqrt{2n+1}}.$$

Following the proofs in [15, 16], it is clear that if a multivariate function $v : \mathbb{R}^M \to W$ admits an analytic extension to the domain $\Sigma_{\boldsymbol{\tau}} = \{\boldsymbol{\zeta} \in \mathbb{C}^M : |\operatorname{Im}\zeta_m| < \tau_m + \epsilon, m = 1, \ldots, M\}$, $\epsilon > 0$, and satisfies

$$\begin{aligned} &|v(\xi_1 + i\eta_1, \ldots, \xi_M + i\eta_M)| \\ &\quad \le C(\beta_1, \ldots, \beta_M) e^{-\sum_{m=1}^{M} |\xi_m| \sqrt{\beta_m^2 - \eta_m^2}}, \qquad \xi_m \in \mathbb{R},\ |\eta_m| \le \beta_m,\ \forall m, \end{aligned}$$

for all $\beta_m \in [0, \tau_m]$, $m = 1, \ldots, M$, then we have

$$v(\boldsymbol{\zeta}) = \sum_{\mathbf{n}} v_{\mathbf{n}} \prod_{m=1}^{M} h_{n_m}(\zeta_m), \quad v_{\mathbf{n}} = \int_{\mathbb{R}^M} v(\boldsymbol{\xi}) \prod_{m=1}^{M} h_{n_m}(\xi_m)\, \mathrm{d}\boldsymbol{\xi},$$

for all $\boldsymbol{\zeta} \in \Sigma_{\boldsymbol{\tau}}$, and, in particular,

$$v_{\mathbf{n}} \le C \exp\left(-\sum_{m=1}^{M} \tau_m \sqrt{2n_m + 1} \right).$$

Thus, we can generalize [2, Lemma 4.6] by an obvious modification to

Lemma 6 (cf. [2, Lemma 4.6]). *Let $v : \mathbb{R}^M \to W$ admit an analytic extension to*

$$\Sigma_{\boldsymbol{\tau}} = \{z \in \mathbb{C}^M : |\operatorname{Im}\zeta_m| < \tau_m + \epsilon, m = 1, \ldots, M\},$$

$\epsilon > 0$, and satisfy

$$\max_{\boldsymbol{\zeta} \in \Sigma_{\boldsymbol{\tau}}} \sigma(\operatorname{Re}\boldsymbol{\zeta}) \|v(\boldsymbol{\zeta})\|_W \le +\infty.$$

Then there holds

$$\min_{w\in\mathcal{P}_{\mathbf{n}}} \max_{\boldsymbol{\xi}\in\mathbb{R}^M} \left| \|v(\boldsymbol{\xi}) - w(\boldsymbol{\xi})\|_W e^{-\|\boldsymbol{\xi}\|^2/8} \right| \le C\Theta(\mathbf{n}) \exp\left(-\frac{1}{\sqrt{2}} \sum_{m=1}^{M} \tau_m \sqrt{n_m} \right),$$

where $\Theta(\mathbf{n}) = C(\boldsymbol{\tau})(n_1 \cdots n_M)^{1/2}$.

References

1. I. Babuška, R. Tempone, G.E. Zouraris, Galerkin finite element approximations of stochastic elliptical partial differential equations. SIAM J. Numer. Anal. **42**(2), 800–825 (2004)
2. I. Babuška, F. Nobile, R. Tempone, A stochastic collocation method for elliptical partial differential equations with random input data. SIAM J. Numer. Anal. **45**(3), 1005–1034 (2007)
3. J. Beck, R. Tempone, F. Nobile, L. Tamellini, On the optimal polynomial approximation of stochastic PDEs by Galerkin and collocation methods. Math. Models Methods Appl. Sci. **22**(9), 1250023.1–1250023.33. doi:10.1142/S0218202512500236 (2012)
4. A. Bespalov, C.E. Powell, D. Silvester, A priori error analysis of stochastic Galerkin mixed approximations of elliptic PDEs with random data. SIAM J. Numer. Anal. **50**(4), 2039–2063 (2012)
5. F. Brezzi, M. Fortin, *Mixed and Hybrid Finite Element Methods* (Springer, Berlin, 1991)
6. J. Charrier, Strong and weak error estimates for elliptic partial differential equations with random coefficients. SIAM J. Numer. Anal. **50**(1), 216–246 (2012)
7. A. Cliffe, O.G. Ernst, B. Sprungk, E. Ullmann, K.G. van den Boogaart, Probabilistic uncertainty quantification using PDEs with random data: a case study. (2014, in preparation)
8. A. Cohen, R. DeVore, C. Schwab, Analytic regularity and polynomial approximation of parametric and stochastic elliptic PDEs. Anal. Appl. **9**, 11–47 (2011)
9. A. Ern, J.-L. Guermond, *Theory and Practice of Finite Elements* (Springer, New York, 2004)
10. O.G. Ernst, A. Mugler, H.-J. Starkloff, E. Ullmann, On the convergence of generalized polynomial chaos expansions. ESAIM Math. Model. Numer. Anal. **46**(2), 317–339 (2012)
11. R.A. Freeze, A stochastic-conceptual analysis of one-dimensional groundwater flow in nonuniform homogeneous media. Water Resour. Res. **11**(5), 725–741 (1975)
12. B. Ganis, H. Klie, M.F. Wheeler, T. Wildey, I. Yotov, D. Zhang, Stochastic collocation and mixed finite elements for flow in porous media. Comput. Methods Appl. Mech. Eng. **197**, 3547–3559 (2008)
13. R.G. Ghanem, P.D. Spanos, *Stochastic Finite Elements: A Spectral Approach* (Springer, New York, 1991)
14. F. Hartogs, Zur Theorie der analytischen Funktionen mehrer Veränderlicher, insbesondere über die Darstellung derselben durch Reihen, welche Potenzen einer Veränderlichen fortschreiten. Math. Ann. **62**, 1–88 (1906)
15. E. Hille, Contributions to the theory of Hermitian series. Duke Math. J. **5**, 875–936 (1939)
16. E. Hille, Contributions to the theory of Hermitian series. II. The representation problem. Trans. Am. Math. Soc. **47**, 80–94 (1940)
17. O. Kallenberg, *Foundations of Modern Probability* (Springer, Berlin, 2002)
18. F. Nobile, R. Tempone, C.G. Webster, An anisotropic sparse grid stochastic collocation method for partial differential equations with random input data. SIAM J. Numer. Anal. **46**(5), 2411–2442 (2008)
19. F. Nobile, R. Tempone, C.G. Webster, A sparse grid stochastic collocation method for partial differential equations with random input data. SIAM J. Numer. Anal. **46**(5), 2309–2345 (2008)

20. C. Schwab, C.J. Gittelson, Sparse tensor discretizations of high-dimensional parametric and stochastic PDEs. Acta Numer. **20**, 291–467 (2011)
21. L.N. Trefethen, *Approximation Theory and Approximation Practice* (SIAM, Philadelphia, 2012)
22. G.W. Wasilkowski, H. Woźniakowski, Explicit cost bounds of algorithms for multivariate tensor product problems. J. Complex. **11**, 1–56 (1995)
23. D. Xiu, J.S. Hesthaven, High-order collocation methods differential equations with random inputs. SIAM J. Sci. Comput. **37**(3), 1118–1139 (2005)

On the Convergence of the Combination Technique

Michael Griebel and Helmut Harbrecht

Abstract Sparse tensor product spaces provide an efficient tool to discretize higher dimensional operator equations. The direct Galerkin method in such ansatz spaces may employ hierarchical bases, interpolets, wavelets or multilevel frames. Besides, an alternative approach is provided by the so-called combination technique. It properly combines the Galerkin solutions of the underlying problem on certain full (but small) tensor product spaces. So far, however, the combination technique has been analyzed only for special model problems. In the present paper, we provide now the analysis of the combination technique for quite general operator equations in sparse tensor product spaces. We prove that the combination technique produces the same order of convergence as the Galerkin approximation with respect to the sparse tensor product space. Furthermore, the order of the cost complexity is the same as for the Galerkin approach in the sparse tensor product space. Our theoretical findings are validated by numerical experiments.

1 Introduction

The discretization in sparse tensor product spaces yields efficient numerical methods to solve higher dimensional operator equations. Nevertheless, a Galerkin discretization in these sparse tensor product spaces requires hierarchical bases, interpolets, wavelets, multilevel frames, or other types of multilevel systems [9, 12, 17] which make a direct Galerkin discretization in sparse tensor product spaces quite involved

M. Griebel
Institut für Numerische Simulation, Universität Bonn, Wegelerstr. 6, 53115 Bonn, Germany
e-mail: griebel@ins.uni-bonn.de

H. Harbrecht (✉)
Mathematisches Institut, Universität Basel, Rheinsprung 21, 4051 Basel, Switzerland
e-mail: helmut.harbrecht@unibas.ch

J. Garcke and D. Pflüger (eds.), *Sparse Grids and Applications - Munich 2012*, Lecture Notes in Computational Science and Engineering 97,
DOI 10.1007/978-3-319-04537-5_3,

and cumbersome in practical applications. To avoid these issues of the Galerkin discretization, the *combination technique* has been introduced in [14]. There, only the Galerkin discretizations and solutions in appropriately chosen, full, but small, tensor product spaces need to be computed and combined.

In [8, 18, 19], it has been shown that, in the special case of operator equations which involve a tensor product operator, the approximation produced by the combination technique indeed coincides exactly with the Galerkin solution in the sparse tensor product space. However, for non-tensor product operators, this is no longer the case. Nevertheless, it is observed in practice that the approximation error is of the same order. But theoretical convergence results are only available for specific applications, see for example [3, 14, 21–23, 25]. Moreover, a general proof of convergence is so far still missing for the combination technique.

In the present paper, we prove optimal convergence rates of the combination technique for elliptic operators acting on arbitrary Gelfand triples. The convergence analysis is based on two compact lemmas (Lemmas 1 and 2) which have basically been proven in [22, 25]. In contrast to these papers, besides considering abstract Gelfand triples, we deal here with the combination technique for the so-called *generalized sparse tensor product spaces* which have been introduced in [10]. Lemma 1 involves a special stability condition for the Galerkin projection (cf. (18)) which, however, holds for certain regularity assumptions on the operator under consideration (see Remark 1).

To keep the notation and the proofs simple, we restrict ourselves to the case of operator equations which are defined on a twofold product domain $\Omega_1 \times \Omega_2$. However, we allow the domains $\Omega_1 \subset \mathbb{R}^{n_1}$ and $\Omega_2 \subset \mathbb{R}^{n_2}$ to be of different spatial dimensions. Our proofs can be generalized without further difficulties to arbitrary L-fold product domains $\Omega_1 \times \Omega_2 \times \cdots \times \Omega_L$ by employing the techniques from [11, 25].

The remainder of this paper is organized as follows. We first present the operator equations under consideration in Sect. 2. Then, in Sect. 3, we specify the requirements of the multiscale hierarchies on each individual subdomain. In Sect. 4, we define the generalized sparse tensor product spaces and recall their basic properties. The combination technique is introduced in Sect. 5 and its convergence is proven in Sect. 6. Section 7 is dedicated to numerical experiments. They are in good agreement with the presented theory. Finally, in Sect. 8, we give some concluding remarks.

Throughout this paper, the notion "essential" in the context of complexity estimates means "up to logarithmic terms". Moreover, to avoid the repeated use of generic but unspecified constants, we signify by $C \lesssim D$ that C is bounded by a multiple of D independently of parameters which C and D may depend on. Obviously, $C \gtrsim D$ is defined as $D \lesssim C$, and $C \sim D$ as $C \lesssim D$ and $C \gtrsim D$.

2 Operator Equations

We consider two sufficiently smooth, bounded domains $\Omega_1 \in \mathbb{R}^{n_1}$ and $\Omega_2 \in \mathbb{R}^{n_2}$, where $n_1, n_2 \in \mathbb{N}$. Moreover, on the product domain $\Omega_1 \times \Omega_2$, let the Hilbert space $\mathcal{H}$ be given such that

$$\mathcal{H} \subset L^2(\Omega_1 \times \Omega_2) \subset \mathcal{H}'$$

forms a Gelfand triple. Thus, the inner product

$$(u, v)_{L^2(\Omega_1 \times \Omega_2)} := \int_{\Omega_1} \int_{\Omega_2} u(\mathbf{x}, \mathbf{y}) v(\mathbf{x}, \mathbf{y}) \, \mathbf{dx} \, \mathbf{dy}$$

in $L^2(\Omega_1 \times \Omega_2)$ can continuously be extended to $\mathcal{H} \times \mathcal{H}'$. For sake of simplicity of presentation, we write $(u, v)_{L^2(\Omega_1 \times \Omega_2)}$ also in the case $u \in \mathcal{H}$ and $v \in \mathcal{H}'$.

Now, let $A : \mathcal{H} \to \mathcal{H}'$ denote a differential or pseudo-differential operator. It is assumed that it maps the Hilbert space $\mathcal{H}$ continuously and bijectively onto its dual $\mathcal{H}'$, i.e.,

$$\|Au\|_{\mathcal{H}'} \sim \|u\|_{\mathcal{H}} \text{ for all } u \in \mathcal{H}.$$

The Hilbert space $\mathcal{H}$ is thus the *energy space* of the operator under consideration. For the sake of simplicity, we further assume that A is $\mathcal{H}$-elliptic. Consequently, the resulting bilinear form

$$a(u, v) := (Au, v)_{L^2(\Omega_1 \times \Omega_2)} : \mathcal{H} \times \mathcal{H} \to \mathbb{R}$$

is continuous

$$a(u, v) \lesssim \|u\|_{\mathcal{H}} \|v\|_{\mathcal{H}} \text{ for all } u, v \in \mathcal{H}$$

and elliptic

$$a(u, u) \gtrsim \|u\|_{\mathcal{H}}^2 \text{ for all } u \in \mathcal{H}.$$

In the following, for given $f \in \mathcal{H}'$, we want to efficiently solve the operator equation $Au = f$ or, equivalently, the variational formulation:

$$\text{find } u \in \mathcal{H} \text{ such that } a(u, v) = (f, v)_{L^2(\Omega_1 \times \Omega_2)} \text{ for all } v \in \mathcal{H}. \tag{1}$$

Of course, since we like to focus on conformal Galerkin discretizations, we should tacitly assume that, for all $j_1, j_2 \geq 0$, the tensor product $V_{j_1}^{(1)} \otimes V_{j_2}^{(2)}$ of the ansatz spaces $V_{j_1}^{(1)}$ and $V_{j_2}^{(2)}$ is contained in the energy space $\mathcal{H}$. Moreover, for the solution $u \in \mathcal{H}$ of (1), we will need a stronger regularity to hold for obtaining decent convergence rates. Therefore, for $s_1, s_2 \geq 0$, we introduce the following Sobolev spaces of dominant mixed derivatives with respect to the underlying space $\mathcal{H}$

$$\mathcal{H}_{mix}^{s_1,s_2} := \left\{ f \in \mathcal{H} : \left\| \frac{\partial^{\alpha+\beta}}{\partial_{\mathbf{x}}^{\alpha}\partial_{\mathbf{y}}^{\beta}} f \right\|_{\mathcal{H}} < \infty \text{ for all } |\boldsymbol{\alpha}| \leq s_1 \text{ and } |\boldsymbol{\beta}| \leq s_2 \right\}.$$

We shall illustrate our setting by the following specific examples.

Example 1. A first simple example is the operator $A : L^2(\Omega_1 \times \Omega_2) \to L^2(\Omega_1 \times \Omega_2)$ which underlies the bilinear form

$$a(u, v) = \int_{\Omega_1} \int_{\Omega_2} \alpha(\mathbf{x}, \mathbf{y}) u(\mathbf{x}, \mathbf{y}) v(\mathbf{x}, \mathbf{y}) \, d\mathbf{x} \, d\mathbf{y},$$

where the coefficient function α satisfies

$$0 < \underline{\alpha} \leq \alpha(\mathbf{x}, \mathbf{y}) \leq \overline{\alpha} \text{ for all } (\mathbf{x}, \mathbf{y}) \in \Omega_1 \times \Omega_2. \tag{2}$$

Here, it holds $\mathcal{H} = L^2(\Omega_1 \times \Omega_2)$. Moreover, our spaces $\mathcal{H}_{mix}^{s_1,s_2}$ of assumed stronger regularity coincide with the standard Sobolev spaces of dominant mixed derivatives, i.e.,

$$\mathcal{H}_{mix}^{s_1,s_2} = H_{mix}^{s_1,s_2}(\Omega_1 \times \Omega_2) := H^{s_1}(\Omega_1) \otimes H^{s_2}(\Omega_2).$$

Example 2. Stationary heat conduction in the product domain $\Omega_1 \times \Omega_2$ yields the bilinear form

$$a(u, v) = \int_{\Omega_1} \int_{\Omega_2} \alpha(\mathbf{x}, \mathbf{y}) \{ \nabla_{\mathbf{x}} u(\mathbf{x}, \mathbf{y}) \nabla_{\mathbf{x}} v(\mathbf{x}, \mathbf{y}) + \nabla_{\mathbf{y}} u(\mathbf{x}, \mathbf{y}) \nabla_{\mathbf{y}} v(\mathbf{x}, \mathbf{y}) \} \, d\mathbf{x} \, d\mathbf{y}.$$

If the coefficient α satisfies (2), then the associated operator A is known to be continuous and elliptic with respect to the space $\mathcal{H} = H_0^1(\Omega_1 \times \Omega_2)$. Moreover, our spaces $\mathcal{H}_{mix}^{s_1,s_2}$ of assumed stronger regularity now coincide with $\mathcal{H}_{mix}^{s_1,s_2} = H_0^1(\Omega_1 \times \Omega_2) \cap H_{mix}^{s_1+1,s_2}(\Omega_1 \times \Omega_2) \cap H_{mix}^{s_1,s_2+1}(\Omega_1 \times \Omega_2)$.

Example 3. Another example appears in two-scale homogenization. Unfolding [4] gives raise to the product of the macroscopic physical domain Ω_1 and the periodic microscopic domain Ω_2 of the cell problem, see [20]. Then, for the first order corrector, one arrives at the bilinear form

$$a(u, v) = \int_{\Omega_1} \int_{\Omega_2} \alpha(\mathbf{x}, \mathbf{y}) \nabla_{\mathbf{y}} u(\mathbf{x}, \mathbf{y}) \nabla_{\mathbf{y}} v(\mathbf{x}, \mathbf{y}) \, d\mathbf{x} \, d\mathbf{y}.$$

The underlying operator A is continuous and elliptic as a operator in the related energy space $\mathcal{H} = L^2(\Omega_1) \otimes H_0^1(\Omega_2)$ provided that the coefficient α satisfies again (2). Furthermore, our spaces $\mathcal{H}_{mix}^{s_1,s_2}$ of assumed stronger regularity coincide with $\mathcal{H}_{mix}^{s_1,s_2} = \big(L^2(\Omega_1) \otimes H_0^1(\Omega_2)\big) \cap H_{mix}^{s_1,s_2+1}(\Omega_1 \times \Omega_2)$.

3 Approximation on the Individual Subdomains

On each domain Ω_i, we consider a nested sequence

$$V_0^{(i)} \subset V_1^{(i)} \subset \cdots \subset V_j^{(i)} \subset \cdots \subset L^2(\Omega_i) \tag{3}$$

of finite dimensional spaces

$$V_j^{(i)} = \operatorname{span}\{\varphi_{j,k}^{(i)} : k \in \Delta_j^{(i)}\}$$

(the set $\Delta_j^{(i)}$ denotes a suitable index set) of piecewise polynomial ansatz functions, such that $\dim V_j^{(i)} \sim 2^{jn_i}$ and

$$L^2(\Omega_i) = \overline{\bigcup_{j \in \mathbb{N}_0} V_j^{(i)}}.$$

We will use the spaces $V_j^{(i)}$ for the approximation of functions. To this end, we assume that the approximation property

$$\inf_{v_j \in V_j^{(i)}} \|u - v_j\|_{H^q(\Omega_i)} \lesssim h_j^{s-q} \|u\|_{H^s(\Omega_i)}, \quad u \in H^s(\Omega_i), \tag{4}$$

holds for $q < \gamma_i$, $q \leq s \leq r_i$ uniformly in j. Here we set $h_j := 2^{-j}$, i.e., h_j corresponds to the width of the mesh associated with the subspace $V_j^{(i)}$ on Ω_i. The parameter $\gamma_i > 0$ refers to the *regularity* of the functions which are contained in $V_j^{(i)}$, i.e.,

$$\gamma_i := \sup\{s \in \mathbb{R} : V_j^{(i)} \subset H^s(\Omega_i)\}.$$

The integer $r_i > 0$ refers to the *polynomial exactness*, that is the maximal order of polynomials which are locally contained in the space $V_j^{(i)}$.

Now, let $Q_j^{(i)} : L^2(\Omega_i) \to V_j^{(i)}$ denote the $L^2(\Omega_i)$-orthogonal projection onto the finite element space $V_j^{(i)}$. Due to the orthogonality, we have $\left(Q_j^{(i)}\right)^\star = Q_j^{(i)}$. Moreover, our regularity assumptions on the ansatz spaces $V_j^{(i)}$ imply the continuity of the related projections relative to the Sobolev space $H^q(\Omega_i)$ for all $|q| < \gamma_i$, i.e., it holds

$$\left\| Q_j^{(i)} u \right\|_{H^q(\Omega_i)} \lesssim \|u\|_{H^q(\Omega_i)}, \quad |q| < \gamma_i, \tag{5}$$

uniformly in $j \geq 0$ provided that $u \in H^q(\Omega_i)$.

By setting $Q_{-1}^{(i)} := 0$, we can define for all $j \geq 0$ the complementary spaces

$$W_j^{(i)} := (Q_j^{(i)} - Q_{j-1}^{(i)})L^2(\Omega_i) \subset V_j^{(i)}.$$

They satisfy

$$V_j^{(i)} = V_{j-1}^{(i)} \oplus W_j^{(i)}, \quad V_{j-1}^{(i)} \cap W_j^{(i)} = \{0\},$$

which recursively yields

$$V_J^{(i)} = \bigoplus_{j=0}^{J} W_j^{(i)}. \tag{6}$$

A given function $f \in H^q(\Omega_i)$, where $|q| < \gamma_i$, admits the unique multiscale decomposition

$$f = \sum_{j=0}^{\infty} f_j \text{ with } f_j := (Q_j^{(i)} - Q_{j-1}^{(i)})f \in W_j^{(i)}.$$

One now has the well-known norm equivalence

$$\|f\|_{H^q(\Omega_i)}^2 \sim \sum_{j=0}^{\infty} 2^{2jq} \|(Q_j^{(i)} - Q_{j-1}^{(i)})f\|_{L^2(\Omega_i)}^2, \quad |q| < \gamma_i,$$

see [5]. Finally, for any $f \in H^s(\Omega_i)$ and $|q| < \gamma_i$, the approximation property (4) induces the estimate

$$\|(Q_j^{(i)} - Q_{j-1}^{(i)})f\|_{H^q(\Omega_i)} \lesssim 2^{-j(s-q)}\|f\|_{H^s(\Omega_i)}, \quad q < s \leq r_i.$$

4 Generalized Sparse Tensor Product Spaces

The canonical approximation method in the Hilbert space $\mathcal{H}$ is the approximation in full tensor product spaces[1]

$$V_{J/\sigma}^{(1)} \otimes V_{J\sigma}^{(2)} = \bigoplus_{\substack{j_1\sigma \leq J \\ j_2/\sigma \leq J}} W_{j_1}^{(1)} \otimes W_{j_2}^{(2)}.$$

[1]Here and in the following, the summation limits are in general no natural numbers and must of course be rounded properly. We leave this to the reader to avoid cumbersome floor/ceil-notations.

Here, $\sigma > 0$ is a given parameter which can be tuned to optimize the cost complexity. There are $2^{Jn_1/\sigma} \cdot 2^{Jn_2\sigma}$ degrees of freedom in the space $V_{J/\sigma}^{(1)} \otimes V_{J\sigma}^{(2)}$. Moreover, for $f \in \mathcal{H}_{mix}^{s_1,0}(\Omega_1 \times \Omega_2) \cap \mathcal{H}_{mix}^{0,s_2}(\Omega_1 \times \Omega_2)$ and $f_J := (Q_{J/\sigma}^{(1)} \otimes Q_{J\sigma}^{(2)})f \in V_{J/\sigma}^{(1)} \otimes V_{J\sigma}^{(2)}$, an error estimate of the type

$$\|f - f_J\|_{\mathcal{H}} \lesssim 2^{-J\min\{s_1/\sigma, s_2\sigma\}} \|f\|_{\mathcal{H}_{mix}^{s_1,0} \cap \mathcal{H}_{mix}^{0,s_2}} \tag{7}$$

holds for all $0 < s_1 \leq p_1$ and $0 < s_2 \leq p_2$. Note that the upper bounds p_1 and p_2 are the largest values such that $\mathcal{H}_{mix}^{p_1,0} \subset H_{mix}^{r_1,r_2}(\Omega_1 \times \Omega_2)$ and $\mathcal{H}_{mix}^{0,p_2} \subset H_{mix}^{r_1,r_2}(\Omega_1 \times \Omega_2)$, respectively.

Alternatively, based on the multiscale decompositions (6) on each individual subdomain, one can define the so-called *generalized sparse tensor product space*, see [2, 10],

$$\hat{V}_J^\sigma := \bigoplus_{j_1\sigma + j_2/\sigma \leq J} W_{j_1}^{(1)} \otimes W_{j_2}^{(2)} = \sum_{j_1\sigma + j_2/\sigma = J} V_{j_1}^{(1)} \otimes V_{j_2}^{(2)}. \tag{8}$$

Thus, a function $f \in \mathcal{H}$ is represented by the Boolean sum

$$\hat{f}_J := \sum_{j_1\sigma + j_2/\sigma \leq J} \Delta_{j_1,j_2}^Q f \in \hat{V}_J^\sigma \tag{9}$$

where, for all $j_1, j_2 \geq 0$, the detail projections Δ_{j_1,j_2}^Q are given by

$$\Delta_{j_1,j_2}^Q := (Q_{j_1}^{(1)} - Q_{j_1-1}^{(1)}) \otimes (Q_{j_2}^{(2)} - Q_{j_2-1}^{(2)}). \tag{10}$$

For further details on sparse grids we refer the reader to the survey [2] and the references therein.

The dimension of the generalized sparse tensor product space $\hat{V}_J^\sigma$ is essentially equal to the dimension of the finest univariate finite element spaces which enter its construction, i.e., it is essentially equal to the value of $\max\{\dim V_{J/\sigma}^{(1)}, \dim V_{J\sigma}^{(2)}\}$. Nevertheless, by considering smoothness in terms of mixed Sobolev spaces, its approximation power is essentially the same as in the full tensor product space. To be precise, we have

Theorem 1 ([10]). *The generalized sparse tensor product space $\hat{V}_J^\sigma$ possesses*

$$\dim \hat{V}_J^\sigma \sim \begin{cases} 2^{J\max\{n_1/\sigma, n_2\sigma\}}, & \text{if } n_1/\sigma \neq n_2\sigma, \\ 2^{Jn_1/\sigma} J, & \text{if } n_1/\sigma = n_2\sigma, \end{cases}$$

degrees of freedom. Moreover, for a given function $f \in \mathcal{H}_{mix}^{s_1,s_2}$ and its L^2-orthonormal projection $\hat{f}_J \in \hat{V}_J^\sigma$, defined by (9)*, where $0 < s_1 \leq p_1$ and $0 < s_2 \leq p_2$, there holds the error estimate*

$$\|f - \hat{f}_J\|_{\mathcal{H}} \lesssim \begin{cases} 2^{-J\min\{s_1/\sigma, s_2\sigma\}}\|f\|_{\mathcal{H}_{mix}^{s_1,s_2}}, & \text{if } s_1/\sigma \neq s_2\sigma, \\ 2^{-Js_1/\sigma}\sqrt{J}\|f\|_{\mathcal{H}_{mix}^{s_1,s_2}}, & \text{if } s_1/\sigma = s_2\sigma. \end{cases}$$

The optimal choice of the parameter σ has been discussed in [10]. It turns out that the best cost complexity rate among all possible values of s_1, s_2 is obtained for the choice $\sigma = \sqrt{n_1/n_2}$. This choice induces an equilibration of the degrees of freedom in the extremal spaces $V_{J/\sigma}^{(1)}$ and $V_{J\sigma}^{(2)}$.

We shall consider the Galerkin discretization of (1) in the generalized sparse tensor product space $\hat{V}_J^\sigma$, that is we want to

$$\text{find } u_J \in \hat{V}_J^\sigma \text{ such that } a(u_J, v_J) = (f, v_J)_{L^2(\Omega_1\times\Omega_2)} \text{ for all } v_J \in \hat{V}_J^\sigma. \tag{11}$$

In view of Theorem 1, we arrive at the following error estimate due to Céa's lemma.

Corollary 1. *The Galerkin solution* (11) *satisfies the error estimate*

$$\|u - u_J\|_{\mathcal{H}} \lesssim \|u - \hat{u}_J\|_{\mathcal{H}} \lesssim \begin{cases} 2^{-J\min\{s_1/\sigma, s_2\sigma\}}\|u\|_{\mathcal{H}_{mix}^{s_1,s_2}}, & \text{if } s_1/\sigma \neq s_2\sigma, \\ 2^{-Js_1/\sigma}\sqrt{J}\|u\|_{\mathcal{H}_{mix}^{s_1,s_2}}, & \text{if } s_1/\sigma = s_2\sigma, \end{cases}$$

for all $0 < s_1 \leq p_1$ *and* $0 < s_2 \leq p_2$ *provided that* $u \in \mathcal{H}_{mix}^{s_1,s_2}(\Omega_1 \times \Omega_2)$.

Nevertheless, for the discretization of (11), hierarchical bases, interpolets, wavelets, multilevel frames, or other types of multilevel systems [2, 9, 12, 13, 15–17, 24, 26] are required which make a direct Galerkin discretization in sparse tensor product spaces quite involved and cumbersome in practical applications.

5 Combination Technique

The combination technique is a different approach for the discretization in sparse tensor product spaces. It avoids the explicit need of hierarchical bases, interpolets, wavelets or frames for the discretization of (11). In fact, one only has to compute the Galerkin solutions with respect to certain full tensor product spaces $V_{j_1}^{(1)} \otimes V_{j_2}^{(2)}$ and to appropriately combine them afterwards. The related Galerkin solutions u_{j_1,j_2} are given by

$$\text{find } u_{j_1,j_2} \in V_{j_1}^{(1)} \otimes V_{j_2}^{(2)} \text{ such that}$$
$$a(u_{j_1,j_2}, v_{j_1,j_2}) = (f, v_{j_1,j_2})_{L^2(\Omega_1\times\Omega_2)} \text{ for all } v_{j_1,j_2} \in V_{j_1}^{(1)} \otimes V_{j_2}^{(2)}.$$

This introduces the Galerkin projection

$$P_{j_1,j_2} : \mathcal{H} \to V_{j_1}^{(1)} \otimes V_{j_2}^{(2)}, \quad P_{j_1,j_2}u := u_{j_1,j_2}$$

which especially satisfies the Galerkin orthogonality

$$a(u - P_{j_1,j_2}u, v_{j_1,j_2}) = 0 \text{ for all } v_{j_1,j_2} \in V_{j_1}^{(1)} \otimes V_{j_2}^{(2)}.$$

The Galerkin projection P_{j_1,j_2} is well defined for all $j_1, j_2 \geq 0$ due to the ellipticity of the bilinear form $a(\cdot,\cdot)$. Moreover, as in (7), we conclude the error estimate

$$\|u - P_{j_1,j_2}u\|_{\mathcal{H}} \lesssim \|u - (Q_{j_1}^{(1)} \otimes Q_{j_2}^{(2)})u\|_{\mathcal{H}} \lesssim 2^{-\min\{j_1 s_1, j_2 s_2\}} \|u\|_{\mathcal{H}_{mix}^{s_1,0} \cap \mathcal{H}_{mix}^{0,s_2}}$$

for all $0 < s_1 \leq p_1$ and $0 < s_2 \leq p_2$ provided that $u \in \mathcal{H}_{mix}^{s_1,0} \cap \mathcal{H}_{mix}^{0,s_2}$. In particular, for fixed $j_1 \geq 0$ and $j_2 \to \infty$, we obtain the Galerkin projection $P_{j_1,\infty}$ onto the space $V_{j_1,\infty} := (Q_{j_1}^{(1)} \otimes I)\mathcal{H} \subset \mathcal{H}$. It satisfies the error estimate

$$\|u - P_{j_1,\infty}u\|_{\mathcal{H}} \lesssim \|u - (Q_{j_1}^{(1)} \otimes I)u\|_{\mathcal{H}} \lesssim 2^{-j_1 s_1} \|u\|_{\mathcal{H}_{mix}^{s_1,0}} \tag{12}$$

for all $0 < s_1 \leq p_1$. Likewise, for fixed $j_2 \geq 0$ and $j_1 \to \infty$, we obtain the Galerkin projection P_{∞,j_2} onto the space $V_{\infty,j_2} := (I \otimes Q_{j_2}^{(2)})\mathcal{H} \subset \mathcal{H}$. Analogously to (12), we find

$$\|u - P_{\infty,j_2}u\|_{\mathcal{H}} \lesssim \|u - (I \otimes Q_{j_2}^{(2)})u\|_{\mathcal{H}} \lesssim 2^{-j_2 s_2} \|u\|_{\mathcal{H}_{mix}^{0,s_2}} \tag{13}$$

for all $0 < s_2 \leq p_2$.

With the help of the Galerkin projections, we can define

$$\Delta_{j_1,j_2}^P u := (P_{j_1,j_2} - P_{j_1-1,j_2} - P_{j_1,j_2-1} + P_{j_1-1,j_2-1})u \tag{14}$$

where we especially set $P_{j_1,-1} := 0$, $P_{-1,j_2} := 0$, and $P_{-1,-1} := 0$. Then, the combination technique is expressed as the Boolean sum (cf. [6–8])

$$\hat{u}_J = \sum_{j_1\sigma + j_2/\sigma \leq J} \Delta_{j_1,j_2}^P u = u - \sum_{j_1\sigma + j_2/\sigma > J} \Delta_{j_1,j_2}^P u. \tag{15}$$

Straightforward calculation shows

$$\hat{u}_J = \sum_{j_1=0}^{\lceil J/\sigma \rceil} (P_{j_1,\lceil J\sigma - j_1\sigma^2 \rceil} - P_{j_1-1,\lceil J\sigma - j_1\sigma^2 \rceil})u \tag{16}$$

if $j_1 \leq j_2\sigma^2$, and

$$\hat{u}_J = \sum_{j_2=0}^{\lceil J\sigma \rceil} (P_{\lceil J/\sigma - j_2/\sigma^2 \rceil, j_2} - P_{\lceil J/\sigma - j_2/\sigma^2 \rceil, j_2-1})u \tag{17}$$

if $j_1 > j_2\sigma^2$. A visualization of the formula (17) is found in Fig. 1.

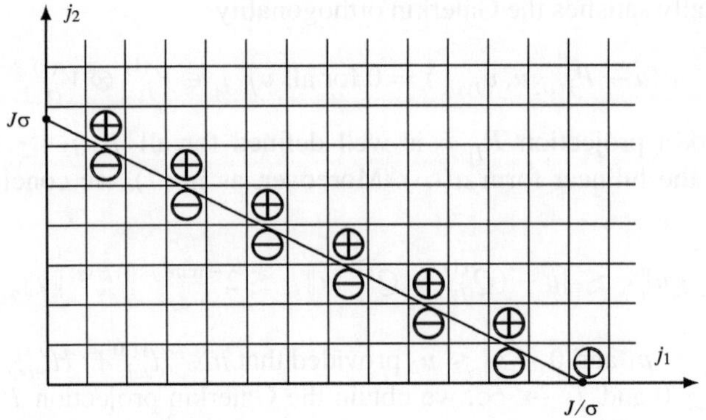

Fig. 1 The combination technique in $\hat{V}_J^\sigma$ combines all the indicated solutions $P_{j_1,j_2}u$ with *positive sign* ("⊕") and *negative sign* ("⊖")

Our goal is now to show that the error $\|u - \hat{u}_J\|_{\mathcal{H}}$ converges as good as the error of the true sparse tensor product Galerkin solution given in Corollary 1.

6 Proof of Convergence

To prove the desired error estimate for the combination technique (16) and (17), respectively, we shall prove first the following two helpful lemmata.

Lemma 1. *For all* $0 < s_1 \leq p_1$ *and* $0 < s_2 \leq p_2$*, it holds*

$$\|(P_{j_1,j_2} - P_{j_1-1,j_2})u\|_{\mathcal{H}} \lesssim 2^{-j_1 s_1} \|u\|_{\mathcal{H}_{mix}^{s_1,0}},$$

$$\|(P_{j_1,j_2} - P_{j_1,j_2-1})u\|_{\mathcal{H}} \lesssim 2^{-j_2 s_2} \|u\|_{\mathcal{H}_{mix}^{0,s_2}},$$

provided that u is sufficiently smooth and provided that the Galerkin projection satisfies

$$\|P_{\infty,j_2}u\|_{\mathcal{H}_{mix}^{s_1,0}} \lesssim \|u\|_{\mathcal{H}_{mix}^{s_1,0}}, \quad \|P_{j_1,\infty}u\|_{\mathcal{H}_{mix}^{0,s_2}} \lesssim \|u\|_{\mathcal{H}_{mix}^{0,s_2}}. \tag{18}$$

Proof. We shall prove only the first estimate, the second one follows in complete analogy. To this end, we split

$$\|(P_{j_1,j_2} - P_{j_1-1,j_2})u\|_{\mathcal{H}} \leq \|(P_{j_1,j_2} - P_{\infty,j_2})u\|_{\mathcal{H}} + \|(P_{\infty,j_2} - P_{j_1-1,j_2})u\|_{\mathcal{H}}.$$

Due to $V_{j_1-1,j_2}, V_{j_1,j_2} \subset V_{\infty,j_2}$, the associated Galerkin projections satisfy the identities $P_{j_1,j_2} = P_{j_1,j_2}P_{\infty,j_2}$ and $P_{j_1-1,j_2} = P_{j_1-1,j_2}P_{\infty,j_2}$. Hence, we obtain

$$\|(P_{j_1,j_2} - P_{j_1-1,j_2})u\|_{\mathcal{H}} \leq \|(P_{j_1,j_2} - I)P_{\infty,j_2}u\|_{\mathcal{H}} + \|(I - P_{j_1-1,j_2})P_{\infty,j_2}u\|_{\mathcal{H}}.$$

By employing now the fact that the Galerkin projections $P_{j_1-1,j_2}u$ and $P_{j_1,j_2}u$ are quasi-optimal, i.e., $\|(I - P_{j_1,j_2})u\|_{\mathcal{H}} \lesssim \|(I - Q_{j_1}^{(1)} \otimes Q_{j_2}^{(2)})u\|_{\mathcal{H}}$ and likewise for $P_{j_1-1,j_2}u$, we arrive at

$$\begin{aligned}&\|(P_{j_1,j_2} - P_{j_1-1,j_2})u\|_{\mathcal{H}}\\&\quad\lesssim \|(Q_{j_1}^{(1)} \otimes Q_{j_2}^{(2)} - I)P_{\infty,j_2}u\|_{\mathcal{H}} + \|(I - Q_{j_1-1}^{(1)} \otimes Q_{j_2}^{(2)})P_{\infty,j_2}u\|_{\mathcal{H}}.\end{aligned}$$

The combination of $Q_{j_1}^{(1)} \otimes Q_{j_2}^{(2)} = (Q_{j_1}^{(1)} \otimes I)(I \otimes Q_{j_2}^{(2)})$ and $(I \otimes Q_{j_2}^{(2)})P_{\infty,j_2} = P_{\infty,j_2}$ yields the operator identity

$$(Q_{j_1}^{(1)} \otimes Q_{j_2}^{(2)})P_{\infty,j_2} = \big(Q_{j_1}^{(1)} \otimes I\big)P_{\infty,j_2},$$

and likewise

$$(Q_{j_1-1}^{(1)} \otimes Q_{j_2}^{(2)})P_{\infty,j_2} = \big(Q_{j_1-1}^{(1)} \otimes I\big)P_{\infty,j_2}.$$

Hence, we conclude

$$\begin{aligned}&\|(P_{j_1,j_2} - P_{j_1-1,j_2})u\|_{\mathcal{H}}\\&\quad\lesssim \big\|\big((I - Q_{j_1}^{(1)}) \otimes I\big)P_{\infty,j_2}u\big\|_{\mathcal{H}} + \big\|\big((I - Q_{j_1-1}^{(1)}) \otimes I\big)P_{\infty,j_2}u\big\|_{\mathcal{H}}\\&\quad\lesssim 2^{-j_1s_1}\|P_{\infty,j_2}u\|_{\mathcal{H}_{mix}^{s_1,0}}.\end{aligned}$$

Using the condition (18) implies finally the desired estimate. ■

Remark 1. Condition (18) holds if $A : \mathcal{H} \to \mathcal{H}'$ is also continuous and bijective as a mapping $A : \mathcal{H}_{mix}^{s_1,0} \to (\mathcal{H}')_{mix}^{s_1,0}$ for all $0 < s_1 \leq p_1$ and also as a mapping $A : \mathcal{H}_{mix}^{0,s_2} \to (\mathcal{H}')_{mix}^{0,s_2}$ for all $0 < s_2 \leq p_2$, respectively. Then, in view of the continuity (5) of the projections $Q_{j_1}^{(1)}$ and $Q_{j_2}^{(2)}$, the Galerkin projections

$$\begin{aligned}P_{\infty,j_2} &= \big((I \otimes Q_{j_2}^{(2)})A(I \otimes Q_{j_2}^{(2)})\big)^{-1}(I \otimes Q_{j_2}^{(2)}) : \mathcal{H} \to V_{\infty,j_2} \subset \mathcal{H},\\P_{j_1,\infty} &= \big((Q_{j_1}^{(1)} \otimes I)A(Q_{j_1}^{(1)} \otimes I)\big)^{-1}(Q_{j_1}^{(1)} \otimes I) : \mathcal{H} \to V_{j_1,\infty} \subset \mathcal{H},\end{aligned}$$

are also continuous as mappings

$$P_{\infty,j_2} : \mathcal{H}_{mix}^{s_1,0} \to V_{\infty,j_2} \subset \mathcal{H}_{mix}^{s_1,0}, \quad P_{j_1,\infty} : \mathcal{H}_{mix}^{0,s_2} \to V_{j_1,\infty} \subset \mathcal{H}_{mix}^{0,s_2},$$

which implies (18).

Lemma 2. *If $u \in \mathcal{H}_{mix}^{s_1,s_2}$, then it holds*

$$\big\|(\Delta_{j_1,j_2}^P - \Delta_{j_1,j_2}^Q)u\big\|_{\mathcal{H}} \lesssim 2^{-j_1s_1-j_2s_2}\|u\|_{\mathcal{H}_{mix}^{s_1,s_2}}$$

for all $0 < s_1 \le p_1$ *and* $0 < s_2 \le p_2$ *where* $\Delta^Q_{j_1,j_2}$ *is given by* (10) *and* $\Delta^P_{j_1,j_2}$ *is given by* (14)*, respectively.*

Proof. Due to $P_{j_1,j_2}(Q^{(1)}_{j_1} \otimes Q^{(2)}_{j_2}) = Q^{(1)}_{j_1} \otimes Q^{(2)}_{j_2}$ for all $j_1, j_2 \ge 0$, we obtain

$$\begin{aligned} \Delta^P_{j_1,j_2} - \Delta^Q_{j_1,j_2} &= P_{j_1,j_2}(I - Q^{(1)}_{j_1} \otimes Q^{(2)}_{j_2}) - P_{j_1-1,j_2}(I - Q^{(1)}_{j_1-1} \otimes Q^{(2)}_{j_2}) \\ &\quad - P_{j_1,j_2-1}(I - Q^{(1)}_{j_1} \otimes Q^{(2)}_{j_2-1}) + P_{i-1,j-1}(I - Q^{(1)}_{j_1-1} \otimes Q^{(2)}_{j_2-1}). \end{aligned} \tag{19}$$

We shall now make use of the identity

$$\begin{aligned} I - Q^{(1)}_{j_1} \otimes Q^{(2)}_{j_2} &= I \otimes I - Q^{(1)}_{j_1} \otimes Q^{(2)}_{j_2} \\ &= I \otimes (I - Q^{(2)}_{j_2}) + (I - Q^{(1)}_{j_1}) \otimes I - (I - Q^{(1)}_{j_1}) \otimes (I - Q^{(2)}_{j_2}). \end{aligned}$$

Inserting this identity into (19) and reordering the terms yields

$$\begin{aligned} \Delta^P_{j_1,j_2} - \Delta^Q_{j_1,j_2} &= (P_{j_1,j_2} - P_{j_1-1,j_2})\big(I \otimes (I - Q^{(2)}_{j_2})\big) \\ &\quad - (P_{j_1,j_2-1} - P_{j_1-1,j_2-1})\big(I \otimes (I - Q^{(2)}_{j_2-1})\big) \\ &\quad + (P_{j_1,j_2} - P_{j_1,j_2-1})\big((I - Q^{(1)}_{j_1}) \otimes I\big) \\ &\quad - (P_{j_1-1,j_2} - P_{j_1-1,j_2-1})\big((I - Q^{(1)}_{j_1-1}) \otimes I\big) \\ &\quad - P_{j_1,j_2}\big((I - Q^{(1)}_{j_1}) \otimes (I - Q^{(2)}_{j_2})\big) \\ &\quad + P_{j_1-1,j_2}\big((I - Q^{(1)}_{j_1-1}) \otimes (I - Q^{(2)}_{j_2})\big) \\ &\quad + P_{j_1,j_2-1}\big((I - Q^{(1)}_{j_1}) \otimes (I - Q^{(2)}_{j_2-1})\big) \\ &\quad - P_{j_1-1,j_2-1}\big((I - Q^{(1)}_{j_1-1}) \otimes (I - Q^{(2)}_{j_2-1})\big). \end{aligned}$$

The combination of the error estimates

$$\begin{aligned} \|(P_{j_1,j_2} - P_{j_1-1,j_2})u\|_{\mathcal{H}} &\lesssim 2^{-j_1 s_1} \|u\|_{\mathcal{H}^{s_1,0}_{mix}}, \\ \|(P_{j_1,j_2} - P_{j_1,j_2-1})u\|_{\mathcal{H}} &\lesssim 2^{-j_2 s_2} \|u\|_{\mathcal{H}^{0,s_2}_{mix}}, \end{aligned}$$

cf. Lemma 1, and

$$\begin{aligned} \big\|\big(I \otimes (I - Q^{(2)}_{j_2})\big)u\big\|_{\mathcal{H}^{s_1,0}_{mix}} &\lesssim 2^{-j_2 s_2} \|u\|_{\mathcal{H}^{s_1,s_2}_{mix}}, \\ \big\|\big((I - Q^{(1)}_{j_1}) \otimes I\big)u\big\|_{\mathcal{H}^{0,s_2}_{mix}} &\lesssim 2^{-j_1 s_1} \|u\|_{\mathcal{H}^{s_1,s_2}_{mix}}, \end{aligned}$$

leads to

$$\begin{aligned}\big\|(P_{j_1,j_2}-P_{j_1-1,j_2})\big(I\otimes(I-Q^{(2)}_{j_2})\big)u\big\|_{\mathcal{H}} &\lesssim 2^{-j_1s_1-j_2s_2}\|u\|_{\mathcal{H}^{s_1,s_2}_{mix}},\\ \big\|(P_{j_1,j_2}-P_{j_1,j_2-1})\big((I-Q^{(1)}_{j_1})\otimes I\big)u\big\|_{\mathcal{H}} &\lesssim 2^{-j_1s_1-j_2s_2}\|u\|_{\mathcal{H}^{s_1,s_2}_{mix}}.\end{aligned} \tag{20}$$

Similarly, from the continuity

$$\|P_{j_1,j_2}u\|_{\mathcal{H}} \lesssim \|u\|_{\mathcal{H}}$$

and

$$\big\|\big((I-Q^{(1)}_{j_1})\otimes(I-Q^{(2)}_{j_2})\big)u\big\|_{\mathcal{H}} \lesssim 2^{-j_1s_1-j_2s_2}\|u\|_{\mathcal{H}^{s_1,s_2}_{mix}},$$

we deduce

$$\big\|P_{j_1,j_2}\big((I-Q^{(1)}_{j_1})\otimes(I-Q^{(2)}_{j_2})\big)u\big\|_{\mathcal{H}} \lesssim 2^{-j_1s_1-j_2s_2}\|u\|_{\mathcal{H}^{s_1,s_2}_{mix}}. \tag{21}$$

With (20) and (21) at hand, we can estimate each of the eight different terms which yields the desired error estimate

$$\big\|(\Delta^P_{j_1,j_2}-\Delta^Q_{j_1,j_2})u\big\|_{\mathcal{H}} \lesssim 2^{-j_1s_1-j_2s_2}\|u\|_{\mathcal{H}^{s_1,s_2}_{mix}}.$$

■

Now, we arrive at our main result which proves optimal convergence rates.

Theorem 2. *The solution* (16) *and* (17)*, respectively, of the combination technique satisfies the error estimate*

$$\|u-\hat{u}_J\|_{\mathcal{H}} \lesssim \begin{cases} 2^{-J\min\{s_1/\sigma,s_2\sigma\}}\|u\|_{\mathcal{H}^{s_1,s_2}_{mix}}, & \text{if } s_1/\sigma \neq s_2\sigma,\\ 2^{-Js_1/\sigma}\sqrt{J}\,\|u\|_{\mathcal{H}^{s_1,s_2}_{mix}}, & \text{if } s_1/\sigma = s_2\sigma,\end{cases}$$

for all $0 < s_1 \leq p_1$ *and* $0 < s_2 \leq p_2$ *provided that* $u \in \mathcal{H}^{s_1,s_2}_{mix}$.

Proof. In view of (15), we have

$$\|u-\hat{u}_J\|^2_{\mathcal{H}} = \Bigg\|\sum_{j_1\sigma+j_2/\sigma>J}\Delta^P_{j_1,j_2}u\Bigg\|^2_{\mathcal{H}}.$$

The Galerkin orthogonality implies the relation

$$\Bigg\|\sum_{j_1\sigma+j_2/\sigma>J}\Delta^P_{j_1,j_2}u\Bigg\|^2_{\mathcal{H}} \sim \sum_{j_1\sigma+j_2/\sigma>J}\big\|\Delta^P_{j_1,j_2}u\big\|^2_{\mathcal{H}}.$$

Thus, we arrive at

$$\|u - \hat{u}_J\|_{\mathcal{H}}^2 \lesssim \sum_{j_1\sigma + j_2/\sigma > J} \|\Delta^Q_{j_1,j_2} u\|_{\mathcal{H}}^2 + \sum_{j_1\sigma + j_2/\sigma > J} \|(\Delta^P_{j_1,j_2} - \Delta^Q_{j_1,j_2})u\|_{\mathcal{H}}^2.$$

We bound the first sum on the right hand side in complete analogy to [10] from above by

$$\sum_{j_1\sigma + j_2/\sigma > J} \|\Delta^Q_{j_1,j_2} u\|_{\mathcal{H}}^2 \lesssim \sum_{j_1\sigma + j_2/\sigma > J} 2^{-2j_1 s_1 - 2j_2 s_2} \|u\|^2_{\mathcal{H}^{s_1,s_2}_{mix}}$$
$$\lesssim \begin{cases} 2^{-2J \min\{s_1/\sigma, s_2\sigma\}} \|u\|^2_{\mathcal{H}^{s_1,s_2}_{mix}}, & \text{if } s_1/\sigma \neq s_2\sigma, \\ 2^{-2J s_1/\sigma} J \|u\|^2_{\mathcal{H}^{s_1,s_2}_{mix}}, & \text{if } s_1/\sigma = s_2\sigma. \end{cases}$$

Likewise, with the help of Lemma 2, the second sum on the right hand side is bounded from above by

$$\sum_{j_1\sigma + j_2/\sigma > J} \|(\Delta^P_{j_1,j_2} - \Delta^Q_{j_1,j_2})u\|_{\mathcal{H}}^2 \lesssim \sum_{j_1\sigma + j_2/\sigma > J} 2^{-2j_1 s_1 - 2j_2 s_2} \|u\|^2_{\mathcal{H}^{s_1,s_2}_{mix}}$$
$$\lesssim \begin{cases} 2^{-2J \min\{s_1/\sigma, s_2\sigma\}} \|u\|^2_{\mathcal{H}^{s_1,s_2}_{mix}}, & \text{if } s_1/\sigma \neq s_2\sigma, \\ 2^{-2J s_1/\sigma} J \|u\|^2_{\mathcal{H}^{s_1,s_2}_{mix}}, & \text{if } s_1/\sigma = s_2\sigma, \end{cases}$$

which, altogether, yields the desired error estimate. ■

7 Numerical Results

We now validate our theoretical findings by numerical experiments. Specifically, we will apply the combination technique for the three examples which were mentioned in Sect. 2. To this end, we consider the most simple case and choose $\Omega_1 = \Omega_2 = (0,1)$, i.e., $n_1 = n_2 = 1$. The ansatz spaces $V_j^{(1)}$ and $V_j^{(2)}$ consist of continuous, piecewise linear ansatz functions on an equidistant subdivision of the interval $(0,1)$ into 2^j subintervals. This yields the polynomial exactnesses $r_1 = r_2 = 2$. For the sake of notational convenience, we set $\square = (0,1) \times (0,1)$.

Example 1. First, we solve the variational problem

$$\text{find } u \in L^2(\square) \text{ such that } a(u,v) = \ell(v) \text{ for all } v \in L^2(\square)$$

where

$$a(u,v) = \int_\square \alpha(x,y) u(x,y) v(x,y) \, \mathrm{d}(x,y)$$

and

$$\ell(v) = \int_{\square} f(x, y)v(x, y)\, \mathrm{d}(x, y). \tag{22}$$

The underlying operator A is the multiplication operator

$$(Au)(x, y) = \alpha(x, y)u(x, y)$$

which is of the order 0. Hence, we have the energy space $\mathcal{H} = L^2(\square)$ and the related spaces of assumed stronger regularity are $\mathcal{H}_{mix}^{s_1,s_2} = H_{mix}^{s_1,s_2}(\square)$. If the multiplier $\alpha(x, y)$ is a smooth function, then A arbitrarily shifts through the Sobolev scales which implies the condition (18) due to Remark 1.

Let the solution u be a smooth function such that $u \in \mathcal{H}_{mix}^{s_1,s_2}$ for given $s_1, s_2 \geq 0$, which holds if the right hand side f is sufficiently regular. Then, the best possible approximation rate for the present discretization with piecewise linear ansatz functions is obtained for $s_1 = r_1 = 2$ and $s_2 = r_2 = 2$, i.e., for $\mathcal{H}_{mix}^{s_1,s_2} = H_{mix}^{2,2}(\square)$. Thus, the *regular sparse tensor product space*

$$\hat{V}_J^1 = \bigoplus_{j_1+j_2 \leq J} W_{j_1}^{(1)} \otimes W_{j_2}^{(2)} = \sum_{j_1+j_2=J} V_{j_1}^{(1)} \otimes V_{j_2}^{(2)}. \tag{23}$$

(cf. (8)) is optimal for the discretization, see [10] for a detailed derivation. In particular, with Theorem 2, the combination technique yields the error estimate

$$\|u - \hat{u}_J\|_{L^2(\square)} \lesssim 4^{-J}\sqrt{J}\,\|u\|_{H_{mix}^{2,2}(\square)}.$$

For our numerical tests, we choose

$$\alpha(x, y) = 1+(x+y)^2, \quad f(x, y) = \alpha(x, y)u(x, y), \quad u(x, y) = \sin(\pi x)\sin(\pi y).$$

The resulting convergence history is plotted as the red curve in Fig. 2. As can be seen there, the convergence rate $4^{-J}\sqrt{J}$, indicated by the dashed red line, is indeed obtained in the numerical experiments.

Example 2. This example concerns the stationary heat conduction in the domain $\square$. In its weak form, it is given by the variational problem

$$\text{find } u \in H_0^1(\square) \text{ such that } a(u, v) = \ell(v) \text{ for all } v \in H_0^1(\square)$$

where

$$a(u, v) = \int_{\square} \alpha(x, y)\left\{\frac{\partial u}{\partial x}(x, y)\frac{\partial v}{\partial x}(x, y) + \frac{\partial u}{\partial y}(x, y)\frac{\partial v}{\partial y}(x, y)\right\} \mathrm{d}(x, y)$$

Fig. 2 Convergence rates in case of the first and second example

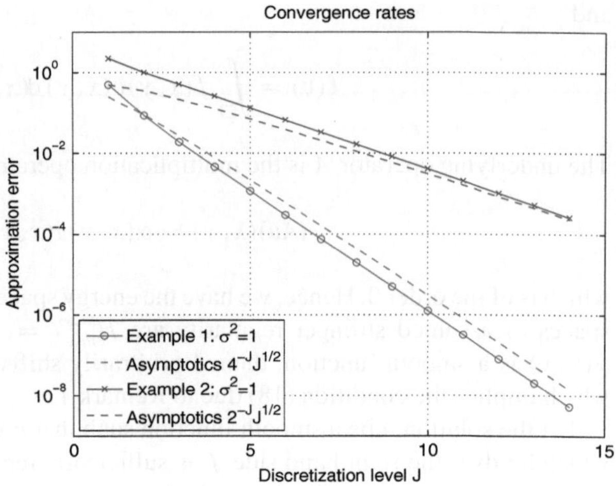

and $\ell(v)$ as in (22). The underlying operator A is the elliptic second order differential operator

$$(Au)(x,y) = -\operatorname{div}_{(x,y)}\big(\alpha(x,y)\nabla_{(x,y)}u(x,y)\big)$$

and maps the energy space $\mathcal{H} = H_0^1(\Box)$ bijectively onto its dual $\mathcal{H}' = H^{-1}(\Box)$. Recall that now the spaces of assumed stronger regularity are $\mathcal{H}_{mix}^{s_1,s_2} = H_0^1(\Box) \cap H_{mix}^{s_1+1,s_2}(\Box) \cap H_{mix}^{s_1,s_2+1}(\Box)$.

Since the domain $\Box$ is convex, the second order boundary value problem under consideration is H^2-regular, which implies that $A : H_0^1(\Box) \cap H^2(\Box) \to L^2(\Box)$ is also bijective. By interpolation arguments, we thus find that $A : \mathcal{H}_{mix}^{1,0} \to (\mathcal{H}')_{mix}^{1,0}$ is continuous and bijective since

$$L^2(\Box) \subset (\mathcal{H}')_{mix}^{1,0} \subset H^{-1}(\Box) \quad \text{and} \quad H_0^1(\Box) \cap H^2(\Box) \subset \mathcal{H}_{mix}^{1,0} \subset H_0^1(\Box).$$

Likewise, $A : \mathcal{H}_{mix}^{0,1} \to (\mathcal{H}')_{mix}^{0,1}$ is continuous and bijective. Hence, the condition (18) holds again due to Remark 1 and Lemma 1 applies.

Again, the regular sparse tensor product space (23) is optimal for the present discretization with piecewise linear ansatz functions. Consequently, Theorem 2 implies as the best possible convergence estimate

$$\|u - \hat{u}_J\|_{H^1(\Box)} \lesssim 2^{-J}\sqrt{J}\,\|u\|_{H_{mix}^{2,1}(\Box)\cap H_{mix}^{1,2}(\Box)}$$

provided that $u \in H_{mix}^{2,1}(\Box) \cap H_{mix}^{1,2}(\Box)$. Here, we exploited that $\mathcal{H}_{mix}^{1,1} = H_0^1(\Box) \cap H_{mix}^{2,1}(\Box) \cap H_{mix}^{1,2}(\Box)$. Nevertheless, in general, we only have $u \in H^2(\Box) \not\subset H_{mix}^{2,1}(\Box) \cap H_{mix}^{1,2}(\Box)$. Thus, due to $H_{mix}^{3/2,1/2}(\Box) \cap H_{mix}^{1/2,3/2}(\Box) \subset H^2(\Box)$, one can only expect the reduced convergence rate

$$\|u - \hat{u}_J\|_{H^1(\square)} \lesssim 2^{-J/2}\sqrt{J}\|u\|_{H^2(\square)}.$$

In our particular numerical computations, we use

$$\alpha(x,y) = 1 + (x+y)^2, \quad u(x,y) = \sin(\pi x)\sin(\pi y),$$
$$f(x,y) = \frac{\partial a}{\partial x}(x,y)\frac{\partial u}{\partial x}(x,y) + \frac{\partial a}{\partial y}(x,y)\frac{\partial u}{\partial y}(x,y) - \alpha(x,y)\Delta u(x,y).$$

Therefore, due to $u \in H^{2,1}_{mix}(\square) \cap H^{1,2}_{mix}(\square)$, we should observe the convergence rate $2^{-J}\sqrt{J}$. The computational approximation errors are plotted as the blue graph in Fig. 2. The dashed blue line corresponds to $2^{-J}\sqrt{J}$ and clearly validates the predicted convergence rate. We even observe the slightly better rate 2^{-J} which can be explained by the fact that the solution u is even in $H^{2,2}_{mix}(\square)$, see [1] for the details.

Example 3. We shall finally consider the variational problem

find $u \in L^2(0,1) \otimes H^1_0(0,1)$ such that $a(u,v) = \ell(v)$ for all $v \in L^2(0,1) \otimes H^1_0(0,1)$

where

$$a(u,v) = \int_\square \alpha(x,y)\frac{\partial u}{\partial y}(x,y)\frac{\partial v}{\partial y}(x,y)\,\mathrm{d}(x,y)$$

and $\ell(v)$ is again given as in (22). The underlying operator A is the elliptic differential operator

$$(Au)(x,y) = -\frac{\partial}{\partial y}\left(\alpha(x,y)\frac{\partial}{\partial y}u(x,y)\right).$$

Its energy space is $\mathcal{H} = L^2(0,1) \otimes H^1_0(0,1) \subset H^{0,1}_{mix}(\square)$ with dual $\mathcal{H}' = L^2(0,1) \otimes H^{-1}(0,1)$. Here, the spaces of assumed stronger regularity coincide with $\mathcal{H}^{s_1,s_2}_{mix} = \big(L^2(0,1) \otimes H^1_0(0,1)\big) \cap H^{s_1,s_2+1}_{mix}(\square)$.

The operator A shifts as a operator $\mathcal{H}^{s_1,s_2+1}_{mix} \to (\mathcal{H}')^{s_1,s_2+1}_{mix}$ for arbitrary $s_1, s_2 \geq 0$ provided that the coefficient α is smooth enough. Thus, Theorem 2 holds and predicts the best possible convergence estimate for our underlying discretization with piecewise linear ansatz functions if u lies in the space $H^{2,2}_{mix}(\square)$.

According to the theory presented in [10], the optimal cost complexity with respect to the generalized sparse tensor product spaces $\hat{V}^\sigma_J$ is obtained for the choice

$$\sigma \in \left[\sqrt{\frac{n_1}{n_2}}, \sqrt{\frac{r_1}{r_2 - 1}}\right] = [1, \sqrt{2}].$$

In order to be able to compare the convergence rates instead of the cost complexities for different choices of σ, we have to consider the generalized sparse tensor product

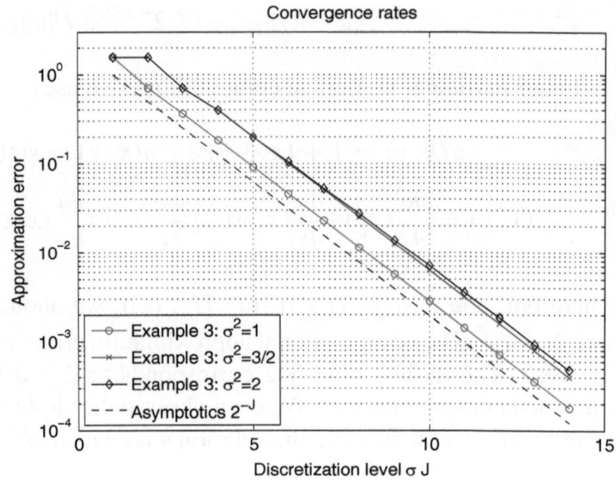

Fig. 3 Convergence rates in case of the third example

spaces $\hat{V}^{\sigma}_{\overline{J}}$, where $\overline{J} := \sigma J$. Then, for all the above choices of σ, we essentially expect the convergence rate

$$\|u - \hat{u}_{\overline{J}}\|_{H^{0,1}_{mix}(\square)} \lesssim 2^{-\overline{J}/\sigma} \|u\|_{H^{2,2}_{mix}(\square)} \sim 2^{-J} \|u\|_{H^{2,2}_{mix}(\square)}$$

while the degrees of freedom of $\hat{V}^{\sigma}_{\overline{J}}$ essentially scale like $2^{\overline{J}/\sigma} \sim 2^{J}$. This setting is employed in our numerical tests, where we further set

$$\alpha(x,y) = 1 + (x+y)^2, \quad u(x,y) = \sin(\pi x)\sin(\pi y),$$

$$f(x,y) = \frac{\partial a}{\partial y}(x,y)\frac{\partial u}{\partial y}(x,y) - \alpha(x,y)\frac{\partial^2 u}{\partial y^2}(x,y).$$

We apply the combination technique for the particular choices

- $\sigma = 1$, which yields an equilibration of the unknowns in all the extremal tensor product spaces $W^{(1)}_{j_1} \otimes W^{(2)}_{\overline{J}-j_1\sigma^2}$,
- $\sigma = \sqrt{2}$, which yields an equilibration of the approximation in all the extremal tensor product spaces $W^{(1)}_{j_1} \otimes W^{(2)}_{\overline{J}-j_1\sigma^2}$, and
- $\sigma = \sqrt{3/2}$, which results in an *equilibrated cost-benefit rate*, see [2, 10] for the details.

The computed approximation errors are found in Fig. 3, where the red curve corresponds to $\sigma = 1$, the black curve corresponds to $\sigma = \sqrt{2}$, and the blue curve corresponds to $\sigma = \sqrt{3/2}$. In the cases $\sigma = 1$ and $\sigma = \sqrt{2}$, we achieve the predicted convergence rate 2^{-J} which is indicated by the dashed black line. In the case $\sigma = \sqrt{2}$ the predicted convergence rate is only $2^{-J}\sqrt{J}$ which is also confirmed by Fig. 3.

8 Conclusion

In the present paper, we proved the convergence of the combination technique in a rather general set-up. Especially, we considered the combination technique in generalized sparse tensor product spaces. We restricted ourselves here to the case of twofold tensor product domains. Nevertheless, all our results can straightforwardly be extended to the case of generalized L-fold sparse tensor product spaces by applying the techniques from [11, 25]. Then, of course, the constants hidden by the "$\sim$"-notation will depend on the given dimension L.

References

1. H.-J. Bungartz, M. Griebel, A note on the complexity of solving Poisson's equation for spaces of bounded mixed derivatives. J. Complex. **15**(2), 167–199 (1999)
2. H.-J. Bungartz, M. Griebel, Sparse grids. Acta Numer. **13**, 147–269 (2004)
3. H. Bungartz, M. Griebel, D. Röschke, C. Zenger, A proof of convergence for the combination technique for the Laplace equation using tools of symbolic computation. Math. Comput. Simul. **42**(4–6), 595–605 (1996)
4. D. Cioranescu, A. Damlamian, G. Griso, The periodic unfolding method in homogenization. SIAM J. Math. Anal. **40**(4), 1585–1620 (2008)
5. W. Dahmen, Wavelet and multiscale methods for operator equations. Acta Numer. **6**, 55–228 (1997)
6. F.-J. Delvos, d-variate Boolean interpolation. J. Approx. Theory **34**(2), 99–114 (1982)
7. F.-J. Delvos, Boolean methods for double integration. Math. Comput. **55**(192), 683–692 (1990)
8. F.-J. Delvos, W. Schempp, *Boolean Methods in Interpolation and Approximation*. Pitman Research Notes in Mathematics Series, vol. 230 (Longman Scientific & Technical, Harlow, 1989)
9. M. Griebel, *Multilevelmethoden als Iterationsverfahren über Erzeugendensystemen*. Teubner Skripten zur Numerik (B.G. Teubner, Stuttgart, 1994)
10. M. Griebel, H. Harbrecht, On the construction of sparse tensor product spaces. Math. Comput. **82**(282), 975–994 (2013)
11. M. Griebel, H. Harbrecht, A note on the construction of L-fold sparse tensor product spaces. Constr. Approx. **38**(2), 235–251 (2013)
12. M. Griebel, P. Oswald, On additive Schwarz preconditioners for sparse grid discretizations. Numer. Math. **66**, 449–463 (1994)
13. M. Griebel, P. Oswald, T. Schiekofer, Sparse grids for boundary integral equations. Numer. Math. **83**(2), 279–312 (1999)
14. M. Griebel, M. Schneider, C. Zenger, A combination technique for the solution of sparse grid problems, in *Iterative Methods in Linear Algebra*, ed. by P. de Groen, R. Beauwens. IMACS (Elsevier, North Holland, 1992), pp. 263–281
15. H. Harbrecht, A finite element method for elliptic problems with stochastic input data. Appl. Numer. Math. **60**(3), 227–244 (2010).
16. H. Harbrecht, R. Schneider, C. Schwab, Sparse second moment analysis for elliptic problems in stochastic domains. Numer. Math. **109**(3), 385–414 (2008)
17. H. Harbrecht, R. Schneider, C. Schwab, Multilevel frames for sparse tensor product spaces. Numer. Math. **110**(2), 199–220 (2008)
18. H. Harbrecht, M. Peters, M. Siebenmorgen, Combination technique based k-th moment analysis of elliptic problems with random diffusion. J. Comput. Phys. **252**, 128–141 (2013)

19. M. Hegland, J. Garcke, V. Challis, The combination technique and some generalisations. Linear Algebra Appl. **420**(2–3), 249–275 (2007)
20. V.H. Hoang, C. Schwab, High-dimensional finite elements for elliptic problems with multiple scales. Multiscale Model. Simul. **3**(1), 168–194 (2004/2005)
21. C. Pflaum, Convergence of the combination technique for second-order elliptic differential equation. SIAM J. Numer. Anal. **34**, 2431–2455 (1997)
22. C. Pflaum, A. Zhou, Error analysis of the combination technique. Numer. Math. **84**, 327–350 (1999)
23. C. Reisinger, Analysis of linear difference schemes in the sparse grid combination technique. IMA J. Numer. Anal. **33**(2), 544–581 (2013)
24. C. Schwab, R.-A. Todor, Sparse finite elements for elliptic problems with stochastic loading. Numer. Math. **95**(4), 707–734 (2003)
25. Y. Xu, A. Zhou, Fast Boolean approximation methods for solving integral equations in high dimensions. J. Integral Equ. Appl. **16**(1), 83–110 (2004)
26. C. Zenger, Sparse grids, in *Parallel Algorithms for Partial Differential Equations. Proceedings of 6th GAMM-Seminar*, Kiel/Germany 1990, ed. by W. Hackbusch. Notes on Numerical Fluid Mechanics, vol. 31 (Vieweg, Braunschweig, 1991), pp. 241–251

Fast Discrete Fourier Transform on Generalized Sparse Grids

Michael Griebel and Jan Hamaekers

Abstract In this paper, we present an algorithm for trigonometric interpolation of multivariate functions on generalized sparse grids and study its application for the approximation of functions in periodic Sobolev spaces of dominating mixed smoothness. In particular, we derive estimates for the error and the cost. We construct interpolants with a computational cost complexity which is substantially lower than for the standard full grid case. The associated generalized sparse grid interpolants have the same approximation order as the standard full grid interpolants, provided that certain additional regularity assumptions on the considered functions are fulfilled. Numerical results validate our theoretical findings.

1 Introduction

In many application areas of numerical simulation, like e.g. physics, chemistry, finance and statistics, high-dimensional approximation problems arise. Here, a conventional numerical approach encounters the so-called curse of dimensionality [4], i.e. the rate of convergence with respect to the number of degrees of freedom deteriorates exponentially with the dimension n. For example a conventional discretization on uniform grids with $\mathcal{O}(2^L)$ points in each direction would involve $M = \mathcal{O}(2^{nL})$ degrees of freedom. Moreover, only a convergence rate of the type

M. Griebel
Institute for Numerical Simulation, University of Bonn, Wegelerstr. 6, 53115 Bonn, Germany
e-mail: griebel@ins.uni-bonn.de

J. Hamaekers (✉)
Fraunhofer Institute for Algorithms and Scientific Computing SCAI, Schloss Birlinghoven, 53754 Sankt Augustin, Germany
e-mail: jan.hamaekers@scai.fraunhofer.de

J. Garcke and D. Pflüger (eds.), *Sparse Grids and Applications - Munich 2012*, Lecture Notes in Computational Science and Engineering 97,
DOI 10.1007/978-3-319-04537-5_4,

$$\|f - f_L^{\mathrm{FG}}\|_{\mathcal{H}^r} \leq c \cdot M^{-\frac{s-r}{n}} \|f\|_{\mathcal{H}^s}$$

can be achieved, where $\|\cdot\|_{\mathcal{H}^r}$ is the usual Sobolev norm in $\mathcal{H}^r$, s denotes the isotropic smoothness of f and c is a constant which may depend on n and the underlying domain Ω but not on the discretization parameter L.

So-called sparse grid based approaches have emerged as useful techniques to tackle higher dimensional problems, since they open the possibility to break the curse of dimensionality under certain conditions. They date back to [47]. For example, if f is in a Sobolev space of bounded mixed smoothness $\mathcal{H}^t_{\mathrm{mix}}(\Omega)$, i.e. if the t-th mixed derivatives of f are bounded, and Ω is a product domain, an error estimate of the type

$$\|f - f_L^{\mathrm{SG}}\|_{\mathcal{H}^r} \leq c \cdot 2^{-(t-r)L} L^{n-1} \|f\|_{\mathcal{H}^t_{\mathrm{mix}}}$$

can be achieved using so-called regular sparse grids where $\mathcal{O}(2^L L^{n-1})$ degrees of freedom are involved.[1] Here, the rate of convergence with respect to the number of degrees of freedom does no longer exponentially deteriorate with the number n of dimensions, except for the logarithmic terms L^{n-1}. Moreover, in specific cases, the use of so-called energy-norm based sparse grids [6] may even result in an error estimate of type

$$\|f - f_L^{\mathrm{ESG}}\|_{\mathcal{H}^r} \leq c \cdot 2^{-(t-r)L} \|f\|_{\mathcal{H}^t_{\mathrm{mix}}},$$

where only $\mathcal{O}(2^L)$ degrees of freedom are involved. Hence, compared to the regular sparse grid case even the logarithmic terms L^{n-1} are eliminated.[2]

For the discretization with sparse grids, Galerkin type methods, finite difference approaches and the so-called combination technique have been developed over the last two decades [6]. Furthermore, these approaches were used in the context of moderate higher-dimensional elliptic, parabolic and hyperbolic differential equations. In addition, sparse grid techniques were successfully applied for the solution of integral equations [25], for quadrature [14], for regression [12] and for time series prediction [5]. Moreover, the sparse grid method was supplemented with adaptive refinement schemes [5, 14, 22], was used for the construction of anisotropic sparse tensor product spaces [20, 21] and was applied in the context of weighted mixed spaces [19, 22]. Sparse grid based collocation schemes were for example discussed in [2, 26, 29, 30, 33, 39]. They recently found widespread use in the important field of uncertainty quantification [38]. On a theoretical level, sparse grids are closely

---

[1] Here, in case of the best linear approximation, estimates of type $\|f - f_L^{\mathrm{SG}}\|_{\mathcal{H}^r} \lesssim 2^{-(t-r)L} L^{(n-1)/2} \|f\|_{\mathcal{H}^t_{\mathrm{mix}}}$ and even of type $\|f - f_L^{\mathrm{SG}}\|_{\mathcal{H}^r} \lesssim 2^{-(t-r)L} \|f\|_{\mathcal{H}^t_{\mathrm{mix}}}$ could be achieved for certain types of basis sets [25, 34, 49]. This holds, e.g. for wavelets and the Fourier basis, respectively.

[2] The constants in the cost and accuracy estimates still depend on n, though.

related to ANOVA-like decompositions [11, 16, 27] which are well-known from statistics. A detailed survey on sparse grids is for example given in [6].

Note that the adaption of the sparse grid techniques to Fourier based methods is done by means of Fourier polynomials from the hyperbolic cross and hence sparse grid methods are also known under the name hyperbolic cross approximation [43, 48]. The properties of such approximations of functions in Sobolev spaces on the n-dimensional torus $\mathbb{T}^n$ have been studied by several authors [7, 9, 15, 31, 32, 34, 36, 41, 44, 45, 48]. In particular, spaces of generalized mixed Sobolev smoothness

$$\mathcal{H}^{t,r}_{\text{mix}}(\mathbb{T}^n) := \left\{ f : \sqrt{\sum_{\mathbf{k}\in\mathbb{Z}^n} \prod_{d=1}^{n} (1 + |k_d|)^{2t} (1 + |\mathbf{k}|_\infty)^{2r} |\hat{f}_{\mathbf{k}}|^2} < \infty \right\}$$

and a specific generalization of the regular sparse grid spaces based on Fourier polynomials $e^{i\mathbf{k}^T\mathbf{x}}$ with frequencies $\mathbf{k}$ from the generalized hyperbolic cross

$$\Gamma_L^T := \left\{ \mathbf{k} \in \mathbb{Z}^n : \prod_{d=1}^{n} (1 + k_d) \cdot (1 + |\mathbf{k}|_\infty) \leq L^{(1-T)} \right\}$$

were introduced in [34], further discussed in [23, 24, 27, 35] and a generalization to Banach spaces is given in [8]. Here, $T \in [-\infty, 1)$ is an additional parameter that controls the mixture of isotropic and mixed smoothness: The case $T = 0$ corresponds to the conventional hyperbolic cross (or regular sparse grid). In that case for example the $\mathcal{H}^r$-error of the best linear approximate f_L^{HC} in the conventional hyperbolic cross discretization space of a function in a periodic Sobolev space of dominated mixed smoothness $\mathcal{H}^{t,r}(\mathbb{T}^n)$ is of order $\mathcal{O}(2^{-tL})$, where $\mathcal{O}(2^L L^{n-1})$ frequencies are involved. Furthermore, the case $T = -\infty$ corresponds to the full grid, the case $T \to 1$ corresponds to a latin hypercube and the case $0 < T < 1$ resembles energy-norm based sparse grids where the order of the amount of included frequencies does not depend on the number of dimensions n, i.e. it is $\mathcal{O}(2^L)$. But let us note here that it is in general not clear, if the approximation error of an *interpolant* exhibits the same convergence rates as that of the best linear approximation.

In this paper, we now mainly deal with trigonometric interpolation on generalized sparse grids and its application for the approximation of multivariate functions in certain periodic Sobolev spaces of bounded mixed smoothness. For functions on the torus, regular sparse grid interpolation methods based on the fast Fourier transform were for example introduced by Hallatschek in [26] and also discussed in [3, 15, 30, 31, 36, 44]. For example, for a function in a Korobov space of mixed smoothness $t > 1$ it is proved in [26] that the approximation error in the maximum norm of its regular sparse grid interpolant is of the order $\mathcal{O}(2^{-L(t-1)} L^{n-1})$ and in [15] a (suboptimal) upper bound estimate of the same order is shown for the approximation error in the $\mathcal{L}^2$ norm. Here, the involved degrees of freedom are of the order

$\mathcal{O}(2^L L^{n-1})$ and the computational cost complexity is of the order[3] $\mathcal{O}(2^L L^n)$. Based on the results of [48] it was furthermore shown in [36, 37] that the approximation error in the $\mathcal{H}^r$-norm of the interpolant associated with a regular sparse grid is of the order $\mathcal{O}(2^{-(t-r)L} L^{n-1})$, if the function is in a periodic Sobolev space $\mathcal{H}^t_{\text{mix}}$ with $t > \frac{1}{2}$.

In this work, we present an extension of the algorithm of Hallatschek given in [26] to the case of interpolation on the generalized sparse grids as introduced in [23]. We will further study its best linear approximation error and will give cost complexity estimates for functions in different variants of periodic Sobolev spaces of dominating mixed smoothness. Moreover, for functions of mixed Sobolev smoothness $\mathcal{H}^t_{\text{mix}}$, we will show estimates for the approximation error of the interpolant in the $\mathcal{H}^r$-norm. To our knowledge this has been done so far only for the regular sparse grid case $T = 0$, but not yet for the case of generalized sparse grids with $0 < T < 1$, which resemble the energy-norm based sparse grids. Note further that the behavior of the approximation error of the interpolant versus the computational complexity of the interpolant is of practical interest. This holds especially with possible applications in the field of uncertainty quantification in mind. Therefore, we give also estimates for its computational complexity. Altogether, it turns out that under specific conditions the order rates of the error complexity and computational complexity are independent of the dimension n of the function. For example, let $f \in \mathcal{H}^2_{\text{mix}}(\mathbb{T}^n)$ and let us measure the approximation error in the $\mathcal{H}^1$-norm, where $\mathbb{T}^n$ denotes the n-dimensional torus. Then, an error of the order $\mathcal{O}(2^{-L})$ and a computational complexity of the order $\mathcal{O}(2^L L)$ can be achieved[4] for interpolants corresponding to a generalized sparse grid with $0 < T < \frac{1}{2}$ for any dimension n.

The remainder of this paper is organized as follows: In Sect. 2 we introduce the fast Fourier transform on general sparse grids with hierarchical bases. In particular, we will recall the conventional Fourier basis representation of periodic functions in Sect. 2.1 and the so-called hierarchical Fourier basis representation in Sect. 2.2. Furthermore, in Sect. 2.3, we will present generalized sparse grids and discuss the construction and application of associated trigonometric interpolation operators and computational complexities. In Sect. 3 we will introduce different variants of periodic Sobolev spaces and will discuss their associated best linear approximation error in Sect. 3.2, the approximation error of the trigonometric general sparse grid interpolants in Sect. 3.3 and its overall complexities in Sect. 3.4. Then, in Sect. 3.5, we will give some short remarks on further generalizations of sparse grids like, e.g. periodic Sobolev spaces with finite-order weights and dimension-adaptive approaches. In Sect. 4 we will apply our approach to some test cases. Finally we give some concluding remarks in Sect. 5.

---

[3] We here do not have $\mathcal{O}(2^L L^{n-1})$ but we have $\mathcal{O}(2^L L^n)$ since one L stems from the computational complexity of the one-dimensional FFT involved.

[4] Here, also the L stems from the involved FFT algorithm.

2 Fourier Transform on General Sparse Grids with Hierarchical Bases

To construct a trigonometric interpolation operator for generalized sparse grids, we will follow the approach of Hallatschek [26]. To this end, we will first recall the conventional Fourier basis and then introduce the so-called hierarchical Fourier basis and its use in the construction of a generalized sparse grid interpolant.

2.1 Fourier Basis Representation

First, let us shortly recall the usual Fourier basis representation of periodic functions. To this end, let $\mathbb{T}^n$ be the n-torus, which is the n-dimensional cube $\mathbb{T}^n \subset \mathbb{R}^n$, $\mathbb{T} = [0, 2\pi]$, where opposite sides are identified. We then have n-dimensional coordinates $\mathbf{x} := (x_1, \ldots, x_n)$, where $x_d \in \mathbb{T}$. We define the basis function associated with a multi index $\mathbf{k} = (k_1, \ldots, k_n) \in \mathbb{Z}^n$ by

$$\omega_{\mathbf{k}}(\mathbf{x}) := \left(\bigotimes_{d=1}^{n} \omega_{k_d}\right)(\mathbf{x}) = \prod_{d=1}^{n} \omega_{k_d}(x_d), \quad \omega_k(x) := e^{ikx}. \tag{1}$$

The set $\{\omega_{\mathbf{k}}\}_{\mathbf{k}\in\mathbb{Z}^n}$ is a complete orthogonal system of the space $\mathcal{L}^2(\mathbb{T}^n)$ and hence every $f \in \mathcal{L}^2(\mathbb{T}^n)$ has the unique expansion

$$f(\mathbf{x}) = \sum_{\mathbf{k}\in\mathbb{Z}^n} \hat{f}_{\mathbf{k}} \omega_{\mathbf{k}}(\mathbf{x}), \tag{2}$$

where the Fourier coefficients are given by

$$\hat{f}_{\mathbf{k}} := \frac{1}{(2\pi)^n} \int_{\mathbb{T}^n} \omega_{\mathbf{k}}^*(\mathbf{x}) f(\mathbf{x})\, d\mathbf{x}. \tag{3}$$

Note that it is common to characterize the smoothness classes of a function f by the decay properties of its Fourier coefficients [28]. In this way, we introduce the periodic Sobolev space of isotropic smoothness as

$$\mathcal{H}^r(\mathbb{T}^n) := \left\{ f(\mathbf{x}) = \hat{f}_{\mathbf{k}} \omega_{\mathbf{k}}(\mathbf{x}) \,:\, \|f\|_{\mathcal{H}^r} := \sqrt{\sum_{\mathbf{k}\in\mathbb{Z}^n} (1 + |\mathbf{k}|_\infty)^{2r} |\hat{f}_{\mathbf{k}}|^2} < \infty \right\}$$

for $r \in \mathbb{R}$.

Let us now define finite-dimensional subspaces of the space $\mathcal{L}^2(\mathbb{T}^n) = \mathcal{H}^0(\mathbb{T}^n)$ for discretization purposes. To this end, we set

$$\sigma : \mathbb{N}_0 \to \mathbb{Z} : j \mapsto \begin{cases} -j/2 & \text{if } j \text{ is even,} \\ (j+1)/2 & \text{if } j \text{ is odd.} \end{cases} \tag{4}$$

For $l \in \mathbb{N}_0$ we introduce the one-dimensional nodal basis

$$\mathcal{B}_l := \{\phi_j\}_{0 \le j \le 2^l - 1} \quad \text{with} \quad \phi_j := \omega_{\sigma(j)} \tag{5}$$

and the corresponding spaces $V_l := \text{span}\{\mathcal{B}_l\}$. For a multi index $\mathbf{l} \in \mathbb{N}_0^n$ we define finite-dimensional spaces by a tensor product construction, i.e. $V_{\mathbf{l}} := \bigotimes_{d=1}^n V_{l_d}$. Finally, we introduce the space[5]

$$V := \sum_{\mathbf{l} \in \mathbb{N}^n} V_{\mathbf{l}}.$$

In the following we will shortly recall the common one-dimensional trigonometric interpolation. Let the Fourier series $\sum_{k \in \mathbb{Z}} \hat{f}_k \omega_k$ be pointwise convergent to f. Then, for interpolation points $\mathcal{S}_l := \{m \frac{2\pi}{2^l} : m = 0, \ldots, 2^l - 1\}$ of level $l \in \mathbb{N}_0$, the interpolation operator can be defined by

$$I_l : V \to V_l : f \mapsto I_l f := \sum_{j \in \mathcal{G}_l} \hat{f}_j^{(l)} \phi_j$$

with indices $\mathcal{G}_l := \{0, \ldots, 2^l - 1\}$ and discrete nodal Fourier coefficients

$$\hat{f}_j^{(l)} := 2^{-l} \sum_{x \in \mathcal{S}_l} f(x) \phi_j^*(x). \tag{6}$$

This way, the 2^l interpolation conditions

$$f(x) = I_l f(x) \quad \text{for all } x \in \mathcal{S}_l$$

are fulfilled. In particular, from (6) and (2) one can deduce the well-known aliasing formula

$$\hat{f}_j^{(l)} = \sum_{k \in \mathbb{Z}} \hat{f}_k 2^{-l} \sum_{x \in \mathcal{S}_l} \omega_{\sigma(j)}^*(x) \omega_k(x) = \sum_{m \in \mathbb{Z}} \hat{f}_{\sigma(j) + m 2^l}. \tag{7}$$

Next, let us consider the case of multivariate functions. To this end, let the Fourier series $\sum_{\mathbf{k} \in \mathbb{Z}^n} \hat{f}_{\mathbf{k}} \omega_{\mathbf{k}}$ be pointwise convergent to f. Then, according to the tensor

[5]Except for the completion with respect to a chosen Sobolev norm, V is just the associated Sobolev space.

product structure of the n-dimensional spaces, we introduce the n-dimensional interpolation operator on full grids as

$$I_{\mathbf{l}} := I_{l_1} \otimes \cdots \otimes I_{l_n} \,:\, V \to V_{\mathbf{l}} \,:\, f \mapsto I_{\mathbf{l}} f = \sum_{\mathbf{j} \in \mathcal{G}_{\mathbf{l}}} \hat{f}_{\mathbf{j}}^{(\mathbf{l})} \phi_{\mathbf{j}},$$

with

$$\mathcal{G}_{\mathbf{l}} := \mathcal{G}_{l_1} \times \cdots \times \mathcal{G}_{l_n} \subset \mathbb{N}_0^n$$

and multi-dimensional discrete nodal Fourier coefficients

$$\hat{f}_{\mathbf{j}}^{(\mathbf{l})} := 2^{-|\mathbf{l}|_1} \sum_{\mathbf{x} \in \mathcal{S}_{\mathbf{l}}} f(\mathbf{x}) \phi_{\mathbf{j}}^*(\mathbf{x}), \tag{8}$$

where

$$\mathcal{S}_{\mathbf{l}} := \mathcal{S}_{l_1} \times \cdots \times \mathcal{S}_{l_n} \subset \mathbb{T}^n.$$

Similar to (7) it holds the multi-dimensional aliasing formula

$$\hat{f}_{\mathbf{j}}^{(\mathbf{l})} = \sum_{\mathbf{m} \in \mathbb{Z}^n} \hat{f}_{\sigma(\mathbf{j}) + \mathbf{m}\, 2^{\mathbf{l}}}, \tag{9}$$

where $\sigma(\mathbf{j}) := (\sigma(j_1), \ldots, \sigma(j_n))$ and $\mathbf{m}\, 2^{\mathbf{l}} := \left(m_1 2^{l_1}, \ldots, m_n 2^{l_n}\right)$.

2.2 One-Dimensional Hierarchical Fourier Basis Representation

Now we discuss a hierarchical variant of the Fourier basis representation. Let us first consider the one-dimensional case. To this end, we introduce an univariate Fourier hierarchical basis function for $j \in \mathbb{N}_0$ by

$$\psi_j := \begin{cases} \phi_0 & \text{for } j = 0, \\ \phi_j - \phi_{2^l - 1 - j} & \text{for } 2^{l-1} \le j \le 2^l - 1,\, l \ge 1, \end{cases} \tag{10}$$

and we define the one-dimensional hierarchical Fourier basis including basis functions up to level $l \in \mathbb{N}_0$ by

$$\mathcal{B}_l^{\mathrm{h}} := \{\psi_j\}_{0 \le j \le 2^l - 1}.$$

Let us further introduce the difference spaces

$$W_l := \begin{cases} \mathrm{span}\{\mathcal{B}_0^{\mathrm{h}}\} & \text{for } l = 0, \\ \mathrm{span}\{\mathcal{B}_l^{\mathrm{h}} \setminus \mathcal{B}_{l-1}^{\mathrm{h}}\} & \text{for } l > 0. \end{cases}$$

Note that it holds the relation $V_l = \mathrm{span}\{\mathcal{B}_l\} = \mathrm{span}\{\mathcal{B}_l^{\mathrm{h}}\}$ for all $l \in \mathbb{N}_0$. Thus, we have the direct sum decomposition $V_l = \bigoplus_{v=0}^{l} W_l$. Now, let $l \in \mathbb{N}_0$ and $u \in V_l$. Then, one can easily switch from the hierarchical representation $u = \sum_{0 \le j \le 2^l - 1} u'_j \psi_j$ to the nodal representation $u = \sum_{0 \le j \le 2^l - 1} u_j \phi_j$ by a linear transform. For example for $l = 0, 1, 2, 3$, the corresponding *de-hierarchization* matrices read as

$$(1), \quad \begin{pmatrix} 1 & -1 \\ 0 & 1 \end{pmatrix}, \quad \begin{pmatrix} 1 & -1 & 0 & -1 \\ 0 & 1 & -1 & 0 \\ 0 & 0 & 1 & 0 \\ 0 & 0 & 0 & 1 \end{pmatrix}, \quad \begin{pmatrix} 1 & -1 & 0 & -1 & 0 & 0 & 0 & -1 \\ 0 & 1 & -1 & 0 & 0 & 0 & -1 & 0 \\ 0 & 0 & 1 & 0 & 0 & -1 & 0 & 0 \\ 0 & 0 & 0 & 1 & -1 & 0 & 0 & 0 \\ 0 & 0 & 0 & 0 & 1 & 0 & 0 & 0 \\ 0 & 0 & 0 & 0 & 0 & 1 & 0 & 0 \\ 0 & 0 & 0 & 0 & 0 & 0 & 1 & 0 \\ 0 & 0 & 0 & 0 & 0 & 0 & 0 & 1 \end{pmatrix},$$

respectively. For all $l \in \mathbb{N}_0$ the de-hierarchization matrix can be easily inverted and its determinant is equal to one. Here, the corresponding *hierarchization* matrices read as

$$(1), \quad \begin{pmatrix} 1 & 1 \\ 0 & 1 \end{pmatrix}, \quad \begin{pmatrix} 1 & 1 & 1 & 1 \\ 0 & 1 & 1 & 0 \\ 0 & 0 & 1 & 0 \\ 0 & 0 & 0 & 1 \end{pmatrix}, \quad \begin{pmatrix} 1 & 1 & 1 & 1 & 1 & 1 & 1 & 1 \\ 0 & 1 & 1 & 0 & 0 & 1 & 1 & 0 \\ 0 & 0 & 1 & 0 & 0 & 1 & 0 & 0 \\ 0 & 0 & 0 & 1 & 1 & 0 & 0 & 0 \\ 0 & 0 & 0 & 0 & 1 & 0 & 0 & 0 \\ 0 & 0 & 0 & 0 & 0 & 1 & 0 & 0 \\ 0 & 0 & 0 & 0 & 0 & 0 & 1 & 0 \\ 0 & 0 & 0 & 0 & 0 & 0 & 0 & 1 \end{pmatrix},$$

for $l = 0, 1, 2, 3$ respectively. Simple algorithms with cost complexity $\mathcal{O}(2^l)$ for hierarchization and de-hierarchization are given in [26].

Let us now define an operator

$$\check{I}_l := (I_l - I_{l-1}) : V \to W_l, \text{ for } l \ge 0,$$

where we set $I_{-1} = 0$. Note that the image of $\check{I}_l$ is a subspace of W_l. Hence, we define the corresponding hierarchical Fourier coefficients $\check{f}_j$ by the unique representation

$$\check{I}_l f = \sum_{0 \le v < l} \sum_{j \in \mathcal{J}_v} (\hat{f}_j^{(l)} - \hat{f}_j^{(l-1)}) \phi_j + \sum_{j \in \mathcal{J}_l} \hat{f}_j^{(l)} \phi_j = \sum_{j \in \mathcal{J}_l} \check{f}_j \psi_j \qquad (11)$$

with

$$\mathcal{J}_v := \begin{cases} \{0\} & \text{for } v = 0, \\ \{2^{v-1}, \ldots, 2^v - 1\} & \text{for } v \geq 1. \end{cases} \tag{12}$$

Moreover, we can write the interpolation operator associated with a level l in the form

$$\begin{aligned} I_l f &= (I_l - I_{l-1} + I_{l-1} - \cdots - I_0 + I_0 - I_{-1}) f \\ &= (\breve{I}_l + \cdots + \breve{I}_0) f \\ &= \sum_{0 \leq v \leq l} \sum_{j \in \mathcal{J}_v} \breve{f}_j \psi_j = \sum_{0 \leq j \leq 2^l - 1} \breve{f}_j \psi_j . \end{aligned}$$

In particular, let us note that the following interpolation relation holds:

$$\breve{I}_l f(x) = f(x) \quad \text{for all } x \in \mathcal{S}_l^{\mathrm{h}}$$

and

$$\breve{I}_l f(x) = 0 \quad \text{for all } x \in \mathcal{S}_{l-1},$$

where

$$\mathcal{S}_l^{\mathrm{h}} := \mathcal{S}_l \setminus \mathcal{S}_{l-1},$$

with $\mathcal{S}_{-1} := \emptyset$.

For $l \in \mathbb{N}_0$ it follows by the definitions (10) and (11) the equation

$$\sum_{0 \leq v < l} \sum_{j \in \mathcal{J}_v} (\hat{f}_j^{(l)} - \hat{f}_j^{(l-1)}) \phi_j + \sum_{j \in \mathcal{J}_l} \hat{f}_j^{(l)} \phi_j = - \sum_{0 \leq v < l} \sum_{j \in \mathcal{J}_v} \breve{f}_{2^l - 1 - j} \phi_j + \sum_{j \in \mathcal{J}_l} \breve{f}_j \phi_j$$

and with it for $j \in \mathcal{J}_l$ that the hierarchical Fourier coefficient $\breve{f}_j$ is equal to the discrete nodal Fourier coefficient $\hat{f}_j^{(l)}$ associated with level l. Hence in the case $l \in \mathbb{N}_0$, $j \in \mathcal{J}_l$ we obtain the relation

$$\breve{f}_j = \hat{f}_j^{(l)} = \sum_{m \in \mathbb{Z}} \hat{f}_{\sigma(j) + m 2^l} \tag{13}$$

with the help of the aliasing formula (7).

Let us remark that the one-dimensional standard Fourier basis representation is sufficient to define multi-dimensional full grids, but the hierarchical Fourier basis representation is indeed necessary for the definition of sparse grids.

2.3 Generalized Sparse Grids

Now we consider the case of multivariate functions. Here, we use a tensor product ansatz to construct n-dimensional basis functions as well as spaces. This way, we set $\psi_{\mathbf{j}} := \bigotimes_{d=1}^{n} \psi_{j_d}$ and $W_{\mathbf{l}} := \bigotimes_{d=1}^{n} W_{l_d}$ for $\mathbf{l} \in \mathbb{N}_0^n$. In particular, we have the direct sum decomposition

$$V = \bigoplus_{\mathbf{l} \in \mathbb{N}^n} W_{\mathbf{l}}.$$

Moreover, we define $W_{\mathcal{I}} := \bigoplus_{\mathbf{l} \in \mathcal{I}} W_{\mathbf{l}}$ for an index set $\mathcal{I} \subset \mathbb{N}_0^n$. For the general sparse grid construction, we restrict ourselves to index sets, which obey the following condition [14, 26]: An index set $\mathcal{I} \subset \mathbb{N}_0^n$ is called admissible if it holds the relation

$$\{\mathbf{v} \in \mathbb{N}_0^n : \mathbf{v} \leq \mathbf{l}\} \subset \mathcal{I}, \tag{14}$$

for all $\mathbf{l} \in \mathcal{I}$. Here, the inequality $\mathbf{v} \leq \mathbf{w}$ is to be understood componentwise, i.e. $\mathbf{v} \leq \mathbf{w} :\Leftrightarrow v_d \leq w_d$ for all $1 \leq d \leq n$. Now, for an admissible index set $\mathcal{I}$, we define generalized sparse grid spaces by

$$V_{\mathcal{I}} := \sum_{\mathbf{l} \in \mathcal{I}} V_{\mathbf{l}} = \bigoplus_{\mathbf{l} \in \mathcal{I}} W_{\mathbf{l}} = W_{\mathcal{I}}. \tag{15}$$

Due to property (14) of $\mathcal{I}$ we are able to introduce the corresponding general sparse grid trigonometric interpolation operator by

$$I_{\mathcal{I}} := \sum_{\mathbf{l} \in \mathcal{I}} \check{I}_{\mathbf{l}} : V \to V_{\mathcal{I}}, \quad \text{where } \check{I}_{\mathbf{l}} := \check{I}_{l_1} \otimes \cdots \otimes \check{I}_{l_n} : V \to W_{\mathbf{l}}.$$

This way, the associated set of interpolation points is given by

$$\mathcal{S}_{\mathcal{I}} := \bigcup_{\mathbf{l} \in \mathcal{I}} \mathcal{S}_{\mathbf{l}}^{\mathrm{h}},$$

where

$$\mathcal{S}_{\mathbf{l}}^{\mathrm{h}} := \mathcal{S}_{l_1}^{\mathrm{h}} \times \cdots \times \mathcal{S}_{l_n}^{\mathrm{h}}.$$

Let us note that this general sparse grid construction includes generalized sparse grids as introduced in [23], i.e. we may employ for $\mathcal{I}$ the index set

$$\mathcal{I}_L^T := \{\mathbf{l} : |\mathbf{l}|_1 - T|\mathbf{l}|_\infty \leq (1 - T)L\}, \quad T < 1. \tag{16}$$

Hence also full grids, i.e. $\mathcal{I}_L^{-\infty} = \{\mathbf{l} : |\mathbf{l}|_\infty \le L\}$ and conventional sparse grids, i.e. $\mathcal{I}_L := \mathcal{I}_L^0 = \{\mathbf{l} : |\mathbf{l}|_1 \le L\}$ of level $L \in \mathbb{N}_0$ are covered as special cases. For a function f with a pointwise convergent Fourier series, the multi-dimensional hierarchical coefficients $\breve{f}_{\mathbf{j}}$ are given by the unique representation

$$\breve{I}_{\mathbf{l}} f = \sum_{\mathbf{j} \in \mathcal{J}_{\mathbf{l}}} \breve{f}_{\mathbf{j}} \psi_{\mathbf{j}},$$

where

$$\mathcal{J}_{\mathbf{l}} := \mathcal{J}_{l_1} \times \cdots \times \mathcal{J}_{l_n}.$$

In particular, the hierarchical Fourier series

$$\sum_{\mathbf{l} \in \mathbb{N}_0^n} \sum_{\mathbf{j} \in \mathcal{J}_{\mathbf{l}}} \breve{f}_{\mathbf{j}} \psi_{\mathbf{j}} \tag{17}$$

converges pointwise to f on all grids $S_{\mathbf{l}}$, $\mathbf{l} \in \mathbb{N}_0^n$. Furthermore, with the help of the multi-dimensional aliasing formula (9) a relation similar to (13) can easily be deduced, that is, for $\mathbf{l} \in \mathbb{N}_0^n$ and $\mathbf{j} \in \mathcal{J}_{\mathbf{l}}$, it holds

$$\breve{f}_{\mathbf{j}}^{(\mathbf{l})} = \hat{f}_{\mathbf{j}}^{(\mathbf{l})} = \sum_{\mathbf{m} \in \mathbb{Z}^n} \hat{f}_{\sigma(\mathbf{j}) + \mathbf{m}\, 2^{\mathbf{l}}}. \tag{18}$$

According to definition (15) of the general sparse grid space $V_{\mathcal{I}}$, we can estimate its number of degrees of freedom by

$$|V_{\mathcal{I}}| \lesssim \sum_{\mathbf{l} \in \mathcal{I}} 2^{|\mathbf{l}|_1}. \tag{19}$$

Starting from relation (19) the following complexity estimate is shown in the case of the general index sets $\mathcal{I}_L^T$ of (16) in [23, 24]:

Lemma 1. *Let $L \in \mathbb{N}_0$ and $T < 1$. The number of degrees of freedom of the general sparse grid spaces $V_{\mathcal{I}_L^T}$ with respect to the discretization parameter L is*

$$\left|V_{\mathcal{I}_L^T}\right| \lesssim \sum_{\mathbf{l} \in \mathcal{I}_L^T} 2^{|\mathbf{l}|_1} \lesssim \begin{cases} 2^L & \text{for } 0 < T < 1, \\ 2^L L^{n-1} & \text{for } T = 0, \\ 2^{L \frac{T-1}{T/n-1}} & \text{for } T < 0, \\ 2^{Ln} & \text{for } T = -\infty. \end{cases} \tag{20}$$

Furthermore, analogously to the well-known case of a multi-dimensional discrete Fourier transform, we can utilize the tensor product structure of the underlying spaces and operators to efficiently compute the general sparse grid interpolant

Algorithm 1 A procedure analog to [26] to apply the general sparse grid interpolation operator $I_{\mathcal{I}}$ for a given admissible index set $\mathcal{I}$ and given interpolation values $\{u_{\mathbf{j}} \in \mathbb{C}\}_{\mathbf{j}\in\mathcal{J}_{\mathbf{l}},\mathbf{l}\in\mathcal{I}}$ associated to the general sparse grid interpolation points $\mathcal{S}_{\mathcal{I}}$. The algorithm works in-place on the given input coefficients, where we use an additional temporary array to perform the involved one-dimensional FFTs

for $d = 1$ to n **do**
  for all $\mathbf{l} \in \mathcal{M}_d$ **do**
    for all $\mathbf{j} \in \mathcal{J}_{l_1} \times \cdots \times \mathcal{J}_{l_{d-1}} \times \mathcal{J}_0 \times \mathcal{J}_{l_{d+1}} \times \cdots \times \mathcal{J}_{l_n}$ **do**
      One-dimensional FFT for $(u_{j_1,\ldots,j_{d-1},0,j_{d+1},\ldots,j_n}, \ldots, u_{j_1,\ldots,j_{d-1},2^{l_d}-1,j_{d+1},\ldots,j_n})$
      Hierarchization for $(u_{j_1,\ldots,j_{d-1},0,j_{d+1},\ldots,j_n}, \ldots, u_{j_1,\ldots,j_{d-1},2^{l_d}-1,j_{d+1},\ldots,j_n})$
    end for
  end for
end for
// At this stage, the hierarchical Fourier coefficients are given in $\{u_{\mathbf{j}}\}_{\mathbf{j}\in\mathcal{J}_{\mathbf{l}},\mathbf{l}\in\mathcal{I}}$.
for $d = n$ to 1 **do**
  for all $\mathbf{l} \in \mathcal{M}_d$ **do**
    for all $\mathbf{j} \in \mathcal{J}_{l_1} \times \cdots \times \mathcal{J}_{l_{d-1}} \times \mathcal{J}_0 \times \mathcal{J}_{l_{d+1}} \times \cdots \times \mathcal{J}_{l_n}$ **do**
      De-hierarchization for $(u_{j_1,\ldots,j_{d-1},0,j_{d+1},\ldots,j_n}, \ldots, u_{j_1,\ldots,j_{d-1},2^{l_d}-1,j_{d+1},\ldots,j_n})$
    end for
  end for
end for
// Finally, the non-hierarchical sparse grid Fourier coefficients are given in $\{u_{\mathbf{j}}\}_{\mathbf{j}\in\mathcal{J}_{\mathbf{l}},\mathbf{l}\in\mathcal{I}}$.

$I_{\mathcal{I}} f$ for a $f \in V$. Here, the multi-dimensional transformation is expressed in terms of one-dimensional discrete Fourier transforms, hierarchizations and de-hierarchizations of different size, cf. [26] and Algorithm 1. Note that the application of a fast Fourier transform algorithm for the computation of a one-dimensional discrete Fourier transform of length 2^l results in a computational complexity of order $\mathcal{O}(l2^l)$. Note furthermore that the complexity for a one-dimensional hierarchization or de-hierarchization of length 2^l is of linear order $\mathcal{O}(2^l)$. In this way, one can give an upper estimate of the order $\mathcal{O}(2^{|\mathbf{l}|_1}|\mathbf{l}|_1)$ for the overall computational complexity in the case of a full grid $V_{\mathbf{l}}$.

In Algorithm 1 we give a procedure to apply the general sparse grid interpolation operator $I_{\mathcal{I}}$ associated to an admissible index set $\mathcal{I}$, where we define for $d \in \{1, \ldots, n\}$ the set

$$\mathcal{M}_d(\mathcal{I}) := \{\mathbf{l} \in \mathcal{I} : \mathbf{l} + \mathbf{e}_d \notin \mathcal{I}\},$$

with the d-th unit vector $\mathbf{e}_d$. Now, an upper estimate for the resulting overall computational complexity $\mathcal{T}[I_{\mathcal{I}}]$ of Algorithm 1 can be easily deduced in the form

$$\begin{aligned}\mathcal{T}[I_{\mathcal{I}}] &\lesssim \sum_{d=1}^{n} \sum_{\mathbf{l}\in\mathcal{M}_d(\mathcal{I})} 2^{l_d} l_d 2^{|\mathbf{l}|_1 - l_d} = \sum_{d=1}^{n} \sum_{\mathbf{l}\in\mathcal{M}_d(\mathcal{I})} 2^{|\mathbf{l}|_1} l_d \le \sum_{d=1}^{n} l_{\max} \sum_{\mathbf{l}\in\mathcal{M}_d(\mathcal{I})} 2^{|\mathbf{l}|_1} \\ &\le n\, l_{\max} \sum_{\mathbf{l}\in\mathcal{I}} 2^{|\mathbf{l}|_1}, \qquad (21)\end{aligned}$$

where $l_{\max} := \max_{\mathbf{l} \in \mathcal{I}} |\mathbf{l}|_\infty$. Note that the inverse operator $I_{\mathcal{I}}^{-1}$ can easily computed by performing the algorithm in a reverse way [26]. In the case of the general sparse grid index sets $\mathcal{I}_L^T$, relation (21) and Lemma 1 lead directly to the following computational complexity estimate:

Lemma 2. *Let $L \in \mathbb{N}_0$ and $T < 1$. An upper estimate for the computational complexity of the general sparse grid interpolation operator $I_{\mathcal{I}_L^T}$ with respect to the discretization parameter L is given by*

$$\mathcal{T}[I_{\mathcal{I}_L^T}] \lesssim L \sum_{\mathbf{l} \in \mathcal{I}_L^T} 2^{|\mathbf{l}|_1} \lesssim \begin{cases} L2^L & \text{for } 0 < T < 1, \\ L2^L L^{n-1} & \text{for } T = 0, \\ L2^{L\frac{T-1}{T/n-1}} & \text{for } T < 0, \\ L2^{Ln} & \text{for } T = -\infty. \end{cases}$$

Let us remark that the case $T = 0$ is already presented in [26], i.e. $\mathcal{T}[I_{\mathcal{I}_L}] = \mathcal{O}(L^n 2^L)$. Note in particular that both, the asymptotic number of degrees of freedom of V_L^T in Lemma 1 and the asymptotic computational complexity of $I_{\mathcal{I}_L^T}$ in Lemma 2, are not exponentially dependent on the dimension n in the case $0 < T < 1$ (see footnote 2).

Let us finally note that, alternatively, the interpolation operator $I_{\mathcal{I}}$ can be applied using the so-called combination technique or the blending scheme [3,13,34]. For an admissible index set $\mathcal{I}$ it holds

$$I_{\mathcal{I}} f = \sum_{\mathbf{l} \in \mathcal{I}} r_{\mathcal{I}}(\mathbf{l}) I_{\mathbf{l}} f, \tag{22}$$

where

$$r_{\mathcal{I}}(\mathbf{l}) := \sum_{\mathbf{v} \in \{0,1\}^n} (-1)^{|\mathbf{v}|} \chi_{\mathcal{I}}(\mathbf{l} + \mathbf{v}),$$

with the characteristic function

$$\chi_{\mathcal{I}}(\mathbf{l}) := \begin{cases} 1 & \text{for } \mathbf{l} \in \mathcal{I}, \\ 0 & \text{otherwise.} \end{cases}$$

The computational complexity of the combination technique (22) can be estimated by

$$\mathcal{T}[I_{\mathcal{I}}] \lesssim \sum_{\mathbf{l} \in \mathcal{I}, r_{\mathcal{I}}(\mathbf{l}) \neq 0} 2^{|\mathbf{l}|_1} |\mathbf{l}|_1.$$

3 Approximation Estimates

In this section, we first define different variants of (periodic) Sobolev spaces on the torus via Fourier series, i.e. we classify functions via the decay of their Fourier coefficients and hence by their smoothness. Then, we give approximation estimates for these spaces. Here, we will first discuss the best linear approximation error and then the approximation error of the interpolant. Based on the derived estimates we further study the resulting error and cost complexities.

3.1 Periodic Sobolev Spaces

As already noted in Sect. 2.1 we characterize the smoothness classes of a function f by the decay properties of its Fourier coefficients [28]. To this end, let $w : \mathbb{Z}^n \to \mathbb{R}_+$ be a continuous and positive weight. Then we define

$$\mathcal{H}_w(\mathbb{T}^n) := \left\{ f(\mathbf{x}) = \sum_{\mathbf{k}\in\mathbb{Z}^n} \hat{f}_{\mathbf{k}}\omega_{\mathbf{k}}(\mathbf{x}) \, : \, \|f\|_w := \sqrt{\sum_{\mathbf{k}\in\mathbb{Z}^n} w(\mathbf{k})^2 |\hat{f}_{\mathbf{k}}|^2} < \infty \right\}. \tag{23}$$

Here, e.g. for $r, t \in \mathbb{R}$ the weights

$$w(\mathbf{k}) = \lambda_{\text{iso}}(\mathbf{k})^r, \quad \text{where} \quad \lambda_{\text{iso}}(\mathbf{k}) := 1 + |\mathbf{k}|_\infty,$$

and

$$w(\mathbf{k}) = \lambda_{\text{mix}}(\mathbf{k})^t, \quad \text{where} \quad \lambda_{\text{mix}}(\mathbf{k}) := \prod_{d=1}^{n} (1 + |k_d|),$$

result in the conventional isotropic Sobolev spaces $\mathcal{H}^r$ [1] and in the standard Sobolev spaces with dominating mixed smoothness $\mathcal{H}^t_{\text{mix}}$ [42], respectively. A further example is the multiplicative combination of these weights, i.e.

$$w(\mathbf{k}) = \lambda_{\text{iso}}(\mathbf{k})^r \lambda_{\text{mix}}(\mathbf{k})^t,$$

which leads to generalized Sobolev spaces of dominating mixed smoothness [23]

$$\mathcal{H}^{t,r}_{\text{mix}}(\mathbb{T}^n) := \left\{ f(\mathbf{x}) = \sum_{\mathbf{k}\in\mathbb{Z}^n} \hat{f}_{\mathbf{k}}\omega_{\mathbf{k}}(\mathbf{x}) \, : \right.$$
$$\left. \|f\|_{\mathcal{H}^{t,r}_{\text{mix}}} := \sqrt{\sum_{\mathbf{k}\in\mathbb{Z}^n} (\lambda_{\text{mix}}(\mathbf{k})^t \lambda_{\text{iso}}(\mathbf{k})^r)^2 \, |\hat{f}_{\mathbf{k}}|^2} < \infty \right\}. \tag{24}$$

In particular, these spaces include the conventional spaces as special cases, i.e.

$$\mathcal{H}^r(\mathbb{T}^n) = \mathcal{H}^{0,r}_{\text{mix}}(\mathbb{T}^n) \quad \text{and} \quad \mathcal{H}^t_{\text{mix}}(\mathbb{T}^n) = \mathcal{H}^{t,0}_{\text{mix}}(\mathbb{T}^n),$$

respectively. Hence, the parameter r from Eq. (24) governs the isotropic smoothness, whereas t governs the mixed smoothness. The spaces $\mathcal{H}^{t,r}_{\text{mix}}$ give us a quite flexible framework for the study of problems in Sobolev spaces.

Moreover, the spaces $\mathcal{H}^{t,r}_{\text{mix}}(\mathbb{T}^n)$ can be generalized to the case of n-dimensional smoothness parameters $\mathbf{t}, \mathbf{r} \in \mathbb{R}^n$ with $\mathbf{r} \geq \mathbf{0}$ [24]. To this end, for $\mathbf{t}, \mathbf{r} \in \mathbb{R}^n$ with $\mathbf{r} \geq \mathbf{0}$ we set $w(\mathbf{k}) = \lambda^{(\mathbf{t})}_{\text{mix}}(\mathbf{k})\lambda^{(\mathbf{r})}_{\text{iso}}(\mathbf{k})$, where

$$\lambda^{(\mathbf{t})}_{\text{mix}}(\mathbf{k}) := \prod_{d=1}^{n}(1+|k_d|)^{t_d} \quad \text{and} \quad \lambda^{(\mathbf{r})}_{\text{iso}}(\mathbf{k}) := \sum_{d=1}^{n}(1+|k_d|)^{r_d}$$

and introduce the spaces

$$\mathcal{H}^{\mathbf{t},\mathbf{r}}_{\text{mix}}(\mathbb{T}^n) := \Bigg\{ f(\mathbf{x}) = \sum_{\mathbf{k}\in\mathbb{Z}^n} \hat{f}_{\mathbf{k}}\omega_{\mathbf{k}}(\mathbf{x}) : \quad \|f\|_{\mathcal{H}^{\mathbf{t},\mathbf{r}}_{\text{mix}}} := \sqrt{\sum_{\mathbf{k}\in\mathbb{Z}^n} \left(\lambda^{(\mathbf{t})}_{\text{mix}}(\mathbf{k})\lambda^{(\mathbf{r})}_{\text{iso}}(\mathbf{k})\right)^2 |\hat{f}_{\mathbf{k}}|^2} < \infty \Bigg\}. \tag{25}$$

In this way, for $r \geq 0$ the spaces $\mathcal{H}^{t,r}_{\text{mix}}$ are up to norm equivalency[6] special cases of the spaces $\mathcal{H}^{\mathbf{t},\mathbf{r}}_{\text{mix}}$, i.e. $\mathcal{H}^{t,r}_{\text{mix}} = \mathcal{H}^{(t,\ldots,t),(r,\ldots,r)}_{\text{mix}}$. We use the short form $\mathcal{H}^{\mathbf{r}} := \mathcal{H}^{\mathbf{0},\mathbf{r}}_{\text{mix}}$ and $\mathcal{H}^{\mathbf{t}}_{\text{mix}} := \mathcal{H}^{\mathbf{t},\mathbf{0}}_{\text{mix}}$.

Furthermore, following [46, 50], for a set of weights $\Gamma := \{\gamma_u\}_{u\subset\{1,\ldots,n\}}$ with $\gamma_u \geq 0$ and a weight function w we introduce a weighted periodic Sobolev space by

$$\mathcal{H}^{\Gamma}_{w}(\mathbb{T}^n) := \Bigg\{ f(\mathbf{x}) = \sum_{\mathbf{k}\in\mathbb{Z}^n} \hat{f}_{\mathbf{k}}\omega_{\mathbf{k}}(\mathbf{x}) : \quad \|f\|_{\mathcal{H}^{\Gamma}_{w}} := \sqrt{\sum_{u\subset\{1,\ldots,n\}} \frac{1}{\gamma_u} \sum_{\mathbf{l}\in\Omega_u} \sum_{\mathbf{j}\in\mathcal{J}_{\mathbf{l}}} w(\sigma(\mathbf{j}))^2 |\hat{f}_{\sigma(\mathbf{j})}|^2} < \infty \Bigg\}, \tag{26}$$

where

[6] For $r \geq 0$ we could also use the weight $\prod_{d=1}^{n}(1+|k_d|)^t \left(\sum_{d=1}^{n}(1+|k_d|)^r\right)$ instead of $\prod_{d=1}^{n}(1+|k_d|)^t\,(1+|\mathbf{k}|_\infty)^r$ to define the space $\mathcal{H}^{t,r}_{\text{mix}}$. This weight is equal to the special case $\lambda^{(t,\ldots,t)}_{\text{mix}}(\mathbf{k})\lambda^{(r,\ldots,r)}_{\text{iso}}(\mathbf{k})$ and hence also the associated spaces $\mathcal{H}^{t,r}_{\text{mix}}$ and $\mathcal{H}^{(t,\ldots,t),(r,\ldots,r)}_{\text{mix}}$ would be equal. However, many of the given proofs would get more technical and thus for reasons of simplicity we restrict ourselves to the definition (24)

$$\Omega_u := \{\mathbf{l} \in \mathbb{N}_0^n : l_d = 0 \text{ for all } d \in \{1,\ldots,n\} \setminus u\}.$$

Let us remark that the orthogonal decomposition

$$f = \sum_{u \subset \{1,\ldots,n\}} f_u, \quad \text{with } f_u := \sum_{\mathbf{l} \in \Omega_u} \sum_{\mathbf{j} \in \mathcal{J}_\mathbf{l}} \hat{f}_{\sigma(\mathbf{j})} \omega_{\sigma(\mathbf{j})} \tag{27}$$

is well-known in statistics under the name ANOVA (analysis of variance) [10], where f_u in particular depends on the coordinates $\{x_d\}_{d \in u}$ only. Note that for $f \in \mathcal{H}_w^\Gamma$ the weight γ_u prescribes the importance of the term f_u and hence the importance of different dimensions and of correlations between dimensions. In particular for a weight $\gamma_u \to 0$ the norm $\|f_u\|_{\mathcal{H}_w}$ is forced to be zero. If the size of terms $\|f_u\|_{\mathcal{H}_w}$ decays fast with e.g. $|u|$, then a proper restriction onto certain lower dimensional functions results in a substantial reduction in computational complexity. For example a set of weights $\Gamma_q := \{\gamma_u\}_{u \subset \{1,\ldots,n\}}$ with $\gamma_u = 0$ for all $u \subset \{1,\ldots,n\}$, $|u| > q$ results in a periodic Sobolev space of finite-order q, cf. [46, 50, 51]. Thus, all terms f_u with $|u| > q$ are either not present at all or can be neglected due to the decay with $|u|$. Then, the problem of approximating a n-dimensional function reduces to the problem of approximating q-dimensional functions.

Note further that the introduced periodic Sobolev spaces, i.e. $\mathcal{H}_{\text{mix}}^{t,r}(\mathbb{T}^n)$, $\mathcal{H}_{\text{mix}}^{\mathbf{t},\mathbf{r}}(\mathbb{T}^n)$ and $\mathcal{H}_w^\Gamma$, can be straightforward generalized to the case of many-particle spaces [17–19, 27].

3.2 Best Linear Approximation Error

In the following, we will consider the error of the best linear approximation in finite-dimensional general sparse grid discretization spaces. Here, we will restrict ourselves to some specific Sobolev spaces of dominating mixed smoothness.

For $\mathbf{l} \in \mathbb{N}_0^n$ we define an approximation operator $Q_\mathbf{l}$ with respect to the $\mathcal{L}^2$-norm by

$$Q_\mathbf{l} := Q_{l_1} \otimes \ldots \otimes Q_{l_n} : \mathcal{L}^2(\mathbb{T}^n) \to V_\mathbf{l},$$

where

$$Q_l : \mathcal{L}^2(\mathbb{T}) \to V_l : f \mapsto \sum_{0 \le j \le 2^l - 1} \hat{f}_{\sigma(j)} \phi_j.$$

For an admissible index set $\mathcal{I}$, as introduced in Sect. 2.3, we define a general sparse grid approximation operator $Q_\mathcal{I} : \mathcal{L}^2(\mathbb{T}^n) \to V_\mathcal{I}$ by

$$Q_\mathcal{I} f := \sum_{\mathbf{l} \in \mathcal{I}} \sum_{\mathbf{j} \in \mathcal{J}_\mathbf{l}} \hat{f}_{\sigma(\mathbf{j})} \omega_{\sigma(\mathbf{j})}.$$

Now let us consider two weight functions w and $\tilde{w}$ with associated Sobolev spaces $\mathcal{H}_w$ and $\mathcal{H}_{\tilde{w}}$ and norms $\|f\|_{\mathcal{H}_w}$ and $\|f\|_{\mathcal{H}_{\tilde{w}}}$, respectively. It should hold $\mathcal{H}_w \subset \mathcal{H}_{\tilde{w}} \subset \mathcal{L}^2$ and thus $w(\mathbf{k}) \lesssim \tilde{w}(\mathbf{k})$. Then, let us consider $f \in \mathcal{H}_w(\mathbb{T}^n) \subset \mathcal{L}^2(\mathbb{T}^n)$ with the unique representation $f = \hat{f}_{\mathbf{k}}\omega_{\mathbf{k}}$. Now, if $\max_{\mathbf{l}\in\mathbb{Z}^n\setminus\mathcal{I},\mathbf{j}\in\mathcal{J}_{\mathbf{l}}} \frac{\tilde{w}(\sigma(\mathbf{j}))^2}{w(\sigma(\mathbf{j}))^2} < \infty$, we obtain for the best linear approximation in $V_{\mathcal{I}}$ the estimate

$$\begin{aligned}
\inf_{\tilde{f}\in V_{\mathcal{I}}} \|f - \tilde{f}\|^2_{\mathcal{H}_{\tilde{w}}} \leq \|f - Q_{\mathcal{I}} f\|^2_{\mathcal{H}_{\tilde{w}}} &= \| \sum_{\mathbf{l}\in\mathbb{Z}^n\setminus\mathcal{I}} \sum_{\mathbf{j}\in\mathcal{J}_{\mathbf{l}}} \hat{f}_{\sigma(\mathbf{j})}\omega_{\sigma(\mathbf{j})} \|^2_{\mathcal{H}_{\tilde{w}}} \\
&= \sum_{\mathbf{l}\in\mathbb{Z}^n\setminus\mathcal{I}} \sum_{\mathbf{j}\in\mathcal{J}_{\mathbf{l}}} \tilde{w}(\sigma(\mathbf{j}))^2 |\hat{f}_{\sigma(\mathbf{j})}|^2 \\
&= \sum_{\mathbf{l}\in\mathbb{Z}^n\setminus\mathcal{I}} \sum_{\mathbf{j}\in\mathcal{J}_{\mathbf{l}}} \frac{\tilde{w}(\sigma(\mathbf{j}))^2}{w(\sigma(\mathbf{j}))^2} |\hat{f}_{\sigma(\mathbf{j})}|^2 w(\sigma(\mathbf{j}))^2 \\
&\leq \left(\max_{\mathbf{l}\in\mathbb{Z}^n\setminus\mathcal{I},\mathbf{j}\in\mathcal{J}_{\mathbf{l}}} \frac{\tilde{w}(\sigma(\mathbf{j}))^2}{w(\sigma(\mathbf{j}))^2} \right) \sum_{\mathbf{l}\in\mathbb{Z}^n\setminus\mathcal{I}} \sum_{\mathbf{j}\in\mathcal{J}_{\mathbf{l}}} |\hat{f}_{\sigma(\mathbf{j})}|^2 w(\sigma(\mathbf{j}))^2 \\
&\leq \left(\max_{\mathbf{l}\in\mathbb{Z}^n\setminus\mathcal{I},\mathbf{j}\in\mathcal{J}_{\mathbf{l}}} \frac{\tilde{w}(\sigma(\mathbf{j}))^2}{w(\sigma(\mathbf{j}))^2} \right) \sum_{\mathbf{l}\in\mathbb{Z}^n} \sum_{\mathbf{j}\in\mathcal{J}_{\mathbf{l}}} |\hat{f}_{\sigma(\mathbf{j})}|^2 w(\sigma(\mathbf{j}))^2 \\
&= \left(\max_{\mathbf{l}\in\mathbb{Z}^n\setminus\mathcal{I},\mathbf{j}\in\mathcal{J}_{\mathbf{l}}} \frac{\tilde{w}(\sigma(\mathbf{j}))^2}{w(\sigma(\mathbf{j}))^2} \right) \|f\|^2_{\mathcal{H}_w}. \qquad (28)
\end{aligned}$$

This general result allows us to derive error estimates for a wide range of situations. We shortly consider two specific cases, namely the pairings $(\mathcal{H}^{t',r'}_{\text{mix}}, \mathcal{H}^{t,r}_{\text{mix}})$ and $(\mathcal{H}^r, \mathcal{H}^{\mathbf{t}}_{\text{mix}})$.

First, for the linear approximation in general sparse grid spaces $V_{\mathcal{I}^T_L}$ with the index $\mathcal{I}^T_L$ set given in (16) the following error estimate for functions in optimized Sobolev spaces of dominating mixed smoothness $\mathcal{H}^{t,r}_{\text{mix}}$ can be derived:

Lemma 3. *For $L \in \mathbb{N}_0$, $T < 1$, $t' + r' < t + r$, $t - t' \geq 0$ and $f \in \mathcal{H}^{t,r}_{\text{mix}}(\mathbb{T}^n)$ it holds:*

$$\begin{aligned}
\inf_{\tilde{f}\in V_{\mathcal{I}^T_L}} \|f - \tilde{f}\|_{\mathcal{H}^{t',r'}_{\text{mix}}} &\leq \|f - Q_{\mathcal{I}^T_L} f\|_{\mathcal{H}^{t',r'}_{\text{mix}}} \\
&\lesssim \begin{cases} 2^{L\left((r'-r)-(t-t')+(T(t-t')-(r'-r))\frac{n-1}{n-T}\right)} \|f\|_{\mathcal{H}^{t,r}_{\text{mix}}} & \text{for } T \geq \frac{r'-r}{t-t'}, \\ 2^{L((r'-r)-(t-t'))} \|f\|_{\mathcal{H}^{t,r}_{\text{mix}}} & \text{for } T \leq \frac{r'-r}{t-t'}. \end{cases}
\end{aligned}$$

Proof. According to (28), the estimation of

$$\max_{\mathbf{l}\in\mathbb{Z}^n\setminus\mathcal{I},\mathbf{j}\in\mathcal{J}_{\mathbf{l}}} \frac{\lambda_{\text{mix}}(\sigma(\mathbf{j}))^{t'}\lambda_{\text{iso}}(\sigma(\mathbf{j}))^{r'}}{\lambda_{\text{mix}}(\sigma(\mathbf{j}))^{t}\lambda_{\text{iso}}(\sigma(\mathbf{j}))^{r}}$$

leads to the desired result, see also [24, 27]. □

Second, for $\mathbf{t} \geq \mathbf{1}$ we can define an anisotropic admissible index set by

$$\mathcal{I}_L^{\mathbf{t}} := \{\mathbf{l} \in \mathbb{N}_0^n : \sum_{d=1}^n t_d l_d \leq L\},$$

see also [20, 21]. Then, the following error estimate can be derived analogously to Lemma 3:

Lemma 4. *For $L \in \mathbb{N}_0$, $\mathbf{t} > \mathbf{0}$, $f \in \mathcal{H}_{\text{mix}}^{\mathbf{t}}$, $|\mathbf{t}|_{min} - r > 0$ and with $\mathbf{t}' := \frac{\mathbf{t}}{|t|_{min}}$, where $|\mathbf{t}|_{min} := \min_{d=1}^n t_d$, it holds:*

$$\inf_{\tilde{f} \in V_{\mathcal{I}_L^{\mathbf{t}'}}} \|f - \tilde{f}\|_{\mathcal{H}^r} \leq \|f - \mathcal{Q}_{\mathcal{I}_L^{\mathbf{t}'}} f\|_{\mathcal{H}_{\text{mix}}^{\mathbf{t}}} \lesssim 2^{-(|\mathbf{t}|_{min} - r)} \|f\|_{\mathcal{H}_{\text{mix}}^{\mathbf{t}}}.$$

Note that, for e.g. $t_1 = |\mathbf{t}|_{min} < t_2 \leq \ldots \leq t_n$, the number of degrees of freedom $|V_{\mathcal{I}_L^{\mathbf{t}'}}|$ is of order $\mathcal{O}(2^L)$, which then results in an overall complexity rate which is independent of the number of dimensions n.

So far we have considered the best linear approximation of a function. However, its coefficients are given by Fourier integrals (3), which can only be evaluated by analytic formulae in special cases. In practice, one possibility to compute an approximation to the best linear approximation, is the numerical calculation of the interpolant of the function. Here, however, it is in general not clear if the associated approximation error exhibits the same convergence rate as that of the best linear approximation. This issue will be discussed in the next section.

3.3 Approximation Error of Interpolant

In the following we will consider the error of the approximation by trigonometric *interpolation*. To this end, let us first recall the following two lemmata:

Lemma 5. *For $L \in \mathbb{N}_0$, $f \in \mathcal{H}^s$, $s > \frac{n}{2}$ and $0 \leq r < s$ it holds:*

$$\|f - I_{\mathcal{I}_L^{-\infty}} f\|_{\mathcal{H}^r} \lesssim 2^{-(s-r)L} \|f\|_{\mathcal{H}^s}. \tag{29}$$

Lemma 6. *For $L \in \mathbb{N}_0$, $f \in \mathcal{H}_{\text{mix}}^t$, $t > \frac{1}{2}$ and $0 \leq r < t$ it holds:*

$$\|f - I_{\mathcal{I}_L^0} f\|_{\mathcal{H}^r} \lesssim 2^{-(t-r)L} L^{n-1} \|f\|_{\mathcal{H}_{\text{mix}}^t}. \tag{30}$$

Let us remark that analogous lemmata are given in [36, 37] based on the works of [40, 48], respectively. We give proofs based on the estimation of the aliasing error in the appendix.

Next, we will extend Lemma 6 to the case of general sparse grids. To this end, we will estimate the hierarchical surplus. Let $f \in \mathcal{H}_w$ obey a pointwise convergent (hierarchical) Fourier series. Then, the relation

$$\begin{aligned}\|f - I_{\mathcal{I}} f\|_{\mathcal{H}_{\tilde{w}}} &= \|\sum_{\mathbf{l}\in\mathbb{N}_0^n}\sum_{\mathbf{j}\in\mathcal{J}_\mathbf{l}} \breve{f}_\mathbf{j}\psi_\mathbf{j} - \sum_{\mathbf{l}\in\mathcal{I}}\sum_{\mathbf{j}\in\mathcal{J}_\mathbf{l}} \breve{f}_\mathbf{j}\psi_\mathbf{j}\|_{\mathcal{H}_{\tilde{w}}} \\ &= \|\sum_{\mathbf{l}\in\mathbb{N}_0^n\setminus\mathcal{I}}\sum_{\mathbf{j}\in\mathcal{J}_\mathbf{l}} \breve{f}_\mathbf{j}\psi_\mathbf{j}\|_{\mathcal{H}_{\tilde{w}}} \\ &\leq \sum_{\mathbf{l}\in\mathbb{N}_0^n\setminus\mathcal{I}} \|\sum_{\mathbf{j}\in\mathcal{J}_\mathbf{l}} \breve{f}_\mathbf{j}\psi_\mathbf{j}\|_{\mathcal{H}_{\tilde{w}}}\end{aligned}$$

holds. By definition of the hierarchical basis we obtain

$$\begin{aligned}\|\sum_{\mathbf{j}\in\mathcal{J}_\mathbf{l}} \breve{f}_\mathbf{j}\psi_\mathbf{j}\|^2_{\mathcal{H}_{\tilde{w}}} &= \|\sum_{\mathbf{j}\in\mathcal{J}_\mathbf{l}}\sum_{\mathbf{v}\in\{0,1\}^n} \breve{f}_\mathbf{j}\bigotimes_{d=1}^{n}\phi_{\mu_{v_d}^{l_d}(j_d)}\|^2_{\mathcal{H}_{\tilde{w}}} \\ &= \sum_{\mathbf{j}\in\mathcal{J}_\mathbf{l}}\sum_{\mathbf{v}\in\{0,1\}^n,\mathbf{l}-\mathbf{v}\geq\mathbf{0}} |\breve{f}_\mathbf{j}|^2\tilde{w}(\sigma(\mu_\mathbf{v}^\mathbf{l}(\mathbf{j})))^2 \\ &= \sum_{\mathbf{j}\in\mathcal{J}_\mathbf{l}}\sum_{\mathbf{v}\in\{0,1\}^n,\mathbf{l}-\mathbf{v}\geq\mathbf{0}} \left|\sum_{\mathbf{m}\in\mathbb{Z}^n}\hat{f}_{\sigma(\mathbf{j})+\mathbf{m}\,2^\mathbf{l}}\right|^2 \tilde{w}(\sigma(\mu_\mathbf{v}^\mathbf{l}(\mathbf{j})))^2 \\ &= \sum_{\mathbf{j}\in\mathcal{J}_\mathbf{l}}\sum_{\mathbf{v}\in\{0,1\}^n,\mathbf{l}-\mathbf{v}\geq\mathbf{0}} \left|\sum_{\mathbf{m}\in\mathbb{Z}^n}\hat{f}_{\sigma(\mathbf{j})+\mathbf{m}\,2^\mathbf{l}}\frac{w(\sigma(\mathbf{j})+\mathbf{m}\,2^\mathbf{l})}{w(\sigma(\mathbf{j})+\mathbf{m}\,2^\mathbf{l})}\right|^2 \tilde{w}(\sigma(\mu_\mathbf{v}^\mathbf{l}(\mathbf{j})))^2,\end{aligned}$$

where $\mu_0^l(j) = j$,

$$\mu_1^l(j) = \begin{cases} -1 & \text{if } l \leq 0, \\ 2^l - 1 - j & \text{if } l \geq 1, \end{cases}$$

$\mu_\mathbf{v}^\mathbf{l} = (\mu_{v_1}^{l_1}, \ldots, \mu_{v_n}^{l_n})$ and $\phi_{-1} = 0$. With Cauchy-Schwarz it follows

$$\begin{aligned}&\left|\sum_{\mathbf{m}\in\mathbb{Z}^n}\hat{f}_{\sigma(\mathbf{j})+\mathbf{m}\,2^\mathbf{l}}\frac{w(\sigma(\mathbf{j})+\mathbf{m}\,2^\mathbf{l})}{w(\sigma(\mathbf{j})+\mathbf{m}\,2^\mathbf{l})}\right|^2 \\ &\quad\leq \left(\sum_{\mathbf{m}\in\mathbb{Z}^n}\left|\hat{f}_{\sigma(\mathbf{j})+\mathbf{m}\,2^\mathbf{l}}w(\sigma(\mathbf{j})+\mathbf{m}\,2^\mathbf{l})\right|^2\right)\left(\sum_{\mathbf{m}\in\mathbb{Z}^n} w(\sigma(\mathbf{j})+\mathbf{m}\,2^\mathbf{l})^{-2}\right) \qquad (31)\end{aligned}$$

and hence it holds

$$\|\sum_{\mathbf{j}\in\mathcal{J}_\mathbf{l}} \breve{f}_\mathbf{j}\psi_\mathbf{j}\|^2_{\mathcal{H}_{\tilde{w}}} \leq \sum_{\mathbf{j}\in\mathcal{J}_\mathbf{l}} \sum_{\mathbf{v}\in\{0,1\}^n,\mathbf{l}-\mathbf{v}\geq\mathbf{0}} \left(\sum_{\mathbf{m}\in\mathbb{Z}^n} \left|\hat{f}_{\sigma(\mathbf{j})+\mathbf{m}2^\mathbf{l}}\right|^2 \left|w(\sigma(\mathbf{j})+\mathbf{m}\,2^\mathbf{l})\right|^2\right) \times$$
$$\times \left(\sum_{\mathbf{m}\in\mathbb{Z}^n} w(\sigma(\mathbf{j})+\mathbf{m}\,2^\mathbf{l})^{-2}\right) \tilde{w}(\sigma(\mu^\mathbf{l}_\mathbf{v}(\mathbf{j})))^2.$$

Now, let us assume that there is a function $g:\mathbb{N}_0^n\to\mathbb{R}$ such that it holds

$$\tilde{w}(\sigma(\mu^\mathbf{l}_\mathbf{v}(\mathbf{j})))^2 \sum_{\mathbf{m}\in\mathbb{Z}^n} \left|w(\sigma(\mathbf{j})+\mathbf{m}\,2^\mathbf{l})\right|^{-2} \leq C^2 g(\mathbf{l})^2 \tag{32}$$

for all $\mathbf{j}\in\mathcal{J}_\mathbf{l}$ and $\mathbf{v}\in\{0,1\}^n, \mathbf{l}-\mathbf{v}\geq\mathbf{0}$ with a constant C independent of $\mathbf{j}$ and $\mathbf{v}$. Then, with $|\{0,1\}^n| = 2^n$, we have

$$\|\sum_{\mathbf{j}\in\mathcal{J}_\mathbf{l}} \breve{f}_\mathbf{j}\psi_\mathbf{j}\|_{\mathcal{H}_{\tilde{w}}} \leq 2^n C g(\mathbf{l}) \left(\sum_{\mathbf{j}\in\mathcal{J}_\mathbf{l}}\sum_{\mathbf{m}\in\mathbb{Z}^n} |\hat{f}_{\sigma(\mathbf{j})+\mathbf{m}2^\mathbf{l}}|^2 w(\sigma(\mathbf{j})+\mathbf{m}\,2^\mathbf{l})^2\right)^{\frac{1}{2}}$$
$$\lesssim g(\mathbf{l})\|f\|_{\mathcal{H}_w}$$

and hence

$$\|f - I_\mathcal{I} f\|_{\mathcal{H}_{\tilde{w}}} \lesssim \sum_{\mathbf{l}\in\mathbb{N}_0^n\setminus\mathcal{I}} g(\mathbf{l})\|f\|_{\mathcal{H}_w}. \tag{33}$$

Let us now consider the approximation error in the $\mathcal{H}^r$-norm for approximating $f\in\mathcal{H}^t_{\text{mix}}$ in the sparse grid space $V_{\mathcal{I}^T_L}$ by interpolation. To this end, let us first recall the following upper bound:

Lemma 7. *For $L\in\mathbb{N}_0$, $T<1$, $r<t$ and $t\geq 0$ it holds:*

$$\sum_{\mathbf{l}\in\mathbb{N}_0^n\setminus\mathcal{I}^T_L} 2^{-t|l|_1+r|l|_\infty} \lesssim \begin{cases} 2^{-((t-r)+(Tt-r)\frac{n-1}{n-T})L} L^{n-1} & \text{for } T\geq\frac{r}{t}, \\ 2^{-(t-r)L}\|f\|_{\mathcal{H}^t_{\text{mix}}} & \text{for } T<\frac{r}{t}. \end{cases}$$

Proof. A proof is given in Theorem 4 in [34]. □

Now, we can give the following lemma:

Lemma 8. *Let $L\in\mathbb{N}_0$, $T<1$, $r<t$, $t>\frac{1}{2}$ and $f\in\mathcal{H}^t_{\text{mix}}$ with a pointwise convergent Fourier series. Then it holds:*

$$\|f - I_{\mathcal{I}^T_L} f\|_{\mathcal{H}^r} \lesssim \begin{cases} 2^{-((t-r)+(Tt-r)\frac{n-1}{n-T})L} L^{n-1}\|f\|_{\mathcal{H}^t_{\text{mix}}} & \text{for } T\geq\frac{r}{t}, \\ 2^{-(t-r)L}\|f\|_{\mathcal{H}^t_{\text{mix}}} & \text{for } T<\frac{r}{t}. \end{cases} \tag{34}$$

Proof. For $t > \frac{1}{2}$, $\mathbf{j} \in \mathcal{J}_{\mathbf{l}}$ and $\mathbf{v} \in \{0,1\}^n$ with $\mathbf{l} - \mathbf{v} \geq \mathbf{0}$ it follows the relation

$$\sum_{\mathbf{m}\in\mathbb{Z}^n} \prod_{d=1}^{n} (1 + |\sigma(j) + m_d 2^{l_d}|)^{-2t} \lesssim \sum_{\mathbf{m}\in\mathbb{Z}^n} \prod_{d=1}^{n} \left(2^{l_d}(1+|m_d|)\right)^{-2t}$$
$$\lesssim 2^{-2t|l|_1} \sum_{\mathbf{m}\in\mathbb{Z}^n} \prod_{d=1}^{n} (1+|m_d|)^{-2t}$$
$$\lesssim 2^{-2t|l|_1}$$

and hence

$$(1 + |\sigma(\mu_{\mathbf{v}}^{\mathbf{l}}(\mathbf{j}))|_\infty)^{2r} \sum_{\mathbf{m}\in\mathbb{Z}^n} \prod_{d=1}^{n} (1 + |\sigma(j) + m_d 2^{l_d}|)^{-2t} \lesssim 2^{-t|l|_1 + r|l|_\infty}.$$

According to (32) and (33) this yields

$$\|f - I_{\mathcal{I}_L^T} f\|_{\mathcal{H}^r} \lesssim \sum_{\mathbf{l}\in\mathbb{N}_0^n \setminus \mathcal{I}_L^T} 2^{-t|l|_1 + r|l|_\infty} \|f\|_{\mathcal{H}^t_{\text{mix}}}$$

and with Lemma 7 we obtain the desired result. □

Let us remark that for regular sparse grids, i.e. $T = 0$, there is a difference in the error behavior between the best approximation and the approximation by interpolation. That is, in the $\mathcal{L}^2$-norm error estimate for the interpolant resulting from Lemma 8 with $t > \frac{1}{2}$, $r = 0$ and $T = 0$, there is a logarithmic factor present, i.e. L^{n-1}. In contrast, for the best linear approximation error in the $\mathcal{L}^2$-norm, there is no logarithmic term L^{n-1} involved according to Lemma 3 with $t > 0, t' = r' = r = 0$ and $T = 0$.

3.4 Convergence Rates with Respect to the Cost

Now, we cast the estimates on the degrees of freedom and the associated error of approximation by interpolation into a form which measures the error with respect to the involved degrees of freedom. In the following, we will restrict ourselves to special cases, where the rates are independent of the dimension:

Lemma 9. *Let $L \in \mathbb{N}_0$, $0 < r < t$, $t > \frac{1}{2}$, $0 < T < \frac{r}{t}$, and $f \in \mathcal{H}^t_{\text{mix}}$ with a pointwise convergent Fourier series. Then it holds:*

$$\|f - I_{\mathcal{I}_L^T} f\|_{\mathcal{H}^r} \lesssim M^{-(t-r)} \|f\|_{\mathcal{H}^t_{\text{mix}}},$$

with respect to the involved number of degrees of freedom $M := |V_{\mathcal{I}_L^T}|$.

Proof. This is a simple consequence of the Lemmata 1 and 8. First, we use the relation (20), that is

$$M = |V_{\mathcal{I}_L^T}| \leq c_1(n) \cdot 2^L$$

for $0 < T < \frac{r}{t}$, which results in $2^{-L} \leq c_1(n) M^{-1}$. We now plug this into (34), i.e. into the relation

$$\|f - I_{\mathcal{I}_L^T} f\|_{\mathcal{H}^r} \leq c_2(n) \cdot 2^{-L(t-r)} \cdot \|f\|_{\mathcal{H}^t_{\text{mix}}}$$

and arrive at the desired result with the order constant $C(n) = c_1(n)^{t-r} \cdot c_2(n)$. □

Analogously, we can measure the error with respect to the computational complexity, which results in the following upper estimate:

Lemma 10. *For $0 < r < t$, $t > \frac{1}{2}$, $0 < T < \frac{r}{t}$, and $f \in \mathcal{H}^t_{\text{mix}}$ with a pointwise convergent Fourier series, it holds:*

$$\|f - I_{\mathcal{I}_L^T} f\|_{\mathcal{H}^r} \lesssim R^{-(t-r)} \log(R)^{t-r} \|f\|_{\mathcal{H}^t_{\text{mix}}},$$

with respect to the involved computational costs $R := \mathcal{T}[I_{\mathcal{I}_L^T}]$.

Proof. This is a simple consequence of the Lemmata 2 and 8. Analogously to the proof of Lemma 9 the relation $\|f - I_{\mathcal{I}_L^T} f\|_{\mathcal{H}^r} \lesssim R^{-(t-r)} L^{t-r} \|f\|_{\mathcal{H}^t_{mix}}$, can been shown, which yields the desired result. □

Note that in [18] a result analogous to Lemma 9 is shown in the case of measuring the best linear approximation error with respect to the involved degrees of freedom.

Finally, let us discuss shortly two cases of regular sparse grids with involved logarithmic terms. First, again a simple consequence of the Lemmata 2 and 8 is that for $L \in \mathbb{N}_0$, $t > \frac{1}{2}$ and $f \in \mathcal{H}^t_{\text{mix}}$ with a pointwise convergent Fourier series it holds the relation

$$\|f - I_{\mathcal{I}_L^0} f\|_{\mathcal{L}^2} \lesssim M^{-t} L^{(t+1)(n-1)} \|f\|_{\mathcal{H}^t_{\text{mix}}} \lesssim M^{-t} \log(M)^{(t+1)(n-1)} \|f\|_{\mathcal{H}^t_{\text{mix}}}. \tag{35}$$

Second, for the case $\frac{1}{2} < t - 1$ the relation

$$\|f - I_{\mathcal{I}_L^0} f\|_{\mathcal{H}^1} \lesssim M^{-(t-1)} L^{(t-1)(n-1)} \|f\|_{\mathcal{H}^t_{\text{mix}}} \lesssim M^{-(t-1)} \log(M)^{(t-1)(n-1)} \|f\|_{\mathcal{H}^t_{\text{mix}}} \tag{36}$$

can be derived.

3.5 Further Generalizations of Sparse Grids

In the following we will give some brief remarks on periodic Sobolev spaces with finite-order weights and dimension-adaptive approaches.

3.5.1 Finite-Order Spaces

First, we consider so-called periodic Sobolev spaces with finite-order weights. These spaces are a special case of the weighted periodic Sobolev space $\mathcal{H}_w^{\Gamma}$, which we introduced in Sect. 3.1 by Definition (26). Let us recall from Sect. 3.1 that a set of weights $\Gamma_q = \{\gamma_u\}_{u \subset \{1,\dots,n\}}$ is denoted to be of finite order $q \in \mathbb{N}_0$, if it holds $\gamma_u = 0$ for all $\gamma_u \in \Gamma$ with $|u| > q$. Note that there are $\binom{n}{q} \lesssim n^q$ possible subsets $u \subset 1, \dots, n$ of order q. Therefore, the problem of the approximation of a n-dimensional function $f \in \mathcal{H}_w^{\Gamma_q}(\mathbb{T}^n)$ is reduced to the problem of the approximation of $\mathcal{O}(n^q)$ functions of dimension q. Hence, in that case the curse of dimensionality can be broken and in particular for $w(\mathbf{k}) = \lambda_{\text{iso}}(\mathbf{k})^r \lambda_{\text{mix}}(\mathbf{k})^t$ the previous lemmata can be straightforwardly adapted, compare also [27].

3.5.2 Dimension-Adaptive Approach

So far we considered admissible index sets $\mathcal{I}_T^L$ and their associated generalized sparse grid spaces $V_{\mathcal{I}_T^L}$, which are chosen such that the corresponding a-priori estimated approximation error for *all* functions in a specific function *class* of dominating mixed smoothness is as small as possible for a given amount of degrees of freedom [6]. The goal of a so-called dimension-adaptive approach is to find an admissible index set such that the corresponding approximation error for a *single* given function is as small as possible for a prescribed amount of degrees of freedom. To this end, a scheme similar to that given in [14] for dimension-adaptive quadrature could be applied.

Such a scheme starts with an a-priori chosen small admissible index set, e.g. $\mathcal{I}^{(0)} = \{\mathbf{0}\}$. The idea is to extend the index set successively such that the index sets remain admissible and that an error reduction as large as possible is achieved. To this end, a so called *error indicator* is computed for each index $\mathbf{l} \in \mathbb{N}_0^n$ and its associated subspace $W_{\mathbf{l}}$. In the case of approximation by interpolation we use the hierarchical surplus to derive an error indicator [29], e.g. $\eta_{\mathbf{l}} = |\sum_{\mathbf{j} \in \mathcal{J}_{\mathbf{l}}} \check{f}_{\mathbf{j}} \Psi_{\mathbf{j}}|$ with appropriate norm. For further details of the dimension-adaptive sparse grid approximation algorithm we refer to [5, 14, 29].

Let us note furthermore that in [16] a relation between the dimension-adaptive sparse grid algorithm and the concept of ANOVA-like decompositions was established. There, it was also shown that general sparse grids correspond to just a hierarchical approximation of the single terms in the ANOVA decomposition (27) see also [11, 22, 38].

Note finally that various locally adaptive sparse grid approaches exist [6, 12, 29] which are based on hierarchical multilevel sparse grid techniques. But, together with fast discrete Fourier transforms, they are not easy to apply at all.

4 Numerical Experiments and Results

We implemented the generalized sparse grid trigonometric interpolation operator $I_{\mathcal{I}}$ for general admissible index sets $\mathcal{I} \subset \mathbb{N}_0^n$ according to Algorithm 1 in a software library called *HCFFT*. This library also includes the functionality for the application of dimension-adaptive approaches. In addition to the fast discrete Fourier transform, which we deal with in this paper, it includes actually the following variants: fast discrete sine transform, fast discrete cosine transform and fast discrete Chebyshev transform.

In the following, we present the results of some numerical calculations performed by the *HCFFT* library. We restrict ourselves to the case of the FFT based application of the interpolation operator $\mathcal{I}_L^T$. Here, we in particular study the dependence of the convergence rates on the number of dimensions for the regular sparse grid case $T = 0$ and the energy-norm like sparse grid case $T > 0$. To this end, we consider the approximation by interpolation of functions in the periodic Sobolev spaces of dominating mixed smoothness $\mathcal{H}^t_{\text{mix}}(\mathbb{T}^n)$. As test cases we use the functions

$$G_p : \mathbb{T}^n \to \mathbb{R} : \mathbf{x} \mapsto \bigotimes_{d=1}^{n} g_p(x_d)$$

with

$$g_p : \mathbb{T} \to \mathbb{R} : x \mapsto N_p \cdot (2 + \operatorname{sgn}(x - \pi) \cdot \sin(x)^p)$$

for $p = 1, 2, 3, 4$. Here, sgn denotes the sign function, i.e.

$$\operatorname{sgn}(x) := \begin{cases} -1 & x < 0, \\ 0 & x = 0, \\ 1 & x > 0 \end{cases}$$

and N_p denotes a normalization constant such that $\|g_p\|_{\mathcal{L}^2} = 1$. Note that for $\epsilon > 0$ we have $g_p \in \mathcal{H}^{\frac{1}{2}+p-\epsilon}(\mathbb{T})$ and thus $G_p \in \mathcal{H}^{\frac{1}{2}+p-\epsilon}_{\text{mix}}(\mathbb{T}^n)$. In particular, the $\mathcal{L}^2$- and $\mathcal{H}^1$-error can be computed by analytic formulae and the relative $\mathcal{L}^2$-error is equal to the absolute $\mathcal{L}^2$-error, i.e. $\|G_p - I_{\mathcal{I}_L^T} G_p\|_{\mathcal{L}^2} / \|G_p\|_{\mathcal{L}^2} = \|G_p - I_{\mathcal{I}_L^T} G_p\|_{\mathcal{L}^2}$. Let us note these test functions are of simple product form, but the decay behavior of its Fourier coefficients reflects that of the considered Sobolev spaces of dominating mixed smoothness. The numerical results for more complicated functions of non-product structure from these Sobolev spaces were basically the same.

For validation we first performed numerical calculations in the one dimensional case for G_p with $p = 1, 2, 3, 4$. We show the measured error versus the number of degrees of freedom in Fig. 1. To estimate the respective convergence rates, we computed a linear least square fit to the results of the three largest levels. This way, we obtained rates of values about 1.50, 2.50, 3.51 and 4.40, respectively,

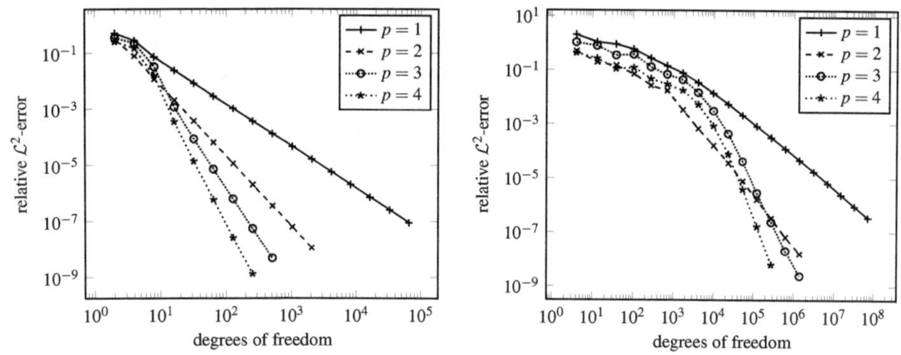

Fig. 1 Convergence behavior for approximating the functions $G_p \in \mathcal{H}_{\text{mix}}^{\frac{1+2p}{2}-\epsilon}$ by trigonometric interpolation on regular spare grids, i.e. $\|G_p - I_{\mathcal{I}_L^0} G_p\|_{\mathcal{L}^2}$ versus $|V_{\mathcal{I}_L^0}|$. *Left*: case of $n = 1$. *Right*: case of $n = 3$

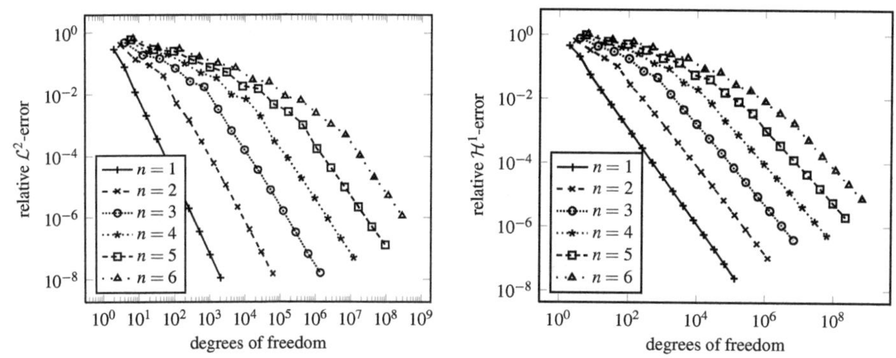

Fig. 2 Convergence behavior for approximating the function $G_2 \in \mathcal{H}_{\text{mix}}^{\frac{5}{2}-\epsilon}$ by trigonometric interpolation on regular spare grids. *Left*: case of relative/absolute $\mathcal{L}^2$-error, i.e. $\|G_2 - I_{\mathcal{I}_L^0} G_2\|_{\mathcal{L}^2}$. *Right*: case of relative $\mathcal{H}^1$-error, i.e. $\|G_2 - I_{\mathcal{I}_L^0} G_2\|_{\mathcal{H}^1} / \|G_2\|_{\mathcal{H}^1}$

which coincide with the theoretically expected rates in the one-dimensional case, cf. Lemmata 1 and 8. Then, we performed calculations for the three dimensional case. The values $p = 1, 2, 3, 4$ result in numerically measured convergence rates of about 1.25, 1.87, 2.83 and 3.90, respectively. Moreover, for the approximation of the test functions G_2 for up to six dimensions by trigonometric interpolation on regular sparse grids, we observe that the rates indeed decrease with the number of dimensions, see Fig. 2. For example in case of the $\mathcal{L}^2$-error the rates deteriorate from a value of 2.50 for $n = 1$ to a value of 1.56 for $n = 6$. All calculated rates are given in Table 1. Note that this decrease in the rates with respect to the number of dimensions is to be expected from theory, since the cost and error estimates in Lemmata 1 and 8 involve dimension-dependent logarithmic terms for the regular sparse grid case, i.e. for $T = 0$. We additionally give in Table 1 the computed rates

Table 1 Numerically measured convergence rates with respect to the number of degrees of freedom according to the relative $\mathcal{L}^2$-norm error and the relative $\mathcal{H}^1$-norm for the approximation of the function $G_2 \in \mathcal{H}_{\text{mix}}^{5/2-\epsilon}$ by trigonometric interpolation on regular spare grids, i.e. $T = 0$

	$n = 1$	$n = 2$	$n = 3$	$n = 4$	$n = 5$	$n = 6$
$\mathcal{L}^2$-norm	2.50	2.17	1.87	1.73	1.60	1.56
$\mathcal{L}^2$-norm / $L^{(\frac{5}{2}+1)(n-1)}$	2.50	2.55	2.49	2.55	2.60	2.76
$\mathcal{H}^1$-norm	1.50	1.38	1.30	1.23	1.19	1.15
$\mathcal{H}^1$-norm / $L^{(\frac{5}{2}-1)(n-1)}$	1.50	1.57	1.51	1.48	1.55	1.44

In addition, we present the rates according to the relative error divided by the respective logarithmic term versus the number of degrees of freedom, see also estimates (35) and (36)

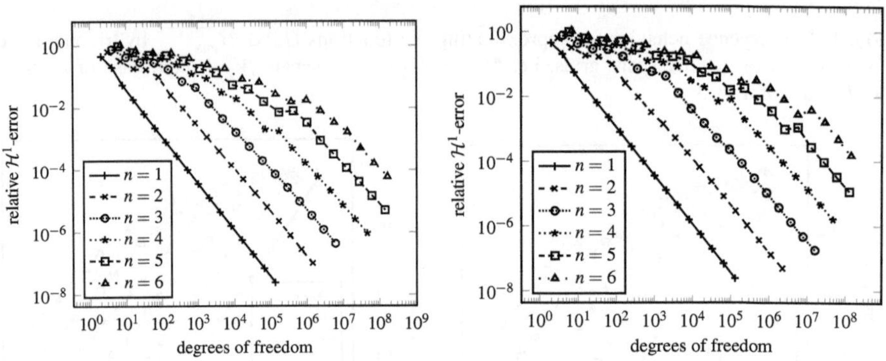

Fig. 3 Convergence behavior for approximating the function $G_2 \in \mathcal{H}_{\text{mix}}^{5/2-\epsilon}$ by the general sparse grid interpolation operator $I_{\mathcal{I}_L^T}$ with respect to the relative $\mathcal{H}^1$-error. *Left*: case of $T = \frac{1}{8}$. *Right*: case of $T = \frac{1}{4}$

associated to the relative errors divided by the respective logarithmic term versus the number of degrees of freedom. Here, the derived values fit quite well to the rates which could be expected from theory, that is, 2.5 and 1.5 for the error measured in the $\mathcal{L}^2$-norm and the $\mathcal{H}^1$-norm, respectively.

Note that according to Lemma 9, we can get rid of the logarithmic terms in some special cases. For example, if we measure the error of the approximation of G_2 by the general sparse grid interpolant $I_{\mathcal{I}_L^T} G_2$ in the $\mathcal{H}^1$-norm, then Lemma 9 leads with $r = 1$, $t = \frac{5}{2} - \epsilon$ to a convergence rate of $\frac{3}{2} - \epsilon$ for $0 < T < \frac{2}{3+2\epsilon}$. Hence, we performed numerical calculations for the generalized sparse grids with $T = \frac{1}{8}$ and $T = \frac{1}{4}$. The obtained errors are plotted in Fig. 3. The results show that the rates are substantially improved compared to the regular sparse grid case. We give all measured rates in Table 2. Note that we still observe a slight decrease of the rates with the number of dimensions. This is surely a consequence of the fact that we are still in the pre-asymptotic regime for the higher-dimensional cases. Note furthermore that the constant involved in the complexity estimate in Lemma 9

Table 2 Numerically measured convergence rates with respect to the number of degrees of freedom for the approximation of the function $G_2 \in \mathcal{H}_{\mathrm{mix}}^{5/2-\epsilon}$ by trigonometric interpolation on generalized spare grids with $0 < T < \frac{2}{3}$

Error	T	$n = 1$	$n = 2$	$n = 3$	$n = 4$	$n = 5$	$n = 6$
$\mathcal{H}^1$	$\frac{1}{8}$	1.50	1.44	1.39	1.29	1.28	1.27
$\mathcal{H}^1$	$\frac{1}{4}$	1.50	1.41	1.36	1.39	1.37	1.49

Table 3 Numerically measured convergence rates with respect to the number of degrees of freedom of the approximation of the function $G_1 \in \mathcal{H}_{\mathrm{mix}}^{3/2-\epsilon}$ by trigonometric interpolation on regular and generalized spare grids

Error	T	$n = 3$	$n = 4$
$\mathcal{H}^1$	0.0	0.45	0.42
$\mathcal{H}^1$	$\frac{1}{8}$	0.47	0.44
$\mathcal{H}^1$	$\frac{1}{4}$	0.49	0.47

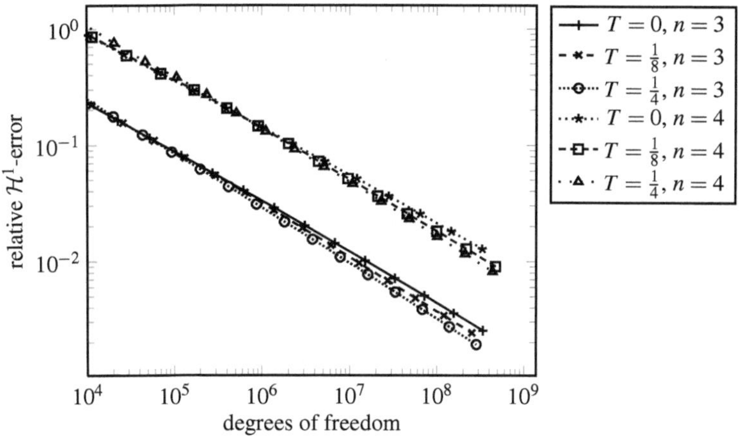

Fig. 4 Convergence behavior for the approximation of the function $G_1 \in \mathcal{H}_{\mathrm{mix}}^{3/2-\epsilon}$ by trigonometric interpolation on generalized spare grids, i.e. $\|G_1 - I_{\mathcal{I}_L^T} G_1\|_{\mathcal{L}^2}$ versus $|V_{\mathcal{I}_L^T}|$, where the error is measured in the relative $\mathcal{H}^1$-error, i.e. $\|G_1 - I_{\mathcal{I}_L^T} G_1\|_{\mathcal{H}^1} / \|G_1\|_{\mathcal{H}^1}$

probably depends exponentially on the number n of dimensions. This explains the offset of the convergence with rising n in Fig. 3.

In [34] it is noted that the involved order constant in the convergence rate estimate for the case $0 < T < 1$ is typically increasing with n and T and it is in particular larger than in the case of regular sparse grids with $T = 0$. In contrast, under certain assumptions, the convergence rate is superior in the case $0 < T < 1$ to that of the regular sparse grid with $T = 0$. Hence, in the pre-asymptotic regime, the effects of constants and order rates counterbalance each other a bit in practice. For example, let us consider the $\mathcal{H}^1$-error of the interpolant $I_{\mathcal{I}_L^T} G_1$ for $T = 0, \frac{1}{8}, \frac{1}{4}$ and $n = 3, 4$. The associated computed rates are given in Table 3. Here, a break-even point can be seen from our numerical results depicted in Fig. 4, i.e. for $n = 4$ the computed $\mathcal{H}^1$-

error is slightly smaller in the case $T = \frac{1}{4}$ than in the cases $T = 0$ and $T = \frac{1}{8}$ for a number of involved degrees of freedom greater than about $|V_{\mathcal{I}_L^T}| \approx 10^6$. A similar effect is also present, albeit barely visible, for $n = 3$ and $|V_{\mathcal{I}_L^T}| \approx 10^5$. Nevertheless, in any case, the various rates are nearly the same anyway and these differences are quite small.

5 Concluding Remarks

In this article, we discussed several variants of periodic Sobolev spaces of dominating mixed smoothness and we constructed the general sparse grid discretization spaces $V_{\mathcal{I}_L^T}$. We gave estimates for their number of degrees of freedom and the best linear approximation error for multivariate functions in $\mathcal{H}^{t,r}_{\text{mix}}(\mathbb{T}^n)$ and $\mathcal{H}^{\mathbf{t}}_{\text{mix}}(\mathbb{T}^n)$. In addition, we presented an algorithm for the general sparse grid interpolation based on the fast discrete Fourier transform and showed its computational cost and the resulting error estimates for the general sparse grid interpolant $I_{\mathcal{I}_L^T}$ of functions in $\mathcal{H}^t_{\text{mix}}(\mathbb{T}^n)$. Specifically, we identified smoothness assumptions that make it possible to choose $I_{\mathcal{I}_L^T}$ in such a way that the number of degrees of freedom is $\mathcal{O}(2^L)$ compared to $\mathcal{O}(2^L L^{n-1})$ and $\mathcal{O}(2^{nL})$ for the regular sparse grid and full grid spaces, respectively, while keeping the optimal order of approximation. For this case, we also showed that the *asymptotic* computational cost complexities rates are independent of the number of dimensions. The constants involved in the $\mathcal{O}$-notation may still depend exponentially on n however.

Let us finally note that we mainly discussed the sparse grid interpolation operator $I_{\mathcal{I}_L^T}$ in the present paper. However, our implemented software library HCFFT allows us to deal with discretization spaces associated with arbitrary admissible index sets and in particular also features dimension-adaptive methods. Furthermore, discrete cosine, discrete sine and discrete Chebyshev based interpolation can be applied. We presently work on its extension to polynomial families which are commonly used in the area of uncertainty quantification [38]. We will discuss these approaches and its applications in a forthcoming paper.

Acknowledgements This work was supported in part by the Collaborative Research Centre 1060 of the Deutsche Forschungsgemeinschaft.

Appendix

In the following, we will give proofs for Lemmata 5 and 6 based on the estimation of the aliasing error.

Let $f \in \mathcal{H}_w$ obey a pointwise convergent Fourier series. Then, it holds the relation

$$\begin{aligned}\|f - I_{\mathcal{I}} f\|_{\mathcal{H}_{\tilde{w}}} &= \|f - I_{\mathcal{I}} f + Q_{\mathcal{I}} f - Q_{\mathcal{I}} f\|_{\mathcal{H}_{\tilde{w}}} \\ &\leq \|f - Q_{\mathcal{I}} f\|_{\mathcal{H}_{\tilde{w}}} + \|I_{\mathcal{I}} f - Q_{\mathcal{I}} f\|_{\mathcal{H}_{\tilde{w}}}.\end{aligned} \tag{37}$$

For the first term of the right hand side, an upper bound can be obtained according to (28). For the second term, it holds with (22) the relation[7]

$$\begin{aligned}\|I_{\mathcal{I}} f - Q_{\mathcal{I}} f\|_{\mathcal{H}_{\tilde{w}}} &= \|\sum_{\mathbf{l}\in\mathcal{I}} r_{\mathcal{I}}(\mathbf{l}) I_{\mathbf{l}} f - Q_{\mathcal{I}} f\|_{\mathcal{H}_{\tilde{w}}} = \|\sum_{\mathbf{l}\in\mathcal{I}} r_{\mathcal{I}}(\mathbf{l})(I_{\mathbf{l}} - Q_{\mathbf{l}}) f\|_{\mathcal{H}_{\tilde{w}}} \\ &\lesssim \sum_{\mathbf{l}\in\mathcal{I}, r_{\mathcal{I}}(\mathbf{l})\neq 0} \|(I_{\mathbf{l}} - Q_{\mathbf{l}}) f\|_{\mathcal{H}_{\tilde{w}}}.\end{aligned}$$

With the help of the aliasing formula (9) and the Cauchy-Schwarz inequality, we obtain for $\mathbf{l} \in \mathbb{N}_0^n$ the relation

$$\begin{aligned}\|I_{\mathbf{l}} f - Q_{\mathbf{l}} f\|^2_{\mathcal{H}_{\tilde{w}}} &= \|\sum_{\mathbf{j}\in\mathcal{J}_{\mathbf{l}}} (\hat{f}^{(\mathbf{l})}_{\mathbf{j}} - \hat{f}_{\sigma(\mathbf{j})})\phi_{\mathbf{j}}\|^2_{\mathcal{H}_{\tilde{w}}} \\ &= \sum_{\mathbf{j}\in\mathcal{J}_{\mathbf{l}}} |\hat{f}^{(\mathbf{l})}_{\mathbf{j}} - \hat{f}_{\sigma(\mathbf{j})}|^2 \tilde{w}(\sigma(\mathbf{j}))^2 \\ &= \sum_{\mathbf{j}\in\mathcal{J}_{\mathbf{l}}} |\sum_{\mathbf{m}\in\mathbb{Z}^n\setminus\{\mathbf{0}\}} \hat{f}_{\sigma(\mathbf{j})+\mathbf{m}\,2^{\mathbf{l}}}|^2 \tilde{w}(\sigma(\mathbf{j}))^2 \\ &= \sum_{\mathbf{j}\in\mathcal{J}_{\mathbf{l}}} |\sum_{\mathbf{m}\in\mathbb{Z}^n\setminus\{\mathbf{0}\}} \hat{f}_{\sigma(\mathbf{j})+\mathbf{m}\,2^{\mathbf{l}}} w(\sigma(\mathbf{j}) + \mathbf{m}\,2^{\mathbf{l}})|^2 \frac{\tilde{w}(\sigma(\mathbf{j}))^2}{w(\sigma(\mathbf{j}) + \mathbf{m}\,2^{\mathbf{l}})^2} \\ &\leq \sum_{\mathbf{j}\in\mathcal{J}_{\mathbf{l}}} \left(\sum_{\mathbf{m}\in\mathbb{Z}^n\setminus\{\mathbf{0}\}} |\hat{f}_{\sigma(\mathbf{j})+\mathbf{m}\,2^{\mathbf{l}}}|^2 |w(\sigma(\mathbf{j}) + \mathbf{m}\,2^{\mathbf{l}})|^2 \right) \times \\ &\qquad \times \left(\sum_{\mathbf{m}\in\mathbb{Z}^n\setminus\{\mathbf{0}\}} |w(\sigma(\mathbf{j}) + \mathbf{m}\,2^{\mathbf{l}})|^{-2} \right) \tilde{w}(\sigma(\mathbf{j}))^2\end{aligned} \tag{38}$$

Let us assume that there is a function $F : \mathbb{N}_0^n \to \mathbb{R}$ such that it holds

$$\tilde{w}(\sigma(\mathbf{j}))^2 \sum_{\mathbf{m}\in\mathbb{Z}^n\setminus\{\mathbf{0}\}} |w(\sigma(\mathbf{j}) + \mathbf{m}\,2^{\mathbf{l}})|^{-2} \leq c F(\mathbf{l})^2 \tag{39}$$

for all $\mathbf{j} \in \mathcal{J}_{\mathbf{l}}$ with a constant c independent of $\mathbf{j}$. Then, (38) yields

[7]Note that analogously to (22) for $I_{\mathcal{I}}$, it holds $Q_{\mathcal{I}} f = \sum_{\mathbf{l}\in\mathcal{I}} r_{\mathcal{I}}(\mathbf{l}) Q_{\mathbf{l}} f$.

$$\|(I_{\mathbf{l}} - Q_{\mathbf{l}})f\|^2_{\mathcal{H}_{\tilde{w}}} \leq cF(\mathbf{l})^2 \sum_{\mathbf{j}\in\mathcal{J}_{\mathbf{l}}} \left(\sum_{\mathbf{m}\in\mathbb{Z}^n\setminus\{\mathbf{0}\}} |\hat{f}_{\sigma(\mathbf{j})+\mathbf{m}\,2^{\mathbf{l}}}|^2 |w(\sigma(\mathbf{j}) + \mathbf{m}\,2^{\mathbf{l}})|^2 \right)$$

$$\lesssim F(\mathbf{l})^2 \|f\|^2_{\mathcal{H}_w},$$

and altogether we obtain

$$\begin{aligned} \|I_{\mathcal{I}} f - Q_{\mathcal{I}} f\|_{\mathcal{H}_{\tilde{w}}} &\lesssim \sum_{\mathbf{l}\in\mathcal{I}, r_{\mathcal{I}}(\mathbf{l})\neq 0} F(\mathbf{l}) \|f\|_{\mathcal{H}_w} \\ &\lesssim \left(\max_{\mathbf{l}\in\mathcal{I}, r_{\mathcal{I}}(\mathbf{l})\neq 0} F(\mathbf{l}) \right) \left(\sum_{\mathbf{l}\in\mathcal{I}, r_{\mathcal{I}}(\mathbf{l})\neq 0} 1 \right) \|f\|_{\mathcal{H}_w}. \end{aligned} \tag{40}$$

Let us now consider the approximation error in the $\mathcal{H}^r$-norm for interpolating $f \in \mathcal{H}^s$, $s > \frac{n}{2}$ in the full grid space $V_{\mathcal{I}_L^{-\infty}}$ and $0 \leq r < s$. According to (39), we may estimate

$$\begin{aligned} \sum_{\mathbf{m}\in\mathbb{Z}^n\setminus\{\mathbf{0}\}} (1 + |\sigma(\mathbf{j}) + \mathbf{m}\,2^{\mathbf{l}})|_\infty)^{-2s} &\lesssim 2^{-2s|\mathbf{l}|_{\min}} \sum_{\mathbf{m}\in\mathbb{Z}^n\setminus\{\mathbf{0}\}} (1 + |\mathbf{m}|_\infty)^{-2s} \\ &\lesssim 2^{-2s|\mathbf{l}|_{\min}} \sum_{m\in\mathbb{N}} ((m^n - (m-1)^n)|m|^{-2s} \\ &\lesssim 2^{-2s|\mathbf{l}|_{\min}} \sum_{m\in\mathbb{N}} m^{-2s+n-1} \\ &\lesssim 2^{-2s|\mathbf{l}|_{\min}} \end{aligned}$$

for all $\mathbf{j} \in \mathcal{J}_{\mathbf{l}}$. With (37), Lemma 3 and (40) we finally obtain

$$\|I_{\mathcal{I}_L^{-\infty}} f - f\|_{\mathcal{H}^r} \lesssim 2^{-(s-r)L} \|f\|_{\mathcal{H}^s}, \tag{41}$$

which proves Lemma 5.

Now, we consider the approximation error in the $\mathcal{H}^r$-norm for interpolating $f \in \mathcal{H}^t_{\text{mix}}$, $t > \frac{1}{2}$ in the sparse grid space $V_{\mathcal{I}_L}$ and $0 \leq r < t$. Here, according to (39), we may estimate

$$\begin{aligned} \sum_{\mathbf{m}\in\mathbb{Z}^n\setminus\{\mathbf{0}\}} \prod_{d=1}^{n} (1 + |\sigma(j_d) + m_d 2^{l_d}|)^{-2t} &\lesssim 2^{-2t|\mathbf{l}|_1} \sum_{\mathbf{m}\in\mathbb{Z}^n\setminus\{\mathbf{0}\}} \prod_{d=1}^{n} (1 + |m_d|)^{-2t} \\ &\lesssim 2^{-2t|\mathbf{l}|_1}. \end{aligned}$$

With (37), Lemma 3, the identity

$$I_{\mathcal{I}_L} = \sum_{d=0}^{n-1} (-1)^d \binom{n-1}{d} \sum_{|\mathbf{l}|_1 = L-d} I_{\mathbf{l}}$$

and (40), we finally obtain

$$\|I_{\mathcal{I}_L} f - f\|_{\mathcal{H}^r} \lesssim 2^{-(t-r)L} \|f\|_{\mathcal{H}^t_{\text{mix}}} \left(\sum_{\mathbf{l} \in \mathcal{I}_L, r_{\mathcal{I}_L}(\mathbf{l}) \neq 0} 1 \right)$$
$$\lesssim 2^{-(t-r)L} L^{n-1} \|f\|_{\mathcal{H}^t_{\text{mix}}}, \quad (42)$$

which proves Lemma 6. This is in particular a special case of Lemma 8. Let us finally remark that the estimates (41) and (42) are also shown in [36, 37] based on the works of [40] and [48], respectively.

References

1. R. Adams, *Sobolev Spaces* (Academic, London, 1975)
2. V. Barthelmann, E. Novak, K. Ritter, High dimensional polynomial interpolation on sparse grids. Adv. Comput. Math. **12**, 273–288 (2000)
3. G. Baszenski, F. Delvos, A discrete Fourier transform scheme for Boolean sums of trigonometric operators. Int. Ser. Numer. Math. **90**, 15–24 (1989)
4. R. Bellmann, *Adaptive Control Processes: A Guided Tour* (Princeton University Press, Princeton, 1961)
5. B. Bohn, M. Griebel, An adaptive sparse grid approach for time series predictions, in *Sparse Grids and Applications*, ed. by J. Garcke, M. Griebel. Lecture Notes in Computational Science and Engineering, vol. 88 (Springer, Berlin, 2012), pp. 1–30
6. H. Bungartz, M. Griebel, Sparse grids. Acta Numer. **13**, 1–123 (2004)
7. D. Dũng, Best multivariate approximations by trigonometric polynomials with frequencies from hyperbolic crosses. J. Approx. Theory **91**(2), 205–225 (1997)
8. D. Dũng, Sampling recovery and cubature on sparse grids. ArXiv e-prints, 2012. arXiv:1211.4319v1 [math.NA]
9. D. Dũng, T. Ullrich, N-widths and ε-dimensions for high-dimensional approximations. Found. Comput. Math. **13**, 965–1003 (2013)
10. B. Efron, C. Stein, The Jackknife estimate of variance. Ann. Stat. **9**(3), 586–596 (1981)
11. C. Feuersänger, Sparse grid methods for higher dimensional approximation. Dissertation, Institute for Numerical Simulation, University of Bonn, 2010
12. C. Feuersänger, M. Griebel, Principal manifold learning by sparse grids. Computing **85**(4), 267–299 (2009)
13. J. Garcke, Maschinelles Lernen durch Funktionsrekonstruktion mit verallgemeinerten dünnen Gittern. Dissertation, Institute for Numerical Simulation, University of Bonn, 2004
14. T. Gerstner, M. Griebel, Dimension-adaptive tensor-product quadrature. Computing **71**(1), 65–87 (2003)
15. V. Gradinaru, Fourier transform on sparse grids: code design and the time dependent Schrödinger equation. Computing **80**, 1–22 (2007)
16. M. Griebel, Sparse grids and related approximation schemes for higher dimensional problems, in *Foundations of Computational Mathematics (FoCM05)*, Santander, ed. by L. Pardo, A. Pinkus, E. Suli, M. Todd (Cambridge University Press, Cambridge, 2006), pp. 106–161

17. M. Griebel, J. Hamaekers, A wavelet based sparse grid method for the electronic Schrödinger equation, in *Proceedings of the International Congress of Mathematicians*, Madrid, 22–30 August, vol. III, ed. by M. Sanz-Solé, J. Soria, J. Varona, J. Verdera (European Mathematical Society, Switzerland, 2006)
18. M. Griebel, J. Hamaekers, Sparse grids for the Schrödinger equation. Math. Model. Numer. Anal. **41**(2), 215–247 (2007)
19. M. Griebel, J. Hamaekers, Tensor product multiscale many-particle spaces with finite-order weights for the electronic Schrödinger equation. Z. Phys. Chem. **224**, 527–543 (2010)
20. M. Griebel, H. Harbrecht, A note on the construction of L-fold sparse tensor product spaces. Constr. Approx. **38**(2), 235–251 (2013)
21. M. Griebel, H. Harbrecht, On the construction of sparse tensor product spaces. Math. Comput. **82**(268), 975–994 (2013)
22. M. Griebel, M. Holtz, Dimension-wise integration of high-dimensional functions with applications to finance. J. Complex. **26**, 455–489 (2010)
23. M. Griebel, S. Knapek, Optimized tensor-product approximation spaces. Constr. Approx. **16**(4), 525–540 (2000)
24. M. Griebel, S. Knapek, Optimized general sparse grid approximation spaces for operator equations. Math. Comput. **78**, 2223–2257 (2009)
25. M. Griebel, P. Oswald, T. Schiekofer, Sparse grids for boundary integral equations. Numer. Math. **83**(2), 279–312 (1999)
26. K. Hallatschek, Fourier-transform on sparse grids with hierarchical bases. Numer. Math. **63**(1), 83–97 (1992)
27. J. Hamaekers, *Sparse Grids for the Electronic Schrödinger Equation: Construction and Application of Sparse Tensor Product Multiscale Many-Particle Spaces with Finite-Order Weights for Schrödinger's Equation* (Südwestdeutscher Verlag für Hochschulschriften, Saarbrücken, 2010)
28. D. Haroske, H. Triebel, *Distributions, Sobolev Spaces, Elliptic Equations* (European Mathematical Society, Zurich, 2007)
29. J. Jakeman, S. Roberts, Local and dimension adaptive sparse grid interpolation and quadrature. ArXiv e-prints, 2011. arXiv:1110.0010v1 [math.NA]
30. Y. Jiang, Y. Xu, Fast discrete algorithms for sparse Fourier expansions of high dimensional functions. J. Complex. **26**(1), 51–81 (2010)
31. L. Kämmerer, S. Kunis, On the stability of the hyperbolic cross discrete Fourier transform. Numer. Math. **117**, 581–600 (2011)
32. L. Kämmerer, S. Kunis, D. Potts, Interpolation lattices for hyperbolic cross trigonometric polynomials. J. Complex. **28**(1), 76–92 (2012)
33. A. Klimke, B. Wohlmuth, Algorithm 847: spinterp: piecewise multilinear hierarchical sparse grid interpolation in MATLAB. ACM Trans. Math. Softw. **31**(4), 561–579 (2005)
34. S. Knapek, Approximation und Kompression mit Tensorprodukt-Multiskalenräumen. Dissertation, University of Bonn, 2000
35. S. Knapek, Hyperbolic cross approximation of integral operators with smooth kernel. Technical Report 665, SFB 256, University of Bonn, 2000
36. F. Kupka, Sparse grid spectral methods for the numerical solution of partial differential equations with periodic boundary conditions. Ph.D. thesis, University of Wien, 1997
37. F. Kupka, Sparse grid spectral methods and some results from approximation theory, in *Proceedings of the 11th International Conference on Domain Decomposition Methods in Greenwich*, England, ed. by C. Lai, P. Bjørstad, M. Cross, O. Widlund, 1999, pp. 57–64
38. O. Le Maître, O. Knio, *Spectral Methods for Uncertainty Quantification*. Scientific Computation, vol. XVI (Springer, Berlin, 2010)
39. X. Ma, N. Zabaras, An adaptive hierarchical sparse grid collocation algorithm for the solution of stochastic differential equations. J. Comput. Phys. **228**(8), 3084–3113 (2009)
40. J. Pasciak, Spectral and pseudospectral methods for advection equations. Math. Comput. **35**, 1081–1092 (1980)

41. H. Schmeisser, W. Sickel, Spaces of functions of mixed smoothness and approximation from hyperbolic crosses. J. Approx. Theory **128**(2), 115–150 (2004)
42. H. Schmeisser, H. Triebel, *Fourier Analysis and Functions Spaces* (Wiley, London, 1987)
43. J. Shen, L. Wang, Sparse spectral approximations of high-dimensional problems based on hyperbolic cross. SIAM J. Numer. Anal. **48**(3), 1087–1109 (2010)
44. W. Sickel, F. Sprengel, Interpolation on sparse grids and tensor products of Nikol'skij-Besov spaces. J. Comput. Anal. Appl. **1**, 263–288 (1999)
45. W. Sickel, T. Ullrich, Tensor products of Sobolev–Besov spaces and applications to approximation from the hyperbolic cross. J. Approx. Theory **161**(2), 748–786 (2009)
46. I. Sloan, X. Wang, H. Woźniakowski, Finite-order weights imply tractability of multivariate integration. J. Complex. **20**(1), 46–74 (2004)
47. S. Smolyak, Quadrature and interpolation formulas for tensor products of certain classes of functions. Soviet Math. Dokl. **4**, 240–243 (1963) [Russian original in Dokl. Akad. Nauk SSSR **148**, 1042–1045 (1963)]
48. V. Temlyakov, *Approximation of Periodic Functions* (Nova Science, New York, 1993)
49. T. von Petersdorff, C. Schwab, Numerical solution of parabolic equations in high dimensions. Math. Model. Numer. Anal. **38**, 93–127 (2004)
50. G. Wasilkowski, H. Woźniakowski, Finite-order weights imply tractability of linear multivariate problems. J. Approx. Theory **130**(1), 57–77 (2004)
51. A. Werschulz, H. Woźniakowski, Tractability of quasilinear problems I: general results. J. Approx. Theory **145**(2), 266–285 (2007)

Dimension-Adaptive Sparse Grid Quadrature for Integrals with Boundary Singularities

Michael Griebel and Jens Oettershagen

Abstract Classical Gaussian quadrature rules achieve exponential convergence for univariate functions that are infinitely smooth and where all derivatives are uniformly bounded. The aim of this paper is to construct generalized Gaussian quadrature rules based on non-polynomial basis functions, which yield exponential convergence even for integrands with (integrable) boundary singularities whose exact type is not a-priori known. Moreover, we use sparse tensor-products of these new formulae to compute d-dimensional integrands with boundary singularities by means of a dimension-adaptive approach. As application, we consider, besides standard model problems, the approximation of multivariate normal probabilities using the Genz-algorithm.

1 Introduction

The approximation of an integral of a function $f : \Omega^{(d)} \to \mathbb{R}, \Omega^{(d)} \subseteq \mathbb{R}^d$ using point-evaluations is an important task in numerical analysis. It is part of numerous methods and algorithms in almost every scientific area where computers are employed. Such a quadrature rule takes the general form

$$\int_{\Omega^{(d)}} f(\mathbf{x})\, \omega(\mathbf{x})\, \mathrm{d}\mathbf{x} \approx \sum_{i=1}^{n} w_i f(\mathbf{x}_i), \tag{1}$$

where $\omega : \Omega^{(d)} \to \mathbb{R}^+$ is a positive weight-function, the $\mathbf{x}_i \in \Omega^{(d)}$ are the quadrature nodes and the $w_i \in \mathbb{R}$ are the quadrature weights. The quality of such

M. Griebel • J. Oettershagen (✉)
Institute for Numerical Simulation, Bonn, Germany
e-mail: griebel@ins.uni-bonn.de; oettershagen@ins.uni-bonn.de

J. Garcke and D. Pflüger (eds.), *Sparse Grids and Applications - Munich 2012*, Lecture Notes in Computational Science and Engineering 97,
DOI 10.1007/978-3-319-04537-5_5,

an approximation depends on the regularity of f and the specific choice of the nodes and weights of the quadrature formula. Most quadrature rules are constructed in such a way that they are exact on a certain finite-dimensional subspace of $L_1(\Omega^{(d)}, \omega)$, e.g., on the polynomials up to a certain degree.

In this paper we will consider integration over the open unit cube $\Omega^{(d)} = (0,1)^d$ with respect to the uniform Lebesgue measure $\omega \equiv 1$. We treat integrands that might possess integrable boundary singularities, whose exact type is not a priori known.[1]

A common approach to deal with boundary singularities in the univariate setting are variable transformations $\tau : \mathbb{R} \supseteq \hat{\Omega} \to (0,1)$, such that

$$\int_0^1 f(x)\,\mathrm{d}x = \int_{\hat{\Omega}} f \circ \tau(y) \cdot \tau'(y)\,\mathrm{d}y.$$

If τ is properly chosen, the transformed integrand $\hat{f}(y) = f \circ \tau(y) \cdot \tau'(y)$ is no longer singular, but decays to zero as y approaches the boundary of $\hat{\Omega}$. Popular examples are the tanh [44] and double exponential [32, 45] transforms, where a truncated trapezoidal rule is applied to $\hat{f}$ on $\hat{\Omega} = \mathbb{R}$. Other examples are the so-called periodization transforms [27, 37, 39, 40], which use a mapping to $\hat{\Omega} = (0,1)$ that makes the resulting integrand periodic. Eventhough these approaches work very well in the univariate setting, their multivariate versions suffer from an exponential blowup of the norm of the transformed integrand $\hat{f}$, as it was pointed out in [27] for the case of periodizing transformations in the context of lattice rules. Thus it takes exponentially long (in d), until the good asymptotic convergence rate kicks in, which makes these approaches very costly for practical problems in higher dimensions.

For this reason we are not going to apply a trapezoidal rule to the transformed integrand $\hat{f}$, but rather use a suitable Gaussian rule tailored to the weight function $\tau'(y)$ on $\hat{\Omega}$ and then map back the associated Gaussian quadrature nodes to the unit interval $(0,1)$. This approach results in a so-called generalized Gaussian quadrature rule on $(0,1)$ that is exact not on a space of polynomials as are conventional Gaussian rules, but on a $2n$-dimensional subspace of univariate *singular* functions from $L_1(0,1)$. Its basis functions are given by powers of certain monotonous singular functions. We will prove exponential convergence for integrands with arbitrary algebraic boundary singularities. Moreover, we explicitly compute error constants for quadrature on $(0,1)$ in the Hardy space of functions that are analytic in the unit disc. In contrast to Gauss-Legendre quadrature, which only achieves an algebraic rate of convergence, our approach shows an exponential decay of the error. For the higher dimensional case we then employ these univariate quadrature rules within a sparse tensor product construction which also exhibits a certain degree of exactness on tensor products of the univariate singular functions. We give numerical

---

[1]Otherwise one could reformulate the problem to integration with respect to a weight function ω that resembles the singularity.

evidence that, for singular problems in the unit cube, our approach significantly outperforms the conventional sparse grid quadrature methods which are based on the Gauss-Legendre or Clenshaw-Curtis rules, respectively. Furthermore, we use our new method in combination with dimension-adaptive sparse grids for various standard model problems and for the computation of multivariate normal probabilities by the Genz-algorithm [12].

The remainder of this article is organized as follows: In Sect. 2 we will shortly revise the classical (univariate) Gaussian quadrature and its generalization to Tschebyscheff-systems. Then we introduce certain classes of Tschebyscheff-systems, whose associated Gaussian quadrature formulae can be described in terms of classical Gaussian formulae on an unbounded domain $\hat{\Omega}$ and a mapping back to $(0, 1)$. This allows for an easy construction of the new generalized Gaussian quadrature. We give a few examples for this approach and prove error bounds for a special case which is related to the classical Gauss-Laguerre formula. In Sect. 3 we will introduce sparse tensor products of the new univariate quadrature to deal with multivariate problems. Here, we specifically employ the dimension-adaptive approach from [17]. In Sect. 4 we give the results of our numerical experiments. First, we demonstrate the quality of our generalized Gaussian quadrature formula in the univariate case for singular problems and quadrature in the Hardy space. Then we apply it within the dimension-adaptive sparse grid algorithm to several model problems and to the computation of multivariate normal probabilities using the algorithm of Genz. We close with some remarks in Sect. 5.

2 A Generalized Gaussian Quadrature Approach

In this section we will introduce a class of generalized Gaussian quadrature rules that are exact on a certain $2n$-dimensional subspace of singular functions on $(0, 1)$. First, we shortly recall the basic properties of classical Gaussian quadrature in the univariate case. Then we generalize it to Tschebyscheff-systems on $(0, 1)$ and introduce a framework for which the generalized Gaussian quadrature formula can be described by a classical Gaussian formula with respect to a certain weight function on $(0, \infty)$ together with a map to $(0, 1)$. We give examples and an error estimate for a class of integrands with algebraic singularities at the boundary.

2.1 Classical Gaussian Quadrature

Given an interval $\Omega \subseteq \mathbb{R}$ and a positive weight function $\omega : \Omega \to \mathbb{R}^+$ the classical n-point Gaussian rule is the standard approach to approximate

$$\int_{\Omega} f(x)\omega(x) \, \mathrm{d}x \approx \sum_{i=1}^{n} w_i f(x_i)$$

for smooth functions $f \in C^{2n}(\Omega)$. Classical Gaussian quadrature rules are defined by the property that polynomials up to degree $2n - 1$ are integrated exactly (with respect to the weight function ω) with only n quadrature points, i.e.

$$\int_{\Omega} p(x)\omega(x)\,\mathrm{d}x = \sum_{i=1}^{n} w_i\, p(x_i) \tag{2}$$

for all $p \in \operatorname{span}\{x^k, k = 0, \ldots, 2n-1\}$. It is well known that there can not exist a quadrature rule that yields a higher degree of polynomial exactness, thus Gaussian quadrature is optimal in this sense.

Note that (2) is a nonlinear system of $2n$ equations which defines the nodes x_i and weights w_i of the Gaussian quadrature rule. In general, the direct solution of a nonlinear system is a difficult task. But here one can resort to orthogonal polynomials and then use their recurrence relation to compute the nodes and weights of Gaussian quadrature formulae. To this end, we consider the set of polynomials which are orthogonal with respect to the weighted L_2 inner product

$$\langle p, q \rangle_{\Omega,\omega} = \int_{\Omega} p(x)q(x)\,\omega(x)\,\mathrm{d}x.$$

It is known that for any domain Ω and weight function ω there exists a polynomial p_n of degree n that fulfills

$$\int_{\Omega} x^k\, p_n(x)\omega(x)\,\mathrm{d}x = 0 \quad \text{for all } k = 0, 1, \ldots, n-1.$$

We call p_n the n-th degree orthogonal polynomial (with respect to Ω and ω). Note that the set of all $p_0, p_1, \ldots$ is a complete orthogonal system of $L_2(\Omega, \omega)$. Moreover, p_n has exactly n distinct simple roots $x_1, \ldots, x_n \in \Omega$, which turn out to be the quadrature nodes of the Gaussian formula. This is important because, for any given weight-function ω on a domain $\Omega \subseteq \mathbb{R}$, there exists a sequence of orthogonal polynomials and thus there exists a uniquely determined corresponding Gaussian quadrature rule. If one knows the coefficients of $p_n(x) := a_n x^n + a_{n-1}x^{n-1} + \ldots + a_0$ and its zeros $x_1, \ldots, x_n$ it is easy to compute the corresponding weights by the formula

$$w_i = \frac{a_n}{a_{n-1}} \frac{\langle p_{n-1}, p_{n-1} \rangle_{\Omega,\omega}}{p_n'(x_i)\, p_{n-1}(x_i)} \geq 0. \tag{3}$$

In order to obtain the coefficients and roots of such orthogonal polynomials one usually employs their recurrence formula to assemble a so-called companion matrix, whose eigenvalues are the roots of p_n. For details see [7, 11].

Now we consider error bounds for classical Gaussian quadrature. First we define the n-th error functional as

$$R_n(f) := \int_{\Omega} f(x)\,\omega(x)\,\mathrm{d}x - \sum_{i=1}^{n} w_i\, f(x_i). \tag{4}$$

Because of the Weierstrass approximation theorem and the positivity of the quadrature weights w_i, Gaussian quadrature rules converge for any continuous function, i.e.

$$f \in C^0(\Omega) \Rightarrow \lim_{n\to\infty} R_n(f) = 0. \tag{5}$$

For integrands which possess at least $2n$ continuous derivatives there is the well-known error-bound

$$|R_n(f)| \leq \frac{f^{(2n)}(\xi)}{a_n^2(2n)!} \quad \text{for some } \xi \in \Omega. \tag{6}$$

Thus, if for all $n \in \mathbb{N}$ it holds $|f^{(2n)}(x)| \leq M_n(f)\ \forall x \in \Omega$ where $\frac{M_n(f)}{a_n^2}$ is bounded by a polynomial, the quantity $|R_n(f)|$ converges even exponentially to zero.

On the other hand, if the derivatives of f are unbounded on Ω the bound (6) is useless. For example, for $\Omega = (-1,1)$ and $\omega(x) = 1$ one obtains the well-known Gauss-Legendre rule, for which the bound (6) takes the form [7]

$$|R_n(f)| \leq f^{(2n)}(\xi) \cdot \frac{(n!)^4}{(2n+1)((2n)!)^3}, \qquad \xi \in (-1,1).$$

Now consider $f(x) = (1-x)^{-\alpha}, \alpha > 0$, which is unbounded on Ω as are all of its derivatives. In [9, 28] it was shown that the rate of convergence substantially deteriorates with α, i.e.

$$|R_n(f)| = \mathcal{O}(n^{-1+\alpha})$$

and we only obtain an algebraic convergence rate of $(1-\alpha)$ for $\alpha < 1$, while for $\alpha \geq 1$ the integral does not exist anyway. This simple example shows that classical Gaussian rules lose their nice convergence properties when it comes to singular integrands. To deal with such integration problems efficiently, the Gaussian approach must be properly generalized.

2.2 *Generalized Gaussian Quadrature*

The aim of this section is to find quadrature rules that achieve a maximum degree of exactness for systems of functions that are not polynomials, but inherently possess integrable singularities. To this end, we will use the notion of so-called *Tschebyscheff systems* (T-systems) [25].

For $n+1$ functions $\varphi_0, \ldots, \varphi_n$ and $n+1$ pairwise distinct points $t_0, \ldots, t_n \in [a,b]$ we define the generalized Vandermonde determinant as

$$\mathcal{D}(\varphi_0,\ldots,\varphi_n;t_0,\ldots,t_n) := \det \begin{pmatrix} \varphi_0(t_0) & \varphi_0(t_1) & \ldots & \varphi_0(t_n) \\ \varphi_1(t_0) & \varphi_1(t_1) & \ldots & \varphi_1(t_n) \\ \vdots & \vdots & \ddots & \vdots \\ \varphi_n(t_0) & \varphi_n(t_1) & \ldots & \varphi_n(t_n) \end{pmatrix} \tag{7}$$

Definition 1. A set of continuous functions $\varphi_0, \varphi_1, \ldots, \varphi_n$ is called *T-system* over a compact interval $[a,b]$ iff

$$\mathcal{D}(\varphi_0,\varphi_1,\ldots,\varphi_n;t_0,t_1,\ldots,t_n) > 0$$

for all pairwise distinct $t_0 < t_1 < \ldots < t_n \in [a,b]$.

This generalizes the concept of polynomials in the sense that it allows unique interpolation by cardinal functions and thus every linear combination $\sum_{i=0}^{n} c_i\varphi_i(x)$ has at most n zeros.

Examples for sets of functions that form a T-system on their respective domains are of course polynomials $\{1,x,x^2,x^3,\ldots,x^n\}$, fractional polynomials like $\{1,\sqrt{x},x,x\sqrt{x},x^2,x^2\sqrt{x},\ldots,x^{n/2}\}$ or certain radial basis functions like, e.g., $\{\exp(-\frac{(c_0-x)^2}{\sigma^2}),\ldots,\exp(-\frac{(c_n-x)^2}{\sigma^2})\}$ for pairwise distinct $c_i \in \mathbb{R}$.

We are now in the position to define the concept of generalized Gaussian quadrature.

Definition 2 (Generalized Gaussian Quadrature Rule). Let $\Phi = \{\varphi_0,\ldots,\varphi_{2n-1}\}$ be a set of integrable functions on a compact interval $[a,b]$, i.e. $\int_a^b \varphi_i\,\omega(x)\,\mathrm{d}x < \infty$. A *generalized Gaussian quadrature rule* on $[a,b]$ is a n-point rule that is exact on Φ, i.e.

$$\sum_{i=1}^{n} w_i\varphi_j(x_i) = \int_a^b \varphi_j(x)\,\mathrm{d}x, \quad \text{for all } j = 0,\ldots,2n-1. \tag{8}$$

We have the following result from [25].

Theorem 1. *Let $\Phi = \{\varphi_0,\ldots,\varphi_{2n-1}\}$ be a T-system of integrable functions on the bounded interval $[a,b] \subsetneq \mathbb{R}$. Then there exists a generalized Gaussian quadrature rule with n nodes $x_1,\ldots,x_n \in (a,b)$ and non-negative weights $w_1,\ldots,w_n$.*

This result was generalized in [29] to the case of (semi-)open intervals.

Theorem 2. *If Φ constitutes a T-system on any closed interval $[\hat{a},\hat{b}] \subset (a,b) \subseteq \mathbb{R}$, then we call Φ a T-system on (a,b) and there exist n nodes $x_1,\ldots,x_n \in (a,b)$ and non-negative weights $w_1,\ldots,w_n$, such that* (8) *holds.*

Moreover, it is also known [19] that generalized Gaussian quadrature formulae are unique. The determination of a specific generalized Gaussian quadrature formula involves the problem of finding n nodes $(x_1,\ldots,x_n)$ and weights $(w_1,\ldots,w_n)$ such that (8) holds. This is a system of $2n$ equations in both $\mathbf{w} \in \mathbb{R}^n$ and $\mathbf{x} \in (a,b)^n$, which is highly non-linear in $x_1, \ldots, x_n$. Once more, its solution is in general a difficult task which is addressed in [29]. To avoid this issue, we will restrict ourselves in the following to a certain class of T-systems which involves powers of singular functions. This will allow us to resort to orthogonal polynomials again.

2.3 A Certain Class of Singular Tschebyscheff-Systems

In the following we will restrict ourselves to $\Omega = (a,b) = (0,1)$ for the sake of simplicity. But all results easily translate to arbitrary finite intervals (a,b) by affine linear dilation and translation.

We propose the following T-system on $\Omega = (0,1)$, which (as we will see) possesses a structure that reduces the solution of (8) to a certain classical Gaussian quadrature rule on $\hat{\Omega} = (0,\infty)$.

To this end, let ψ be a $C^1(0,1)$ function that fulfills the following conditions:

1. $\psi(0) = 0$ and $\lim_{x\to 1} \psi(x) = \infty$.
2. $\int_0^1 \psi(x)^j \,\mathrm{d}x < \infty$, for all $j \in \mathbb{N}_0$.
3. ψ is strictly monotonous increasing, i.e. $\psi' > 0$.

Remark 1. From the conditions (1)–(3) the following results can be derived:

- $\psi : [0,1) \leftrightarrow [0,\infty)$ is a C^1-bijection.
- $\lim_{y\to\infty} \psi^{-1}(y) = 1$.
- $\int_0^\infty \psi'(\psi^{-1}(y))^{-1} \,\mathrm{d}y = 1$.
- ψ is the inverse of a cumulative distribution function whose underlying distribution has finite moments on $[0,\infty)$.

Since polynomials form a T-system over any subset of $\mathbb{R}$, the following lemma proves, that

$$\varphi_j(x) := \psi(x)^j, \quad j = 0,1,2,\ldots \tag{9}$$

is a complete T-system over $(0,1)$, if ψ fulfills the conditions (1)–(3)

Lemma 1. *If $\hat{\varphi}_0, \ldots, \hat{\varphi}_n$ is a T-system on some domain $\hat{\Omega} \subseteq \mathbb{R}$ and $\psi : \Omega \leftrightarrow \hat{\Omega}$ is a bijection between $\Omega \subseteq \mathbb{R}$ and $\hat{\Omega}$, then the set*

$$\varphi_j := \hat{\varphi}_j \circ \psi : \ \Omega \to \mathbb{R}, \quad j = 0,1,\ldots,n$$

is a T-system on Ω.

Proof. *Suppose that there exist pairwise distinct points $t_0, \ldots, t_n \in \Omega$, such that $\mathcal{D}(\varphi_0, \ldots, \varphi_n; t_0, \ldots, t_n) \leq 0$ holds. Because of the bijectivity of ψ there are points $\hat{t}_j := \psi(t_j) \in \hat{\Omega}$, such that $\mathcal{D}(\hat{\varphi}_0, \ldots, \hat{\varphi}_n; \hat{t}_0, \ldots, \hat{t}_n) \leq 0$, which is a contradiction to the assumption that $\hat{\varphi}_0, \ldots, \hat{\varphi}_n$ is a T-system on $\hat{\Omega}$.* □

Next we will describe the relationship between the generalized Gaussian quadrature with respect to the system $\Phi = \{\psi(\cdot)^j\}_{j=0}^{\infty}$ on $\Omega = (0,1)$ and classical Gaussian quadrature on the unbounded domain $\hat{\Omega} = (0,\infty)$. To this end, we set

$$\omega(y) := \frac{d}{dy}\psi^{-1}(y) = \frac{1}{\psi'(\psi^{-1}(y))},$$

which is non-negative on $\hat{\Omega}$. We know that there exists a sequence of polynomials $p_0, p_1, p_2, \ldots$ on $\hat{\Omega}$ which are orthogonal with respect to ω, i.e. $\langle p_i, p_j\rangle_{\hat{\Omega},\omega} = \delta_{i,j}$.

Remark 2. The orthogonality of the p_i translates to the set of functions $q_j : (0,1) \to \mathbb{R}$

$$q_j(x) := p_j \circ \psi(x), \quad j = 0,1,2,\ldots,$$

i.e.

$$\int_0^1 q_i(x)q_j(x)\,\mathrm{d}x = \int_0^\infty p_i(y)p_j(y)\,\omega(y)dy = \delta_{i,j}.$$

Analogously, the n distinct zeros $y_j \in \hat{\Omega}$ of p_n carry over to the zeros $x_j \in (0,1)$ of q_n as

$$x_j := \psi^{-1}(y_j), \tag{10}$$

since $q_n(x_j) = p_n(y_j) = 0$.

We finally arrive at the following result:

Theorem 3. *With $\Omega = (0,1)$ and $\hat{\Omega} = (0,\infty)$ let $\psi : \Omega \to \hat{\Omega}$ fulfill the conditions (1)–(3), and let $y_j \in \hat{\Omega}$, $w_j \in \mathbb{R}^+$ $(j = 1,\ldots,n)$ be the nodes and weights of the classical Gaussian quadrature rule on $\hat{\Omega}$ with respect to the weight function $\omega(y) = \frac{d}{dy}\psi^{-1}(y)$. Then the quadrature rule*

$$Q_n(f) := \sum_{j=1}^n w_j f(x_j), \quad \textit{where } x_j = \psi^{-1}(y_j) \in (0,1), \tag{11}$$

is exact on the span of $\Phi = \{\varphi_0,\ldots,\varphi_{2n-1}\}$ defined in (9), i.e.

$$Q_n(\varphi_j) = \int_0^1 \varphi_j(x)\,\mathrm{d}x, \quad \textit{for } j = 0,\ldots,2n-1.$$

Proof. Because of $\varphi_k(x_j) = y_j^k$ and

$$\sum_{j=1}^n w_j y_j^k = \int_0^\infty y^k\omega(y)\,dy = \int_0^1 \varphi_k(x)\,\mathrm{d}x$$

it holds for $k = 0, \dots, 2n-1$ *that*

$$Q_n(\varphi_k) = \sum_{j=1}^{n} w_j \varphi_k(x_j) = \int_0^1 \varphi_k(x)\, \mathrm{d}x. \qquad \square$$

By now we have shown that the generalized n-point Gaussian quadrature with respect to the set of singular functions $\varphi_0, \varphi_1, \dots, \varphi_{2n-1}$ from (9) can simply be computed by mapping the nodes of a certain classical Gaussian quadrature in $(0, \infty)$ back to $(0, 1)$. Moreover, because of (5) the quadrature rule (11) converges for any continuous function on $(0, 1)$. This kind of quadrature rule is especially suited for integrands on $(0, 1)$ with a boundary singularity located at $x = 1$. It is possible to extend this approach to integrands with singularities at both endpoints of the domain $(-1, 1)$.

Remark 3. If we change the first and second condition to

1′. $\lim_{x \to \pm 1} \psi(x) = \pm\infty$.
2′. $\int_{-1}^{1} \psi(x)^j\, \mathrm{d}x < \infty$, for all $j \in \mathbb{N}_0$.

we obtain a T-system of functions which have singularities at both $x = -1$ and $x = +1$. Moreover, it follows that

- $\psi : (-1, 1) \leftrightarrow (-\infty, \infty)$ is a C^1 bijection.
- $\lim_{x \to \pm\infty} \psi^{-1}(x) = \pm 1$.
- $\int_{-\infty}^{\infty} \psi'(\psi^{-1}(y))^{-1}\, \mathrm{d}y = 2$.
- ψ is the inverse of a cumulative distribution function whose underlying distribution has finite moments on $(-\infty, \infty)$.

All previous statements also hold true for this case, with obvious modifications in their proofs.

2.4 Examples

We now give three examples for the choice of the function ψ, where the first one relates to the classical Gauss-Laguerre quadrature on $(0, \infty)$ with respect to the weight-function $\omega(y) = e^{-y}$ and the second one to a non-classical rule with respect to $\omega(y) = \cosh(y) \cdot \cosh(\sinh(y))^{-2}$ on $(0, \infty)$ as well. The third example refers to Remark 3 and is related to the classical Gauss-Hermite quadrature on the double-infinite interval $(-\infty, \infty)$ with respect to the weight function $\omega(y) = \exp(-y^2)$.

2.4.1 Powers of Logarithms

Our first example involves the classical Gauss-Laguerre quadrature rule, which is well-known and its nodes and weights are available from numerous libraries.

We choose $\psi : (0, 1) \to (0, \infty)$ as

$$\psi_{\log}(x) := -\log(1-x).$$

With (9), this results in the T-system defined by

$$\varphi_k(x) := (-\log(1-x))^k, \quad k = 0, 1, 2, \ldots. \tag{12}$$

The inverse is given as $\psi^{-1}(y) = 1 - \exp(-y)$ and its derivative is

$$\omega(y) := \frac{d}{dy}\psi^{-1}(y) = \exp(-y). \tag{13}$$

Because of $\int_0^1 (-\log(1-x))^k = k!$ all φ_k are integrable. Thus the conditions (1)–(3) are fulfilled and Theorem 3 relates to the well-known Laguerre polynomials which define the Gauss-Laguerre quadrature rules on $(0, \infty)$ with respect to the weight-function $\omega(y) = e^{-y}$. Let w_i and y_i denote the n quadrature weights and nodes of the Gauss-Laguerre formula on $(0, \infty)$. Then, by mapping the Gaussian nodes y_i back to $(0, 1)$, we obtain the quadrature rule

$$\int_0^1 f(x)\,\mathrm{d}x \approx \sum_{i=1}^n w_i f(x_i), \quad \text{with } x_i = 1 - \exp(-y_i) \tag{14}$$

which is exact for $\varphi_0, \varphi_1, \ldots, \varphi_{2n-1}$ defined in (12) on the interval $(0, 1)$. Moreover, one can prove that the generalized Gaussian quadrature (11) on $(0, 1)$ with φ_i given by (12) achieves basically exponential convergence for certain integrands with algebraic singularities.

Theorem 4. *Let $f(x) = (1-x)^{-\alpha}$ with $\alpha < 1$. Then, the generalized Gaussian formula with respect to the T-system* (12) *converges faster than any polynomial, i.e.*

$$\left| \int_0^1 f(x)\,\mathrm{d}x - \sum_{i=1}^n w_i f(x_i) \right| \le c_{\alpha,k} n^{-k},$$

where $c_{\alpha,k}$ is a constant depending on both α and k, but not on n.

Proof. *By Theorem 1 from [30] the quadrature error of the Gauss-Laguerre formula for a k-times continuous differentiable function g is bounded by*

$$\left| \int_0^\infty g(y)\, e^{-y}\, dy - \sum_{i=1}^n w_i g(y_i) \right| \le c n^{-k} \cdot \int_0^\infty |y^{\frac{k}{2}} g^{(k)}(y)| e^{-y}\, dy. \tag{15}$$

Mapping this result to the unit interval yields

$$\begin{aligned}
\left| \int_0^1 f(x)\, \mathrm{d}x - \sum_{i=1}^n w_i f(x_i) \right| &= \left| \int_0^\infty f(1-e^{-y})\, e^{-y}\, dy - \sum_{i=1}^n w_i f(1-e^{-y_i}) \right| \\
&= \left| \int_0^\infty e^{\alpha y}\, e^{-y}\, dy - \sum_{i=1}^n w_i e^{\alpha y_i} \right| \\
&\overset{(15)}{\leq} cn^{-k} \cdot \alpha^k \int_0^\infty |y^{\frac{k}{2}} e^{\alpha y}| e^{-y}\, dy \\
&= cn^{-k} \cdot \frac{\alpha^k}{(1-\alpha)^{\frac{k}{2}+1}} \Gamma\left(\frac{k}{2}+1\right)
\end{aligned}$$

which holds for any $k = 1, 2, \ldots$. □

Recall that for this class of integrands the Gauss-Legendre quadrature only achieves an algebraic rate of convergence of $n^{-1+\alpha}$.

2.4.2 Powers of Inverse Hyperbolic Functions

Next, we consider a choice for ψ that consists of certain hyperbolic functions. It is given by

$$\psi_{\text{hyp}}(x) := \text{arc sinh}\left(\frac{2}{\pi}\text{arc tanh}(x)\right), \quad x \in (0,1), \tag{16}$$

which is inspired by the so-called double exponential (DE) quadrature [32, 45] that has gained substantial interest within the last years. It leads to the problem of constructing a Gaussian quadrature rule for the weight function

$$\omega(y) = \frac{\cosh(y)}{\cosh(\frac{\pi}{2}\sinh(y))^2}$$

on the infinite interval $(0, \infty)$. For this purpose we use the algorithm proposed in [11] and map the resulting Gaussian nodes $y_1, \ldots, y_n$ back to $(0, 1)$. This results in the quadrature formula

$$\int_0^1 f(x)\, \mathrm{d}x \approx \sum_{i=1}^n w_i f(x_i), \quad \text{with } x_i = \tanh\left(\frac{\pi}{2}\sinh(y_i)\right) \tag{17}$$

which is exact on the T-system

$$\varphi_k(x) = \text{arc sinh}\left(\frac{2}{\pi}\text{arc tanh}(x)\right)^k, \quad k = 0, 1, \ldots 2n-1.$$

Note that the predecessor of the DE-rule was the Tanh-quadrature which was introduced in [44]. In our setting it relates to

$$\psi(x) = \text{arc}\tanh(x)$$

which also fulfills the conditions (1)–(3) and leads to orthogonal polynomials with respect to

$$\omega(y) = \cosh(y)^{-2}.$$

Remark 4. Both the double exponential and the tanh approach rely on the quick decay of $f \circ \psi^{-1}(y) \cdot D\psi^{-1}(y)$ as $y \to \pm\infty$, which allows for an exponential rate of convergence for the trapezoidal rule. This approach does not work well in a multivariate setting, because the factor $\prod_{j=1}^{d} D\psi^{-1}(y_j)$ exponentially blows up the norm of the transformed integrand [27]. Our approach is different in the sense that we do not apply a trapezoidal rule directly to $f \circ \psi^{-1}(y) \cdot D\psi^{-1}(y)$ but use a Gaussian rule tailored to $D\psi^{-1}$, which is applied to $f \circ \psi^{-1}$ only.

2.4.3 Powers of the Inverse Error Function

Our third example illustrates Remark 3 and relates to Gauss-Hermite quadrature on the whole $\mathbb{R}$. We choose $\psi : (-1, 1) \to (-\infty, \infty)$ as

$$\psi_{\text{erf}}(x) := \text{erf}^{-1}(x),$$

where $\text{erf}(x)$ denotes the error-function.[2] This leads to the T-system of functions

$$\varphi_k(x) := \left(\text{erf}^{-1}(x)\right)^k, \quad k = 0, 1, 2, \ldots \tag{18}$$

which have singularities at both $x = -1$ and $x = 1$. Since $\psi(x)^k$ is integrable on $(-1, 1)$ for all $k = 0, 1, \ldots$, the conditions (1′)–(3′) are fulfilled and Theorem 3 relates to the well-known Hermite polynomials which define the Gauss-Hermite quadrature rules on $(-\infty, \infty)$ with respect to $\omega(y) = \frac{2}{\sqrt{\pi}} e^{-y^2}$. Then, if w_i and y_i denote the weights and nodes of the n-point Gauss-Hermite quadrature on $\mathbb{R}$, the resulting quadrature rule on $(-1, 1)$ with respect to the T-system (18) is given by

$$\int_{-1}^{1} f(x)\, \mathrm{d}x \approx \sum_{i=1}^{n} w_i f(x_i), \quad \text{with } x_i = \text{erf}(y_i).$$

Note that it is possible to derive an analogous error bound as in Theorem 4 by an estimate for Gauss-Hermite quadrature which is given in [30].

[2]The error-function $\text{erf} : \mathbb{R} \to (-1, 1)$ is defined as $\text{erf}(x) = \frac{2}{\sqrt{\pi}} \int_0^x e^{-t^2}\, \mathrm{d}t$.

3 Dimension-Adaptive Sparse Grids

In this section we recall the sparse grid method for the integration of multivariate functions. Sparse grid quadrature formulae are constructed using certain combinations of tensor products of one-dimensional quadrature rules, see, e.g., [6, 15, 16, 20, 21,36,43,46]. This way, sparse grid methods can exploit the mixed smoothness of f, if present, and may overcome the curse of dimension to a certain extent. Moreover, they can be employed in a dimension-adaptive fashion.

3.1 *Classical Construction*

For a continuous univariate function $g : (0,1) \to \mathbb{R}$ let

$$Q_{n_k} g := \sum_{i=1}^{n_k} w_{i,k}\, g(x_{i,k})$$

denote a sequence of univariate quadrature rules with $n_{k+1} > n_k$, $Q_0 f = 0$ and $Q_{n_k} g \to \int_0^1 g(x)\, \mathrm{d}x$ for $k \to \infty$. Using the difference quadrature formulae

$$\Delta_k = Q_{n_k} - Q_{n_{k-1}}$$

one has

$$\sum_{k=1}^{\infty} \Delta_k g = \int_0^1 g(x)\, \mathrm{d}x.$$

Then, for a d-variate function $f : (0,1)^d \to \mathbb{R}$, its integral can be represented by the infinite telescoping series

$$\int_{(0,1)^d} g(\mathbf{x})\, d\mathbf{x} = \sum_{\mathbf{k} \in \mathbb{N}^d} \Delta_{\mathbf{k}} f \tag{19}$$

which collects the products of each possible combination of the univariate difference formula. Here, $\mathbf{k} \in \mathbb{N}^d$ denotes a multi-index with $k_j > 0$ and

$$\Delta_{\mathbf{k}} f := \left(\Delta_{k_1} \otimes \ldots \otimes \Delta_{k_d}\right) f.$$

For a given level $l \in \mathbb{N}$ and the choice $n_k = 2^k - 1$ the classical sparse grid method,[3] see, e.g., [16, 36, 43], is then defined by

[3]Often denoted as Smolyak's method, see [43].

$$\mathrm{SG}_l f := \sum_{|\mathbf{k}|_1 \le l+d-1} \Delta_{\mathbf{k}} f \tag{20}$$

where $|\mathbf{k}|_1 := \sum_{j=1}^{d} k_j$. Here, from the set of all possible indices $\mathbf{k} \in \mathbb{N}^d$, only those are considered whose $|\cdot|_1$-norm is smaller than a certain value. Note that the product integration rule is recovered if the norm $|\cdot|_\infty := \max\{k_j : j = 1,\ldots,d\}$ is used for the selection of indices instead of the $|\cdot|_1$-norm in (20).

3.2 Generalized Sparse Grids

The sparse grid construction can be tailored to certain classes of integrands if some information on the importance of the dimensions or the importance of the interactions between the dimensions is a priori known. This is achieved by choosing appropriate finite index sets $\mathcal{I} \subset \mathbb{N}^d$ in the representation (19) such that a given accuracy is attained with as few as possible function evaluations.

To ensure the validity of the hierarchical expansion the index set $\mathcal{I}$ has to satisfy the admissibility condition

$$\mathbf{k} \in \mathcal{I} \text{ and } \mathbf{l} \le \mathbf{k} \Rightarrow \mathbf{l} \in \mathcal{I}.$$

In this way, the generalized sparse grid method

$$\mathrm{SG}_{\mathcal{I}} f := \sum_{\mathbf{k} \in \mathcal{I}} \Delta_{\mathbf{k}} f \tag{21}$$

is defined, see, e.g., [16]. Note that the product rule, the classical sparse grid construction (20), sparse grids with delayed basis sequences [38] or nonisotropic sparse grids based on the weighted norms $|\mathbf{k}|_{1,\mathbf{a}} := \sum_{j=1}^{d} a_j k_j$ with weight factor $\mathbf{a} \in \mathbb{R}_+^d$ for the different coordinate directions [16, 21] are just special cases of this general approach.

3.3 Dimension-Adaptive Sparse Grids

In practice, usually no a priori information on the dimension structure of the integrand is available. In this case algorithms are required which can construct appropriate index sets $\mathcal{I}$ automatically during the actual computation. Such algorithms were presented in [17, 23] where the index sets are found in a dimension-adaptive way by the use of suitable error indicators. The adaptive methods start with the smallest index set $\mathcal{I} = \{(1,\ldots,1)\}$. Then, step-by-step the index $\mathbf{k}$ from the set of all admissible indices is added which has the largest value $|\Delta_{\mathbf{k}} f|$ and is therefore expected to provide the largest error reduction, see [15, 17, 21, 33] for details. Altogether, the algorithm allows for an adaptive detection of the important

Algorithm 1 Dimension-adaptive construction of the index set $\mathcal{I}$

Initialize:

1. set of active indices: $\mathcal{I} = (1, \ldots, 1)$.
2. $s = \Delta_{(1,\ldots,1)} f$.

repeat

1. Determine the set of admissible indices $\mathcal{A} = \{\mathcal{I} + \mathbf{e}_i : i = 1, \ldots, d\}$.
2. For all $\mathbf{k} \in \mathcal{A}$ compute $\Delta_{\mathbf{k}} f$.
3. Determine (some) $\hat{\mathbf{k}} = \arg\max_{\mathbf{k} \in \mathcal{A}} \Delta_{\mathbf{k}} f$.
4. Add the index $\hat{\mathbf{k}}$ to $\mathcal{I}$.
5. Update the sum $s = s + \Delta_{\hat{\mathbf{k}}} f$.

until $|\Delta_{\hat{\mathbf{k}}} f| < \varepsilon$;

Output: $\mathrm{SG}_{\mathcal{I}} f = s$.

dimensions and heuristically constructs optimal index sets $\mathcal{I}$ in the sense of [5, 22] which is closely related to best N-term approximation [8].

For the sake of completeness, we give a simplified version[4] of the dimension-adaptive algorithm from [17, 21] in Algorithm 1. In our numerical experiments of Sect. 4 we will use this approach with the three generalized Gaussian quadrature rules from Sect. 2.4.

Note finally that, besides the dimension-wise adaption, also a purely local adaptivity based on the trapezoidal rule or higher order composite Newton-Cotes formulae is possible [2, 4], which leads to algebraic convergence. However, since our aim in this paper is to explicitly deal with boundary singularities by means of a special generalized Gaussian approach that allows for exponential convergence, we will stick to the dimension-adaptive approach here.

3.4 Degree of Exactness

Now we have a look at the degree of exactness of the sparse grid method. As before, let $\Phi = \{\varphi_j\}_{j=0}^{\infty}$ be a complete Tschebyscheff-system on $(0, 1)$ and let the univariate quadrature rule, on which a particular sparse grid algorithm is based, have a certain degree of exactness $\deg(n_k)$ with respect to Φ, i.e.

$$Q_{n_k} \varphi_j = \int_0^1 \varphi_j(x)\, \mathrm{d}x, \quad \text{for all } j = 0, \ldots, \deg(n_k).$$

[4]The original algorithm from [17] which we employed in our computations in Sect. 4 uses a more sophisticated error criterion than the one described in Algorithm 1.

If one defines $P_k = \text{span}\{\varphi_0, \dots, \varphi_{\deg(n_k)}\}$, the sparse grid algorithm $\text{SG}_{\mathcal{I}}$ is exact on the space

$$\{P_{k_1} \otimes \dots \otimes P_{k_d} : \mathbf{k} \in \mathcal{I}\}.$$

This is similar to a result from [36]. There it was shown that a regular Clenshaw-Curtis sparse grid (i.e. $\mathcal{I}_\ell = \{\mathbf{k} \in \mathbb{N}_+^d : |\mathbf{k}|_1 \leq d + \ell - 1\}$) is exact on

$$\{P_{k_1} \otimes \dots \otimes P_{k_d} : |\mathbf{k}|_1 = d + \ell - 1\},$$

where the φ_i are polynomials of degree i and $\deg(n_k) = n_k$. In our case, we have $\deg(n_k) = 2n - 1$ since we are in a (generalized) Gaussian setting.

4 Numerical Results

In this section we give results for several numerical experiments. First, we study the behaviour of our generalized Gaussian quadrature formulae in the univariate case for both, smooth and singular integrands, as well as the worst-case error in the Hardy space. Then, we deal with the higher-dimensional case where we employ the dimension-adaptive sparse grid approach that is based on our new univariate approach and compare it with dimension-adaptive sparse grids based on classical univariate rules like Gauss-Legendre or Clenshaw-Curtis quadrature. For the sake of completeness, we also compare with plain Monte Carlo and the Quasi-Monte Carlo method that is based on the Sobol sequence. Note that in all experiments the term $\psi_{\log}$ refers to the generalized Gaussian quadrature-formula with (12), while ψ_{hyp} refers to the construction with (16). ψ_{erf} refers to the construction from (18) with the additional linear transformation to the unit-interval $(0, 1)$.

Note that, because of $\int_0^1 f(x)\,dx = \int_0^1 f(1-x)\,dx$, it is advantageous to transform integrands $f(x)$ with a singularity located at $x = +1$ to $f(1-x)$, which has the same singularity at $x = 0$. Since double floating point arithmetic is more precise in a neighbourhood of 0 than in a neighbourhood of ± 1 it is possible to resolve the singularity up to a higher precision, after this transformation is employed. Of course the quadrature nodes have to undergo the same transformation, i.e. $x_i \mapsto 1 - x_i$.

4.1 Univariate Results

As our first univariate test case we consider functions with algebraic singularity of the order $0 < \alpha < 1$ at $x = 1$, i.e. we consider the problem

$$\int_0^1 \frac{1}{(1-x)^\alpha}\,dx, \tag{22}$$

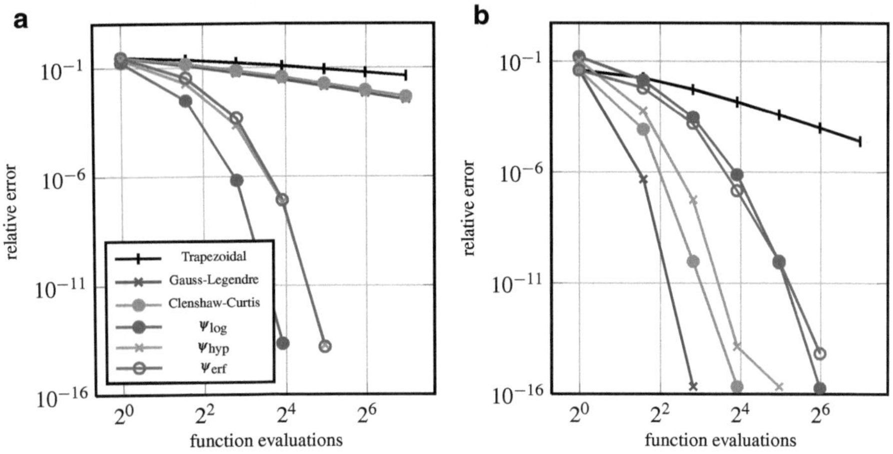

Fig. 1 Convergence behaviour for the singular integrand $f(x) = (1-x)^{-1/2}$ (**a**) and smooth integrand $f(x) = \exp(x)$ (**b**)

for various values of $0 < \alpha < 1$. As can be seen in Fig. 1, the generalized Gaussian approaches achieve exponential convergence not only for the smooth example with $f(x) = \exp(x)$ but also for the singular integrand (22) with $\alpha = 1/2$. Furthermore it can be observed that the rate of convergence for the trapezoidal, the Clenshaw-Curtis and the classical Gauss-Legendre quadrature are only algebraic. We clearly see that the Gauss-Legendre rule loses its exponential convergence when it comes to the treatment of singular functions. This is however not the case for our new generalized Gaussian approach which exhibits exponential convergence for both, the smooth and the singular test function.

Next, we consider how the type of singularity of the integrand (22) affects the convergence. In Fig. 2 one can see the performance of the univariate generalized Gaussian approach based on $\psi_{\log} = -\log(1-x)$ for several values of $\alpha \in (0,1)$. The convergence is always exponential, even though it takes longer to reach the asymptotic exponential regime for big values of α. Note that the integrands are not in $L^2(0,1)$ anymore for $\alpha \geq 1/2$. For the actual computation of the results displayed in Fig. 2 we used long double[5] floating point arithmetic for the evaluation of the integrand. We stress that the quadrature nodes and weights were only stored in standard double[6] precision.

Another example concerns the quadrature error in a Hardy space. It consists of functions that are analytic in the unit-circle $\mathbb{D} = \{z \in \mathbb{C} : |z| < 1\}$ and is given by

$$H^2 = \{f : \mathbb{D} \mapsto \mathbb{C} : \|f\|_{H^2}^2 < \infty\}, \quad \text{where } \|f\|_{H^2}^2 = \int_{\mathbb{D}} |f(z)|^2 \, |dz|.$$

[5]64 bit significant precision and 14 bit exponent precision.

[6]53 bit significant precision and 10 bit exponent precision.

Fig. 2 Convergence behaviour of the generalized Gaussian quadrature w.r.t. $\psi_{\log}(x)$ for the integrand $f(x) = (1-x)^{\alpha}$ with algebraic singularities for different value of α

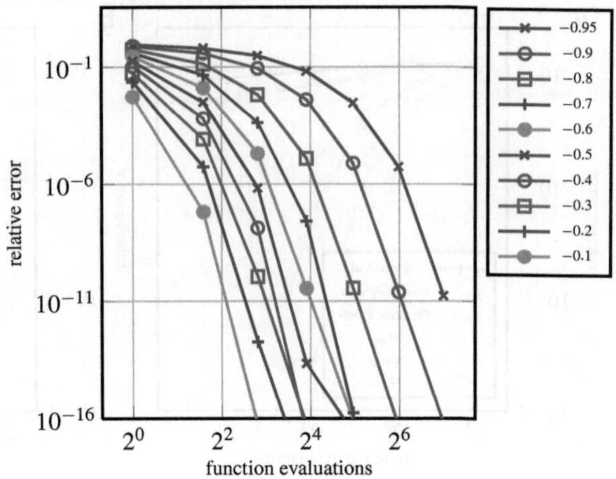

Since H^2 is a reproducing kernel Hilbert space with kernel $K(x,y) = \frac{1}{1-xy}$, see e.g. [47], the quadrature error in this space can be estimated by standard functional analysis as

$$\left| \int_0^1 f(x)\,\mathrm{d}x - \sum_{i=1}^n w_i f(x_i) \right| \leq R_n(\mathbf{x},\mathbf{w}) \cdot \|f\|_{H^2},$$

where $R_n(\mathbf{x},\mathbf{w})$ is the so-called worst-case error which depends on the nodes $\mathbf{x}$ and weights $\mathbf{w}$. It is explicitly given by

$$\begin{aligned} R_n(\mathbf{x},\mathbf{w})^2 &= \int_0^1 \int_0^1 K(s,t)\,\mathrm{d}s\,\mathrm{d}t - 2\sum_{i=1}^n w_i \int_0^1 K(t,x_i)\,\mathrm{d}t \\ &\quad + \sum_{i=1}^n \sum_{j=1}^n w_i w_j K(x_i,x_j) \\ &= \frac{\pi^2}{6} + 2\sum_{i=1}^n w_i \frac{\log(1-x_i)}{x_i} + \sum_{i=1}^n \sum_{j=1}^n w_i w_j \frac{1}{1-x_i x_j}. \end{aligned}$$

In Fig. 3 one can see that all three new methods, i.e. the generalized Gaussian quadrature with respect to $\psi_{\log}$, ψ_{hyp} and ψ_{erf}, significantly outperform polynomial-based methods like Gauss-Legendre as well as Clenshaw-Curtis and the trapezoidal rule in the Hardy space.

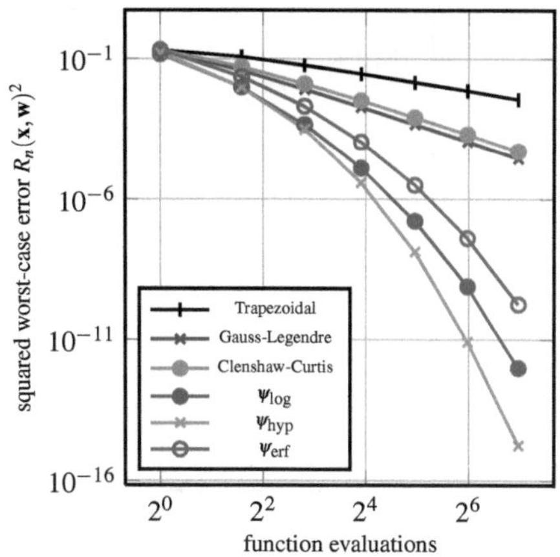

Fig. 3 Comparison of several methods for the worst-case error in the Hardy space H^2

4.2 *Multivariate Results for Standard Test Cases*

Now we consider simple higher-dimensional model problems with a certain multiplicative and additive structure. To this end, let

$$f(\mathbf{x}) = \sum_{\mathfrak{u}\subseteq\{1,\ldots,d\}} \gamma_{\mathfrak{u}} \prod_{i\in\mathfrak{u}} \frac{1}{\sqrt[3]{1-x_i}}, \tag{23}$$

where the so-called product-weights $\gamma_{\mathfrak{u}}$ [42, 46] are defined by

$$\gamma_{\mathfrak{u}} := \prod_{i\in\mathfrak{u}} \gamma_i .$$

Here, the sequence $\gamma_1 \geq \gamma_2 \geq \ldots \geq \gamma_d > 0$ moderates the contribution of the different coordinate directions to the value of the integral.

For the special case $\gamma_i = 2^{-i}$, the resulting convergence behaviour can be seen in Fig. 4 for conventional integration methods like MC and QMC and the dimension-adaptive sparse grid approach based on both, the classical Clenshaw-Curtis and Gauss-Legendre formulae and our new quadrature formula with $\psi_{\log}$ chosen as (12) and ψ_{hyp} chosen as (16), respectively. We clearly see again that the conventional methods only possess an algebraic rate of convergence, whereas our new approach with $\psi_{\log}$ and ψ_{hyp} indeed shows exponential convergence. We furthermore see that, for this particular model problem, we obtain somewhat better results for $\psi_{\log}$ than for ψ_{hyp}. For higher dimensions this exponential behaviour gets less prominent. This stems from a delay in the onset of the asymptotic regime due to the dimension-dependent constants involved here [34]. But the superiority over the conventional methods still can be seen clearly.

As another example we will now deal with multivariate functions that can be represented as a superposition of q-dimensional functions. They are just a special

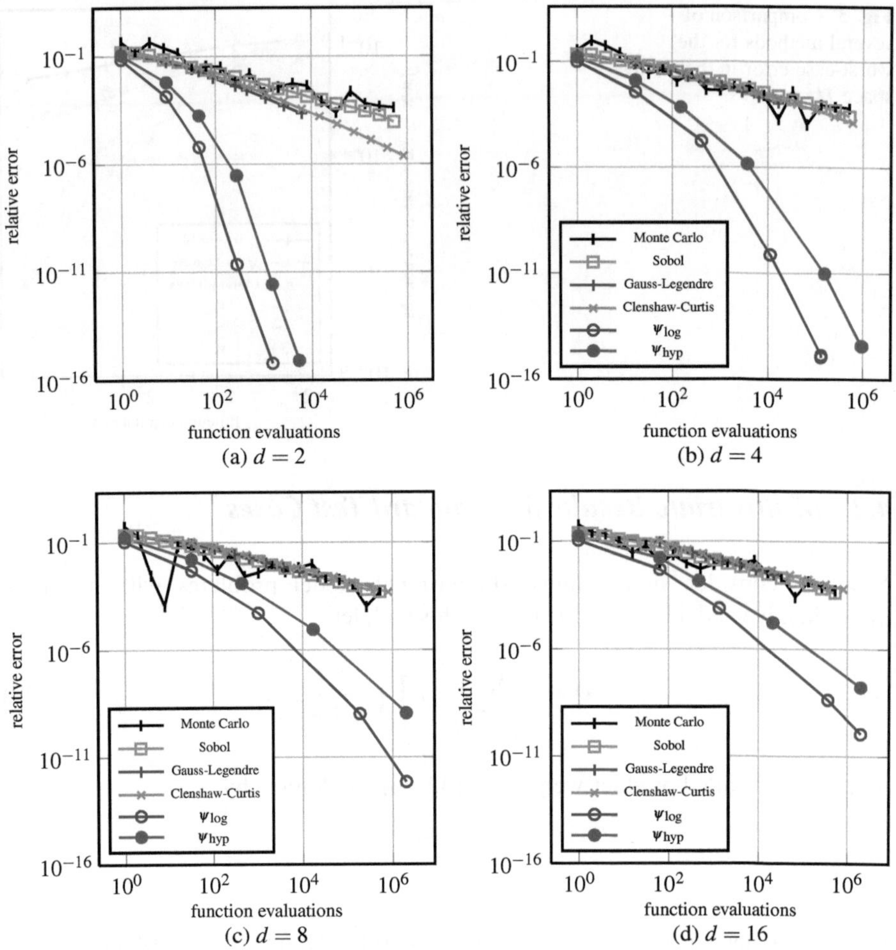

(a) $d = 2$ (b) $d = 4$ (c) $d = 8$ (d) $d = 16$

Fig. 4 Convergence behaviour for the test function with product-weights (23) with $\gamma_i = 2^{-i}$ for dimensions $d = 2, 4, 8, 16$

case of the above framework where all $\gamma_{\mathfrak{u}} = 0$ for sets $\mathfrak{u}$ with cardinality bigger than q, i.e. $|\mathfrak{u}| > q$. This type of weights is often referred to as *finite-order weights*, see e.g. [41]. Here, we consider the test-function

$$f_{d,q}(\mathbf{x}) := \sum_{\substack{\mathfrak{u} \subset \mathcal{D} \\ |\mathfrak{u}| \leq q}} \frac{1}{\sqrt{\sum_{j \in \mathfrak{u}} (1 - x_j)}}. \tag{24}$$

The results are displayed in Fig. 5. We basically observe a similar convergence behaviour as in the previous example. Now, however, ψ_{hyp} is slightly superior to $\psi_{\log}$. Moreover, the offset of the asymptotic convergence behaviour with respect to the dimension is more specific. The number of function evaluations, needed until the

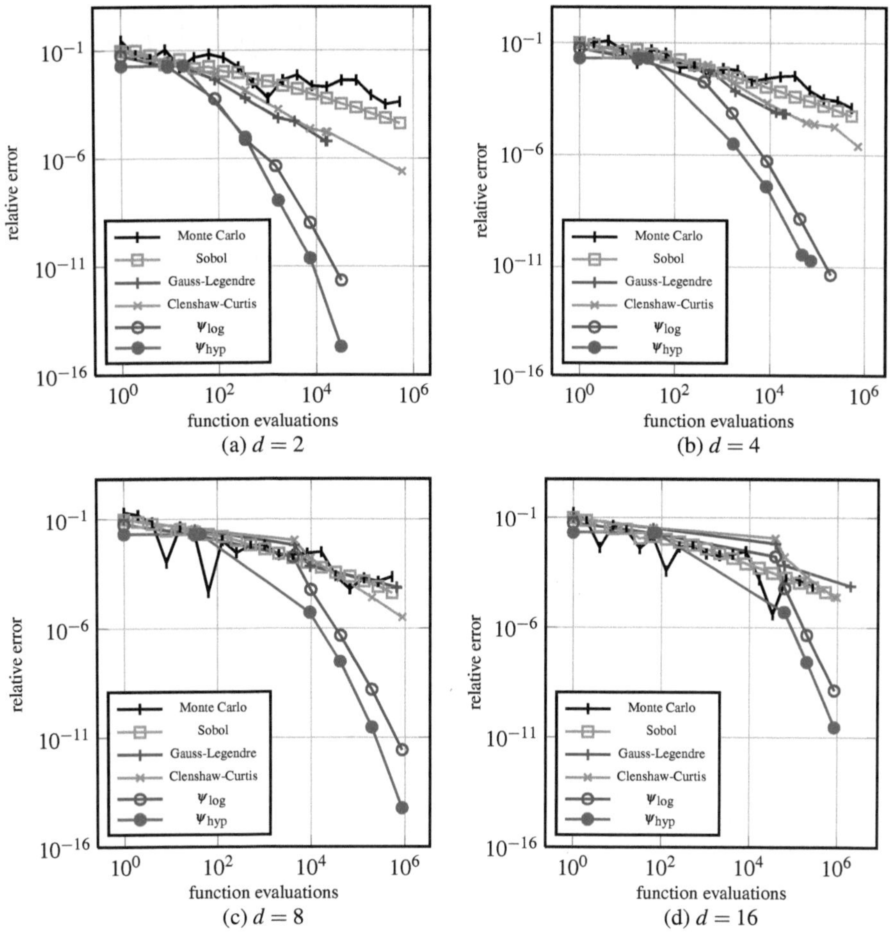

Fig. 5 Convergence behaviour of the finite-order test function (24) with $q = 2$ and $d = 2, 4, 8, 16$

convergence rate for the dimension-adaptive sparse grid methods kicks in, depends quadratically on d. This is due to the fact that for $q = 2$ the sum in (24) consists of $\binom{d}{2}$ parts.

4.3 Computing Multivariate Normal Probabilities by the Genz-Algorithm

As a final example we consider the evaluation of multivariate normal probabilities which is an important part of many numerical methods in statistics [14], financial engineering [15], physics [31] and econometrics [1, 24]. It is defined by the integral

$$F(\mathbf{b}) := \frac{1}{\sqrt{\det(\Sigma)(2\pi)^d}} \int_{-\infty}^{b_1} \cdots \int_{-\infty}^{b_d} \exp\left(-\frac{1}{2}\mathbf{x}^t \Sigma^{-1}\mathbf{x}\right) d\mathbf{x} \tag{25}$$

where $\Sigma \in \mathbb{R}^{d\times d}$ is a covariance matrix which depends on the specific problem under consideration.

For (25) it is common to use a sequence of variable transformations to obtain an integration problem that is defined on the open unit cube. This approach was independently developed by Genz [12], Geweke and Hajivassiliou [3, 18] and Keane [26]. In statistics, this method is often referred to as Genz-algorithm, while in econometrics it is called GHK-simulator. Regular sparse grids based on the Gauss-Legendre quadrature were utilized in [24] for the first time in this setting. In the following we will demonstrate that the sparse grid approach can benefit from our new univariate quadrature formulae. We remark that this approach can also be applied to the computation of other probabilities, e.g. the t-distribution [14].

The Genz-algorithm [12] consists of several transformations and finally leads to the integral

$$F(\mathbf{b}) = \hat{b}_1 \int_{(0,1)^{d-1}} \prod_{i=2}^{d} \hat{b}_i(w_1,\ldots,w_{i-1})\, \mathrm{d}\boldsymbol{w} \tag{26}$$

where the $\hat{b}_i$ are recursively given by

$$\hat{b}_i(w_1,\ldots,w_{i-1}) = \Phi\left(C_{i,i}^{-1} \cdot \left(b_i - \sum_{j=1}^{i-1} C_{i,j} \cdot \Phi^{-1}(w_j \cdot \hat{b}_j(w_1,\ldots,w_{j-1})\right)\right).$$

Here, the matrix $\mathbf{C} \in \mathbb{R}^{d\times d}$ denotes a Cholesky factor[7] of the covariance-matrix, i.e. $\mathbf{C}\mathbf{C}^T = \Sigma$, and $\Phi : \mathbb{R} \to (0,1)$ is the cumulative Gaussian distribution function.

The main advantage of the Genz-algorithm in a dimension-adaptive sparse grid setting stems from the fact that it enforces a priority ordering onto the variables $w_1,\ldots,w_{d-1}$, where w_1 contributes the most and w_{d-1} contributes the fewest to the value of $F(\mathbf{b})$. Furthermore, the dimensionality of the original integration problem is reduced by one. A disadvantage is of course the increased cost for the evaluation of the transformed integrand in formula (26). Moreover, while the original integrand was analytic in the whole complex plane, the new integrand is only analytic within the open disc $\{z \in \mathbb{C} : |z-\frac{1}{2}| < \frac{1}{2}\}$. This is due to the inverse cumulative distribution function Φ^{-1} that introduces a singularity at the origin and in some dimensions a fast growth of the integrand for arguments close to one. This is the reason why we now also included the ψ_{erf} method in our experiments for the Genz-integrand.

[7] Cholesky factorization is here only unique modulo row and column permutation.

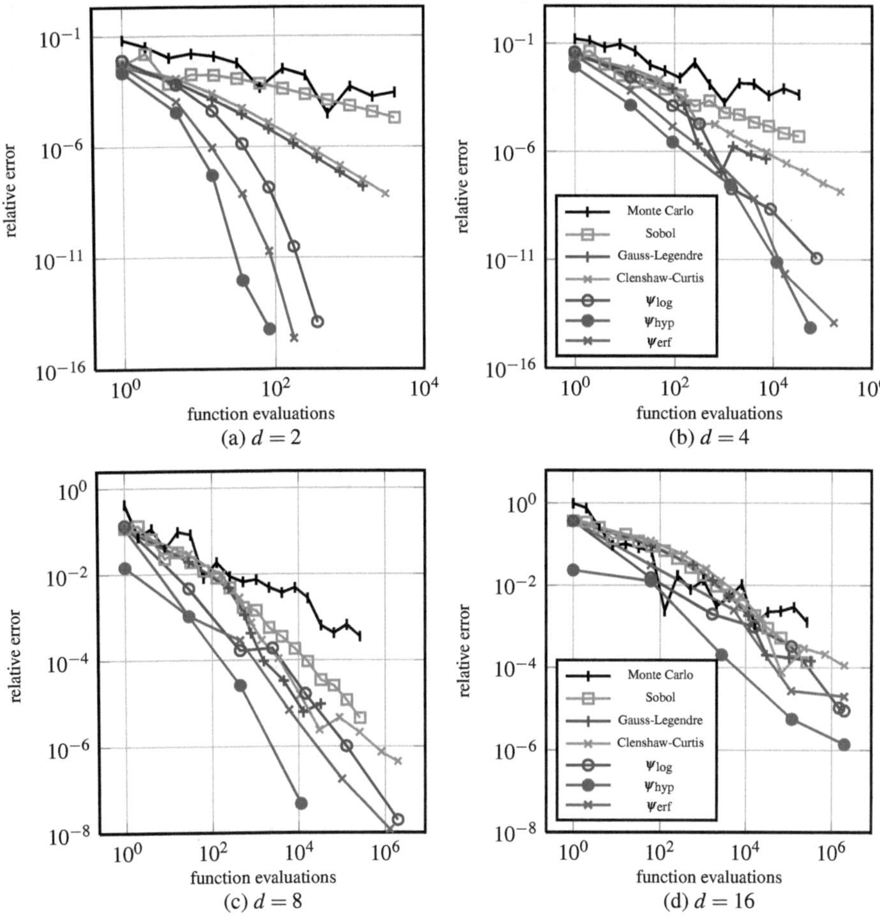

Fig. 6 Convergence behaviour for the Genz-integrand with $\rho_0 = 0.1$ and $b_i = \frac{1}{2}$

In our numerical experiments we will consider a special covariance structure for which the integral in (25) has a closed form solution [10,12]. Namely we assume that the covariance matrix Σ has constant variance $\Sigma_{i,i} = 1$ and covariance $\Sigma_{i,j} = v_i \cdot v_j$ for $i \neq j$, where $v_i \in (-1, 1), i = 1, \ldots, d$. We remark that the normalization of the variance to one is not a restriction because it is always possible to shift the variance via a diagonal transformation to the boundaries of integration $b_1, \ldots, b_d$.

In our first example we choose constant correlation $\Sigma_{i,j} = \rho_0 = 0.1$ and all $b_i = \frac{1}{2}$. In Fig. 6 it can be observed that the dimension-adaptive sparse grid approach is superior to (Q)MC for small values of d. Especially if it is based on the new generalized Gaussian formulae (11) with $\psi_{\log}$, ψ_{hyp} and ψ_{erf}, it performs very well and even achieves exponential convergence for small d. For higher dimensions the convergence should still be exponential, but it takes quite long until this asymptotic

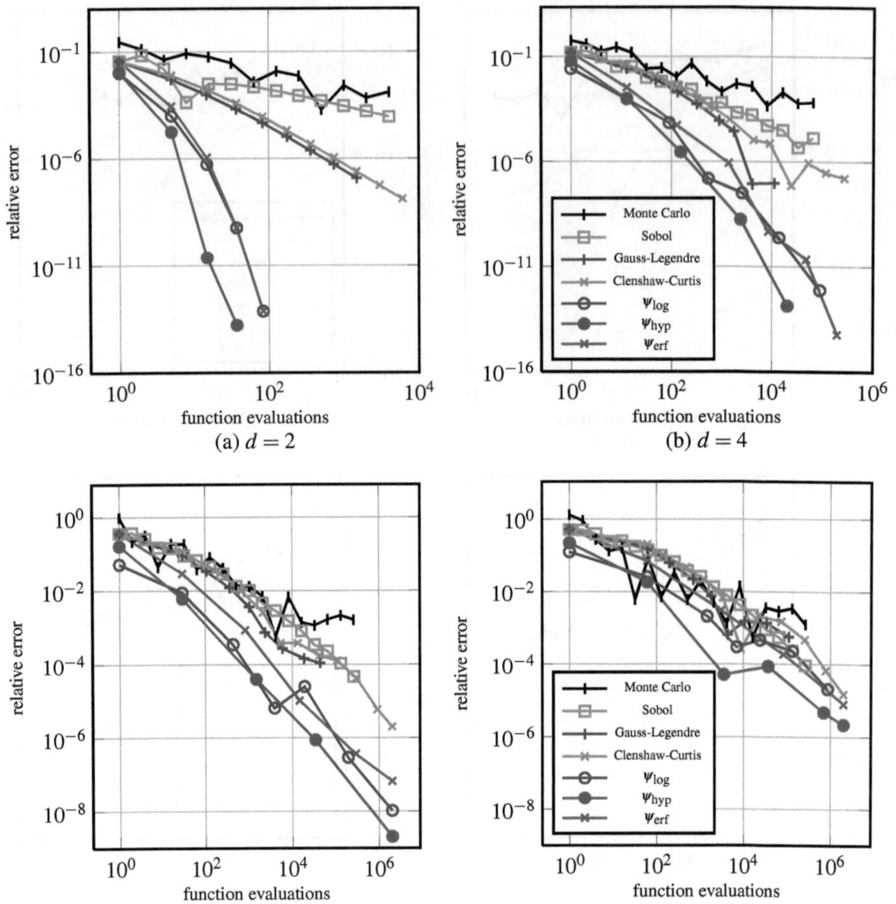

(a) $d = 2$ (b) $d = 4$ (c) $d = 8$ (d) $d = 16$

Fig. 7 Convergence behaviour of the Genz-integrand with $\rho_0 = 0.25$ and $b_i = -1 + \frac{i}{10}$

rate sets in. Thus exponential convergence is not visible in the case $d = 16$. This situation is common for sparse grids, as pointed out in [34, 35].

In our second example we use different values for the boundaries, namely $b_i = -1 + \frac{i}{10}$ and a bigger, but still constant correlation $\Sigma_{i,j} = \rho_0 = 0.25$. The convergence behaviour is similar to the first example, as can be seen from Fig. 7. This demonstrates that our new approach indeed allows to deal with varying boundary values as it is needed in most practical applications.

In the third example we look at a truly high-dimensional example with d up to 256. Here we set $\Sigma_{i,j} = 2^{-(i+j)}$ and $b_i = -1 + \frac{i}{10}$. This enforces an exponential decay of the correlation coefficients which weakens the interaction between the coordinates of the underlying integrand. Albeit involving a somewhat strong restriction, such situations appear often in practical problems with high

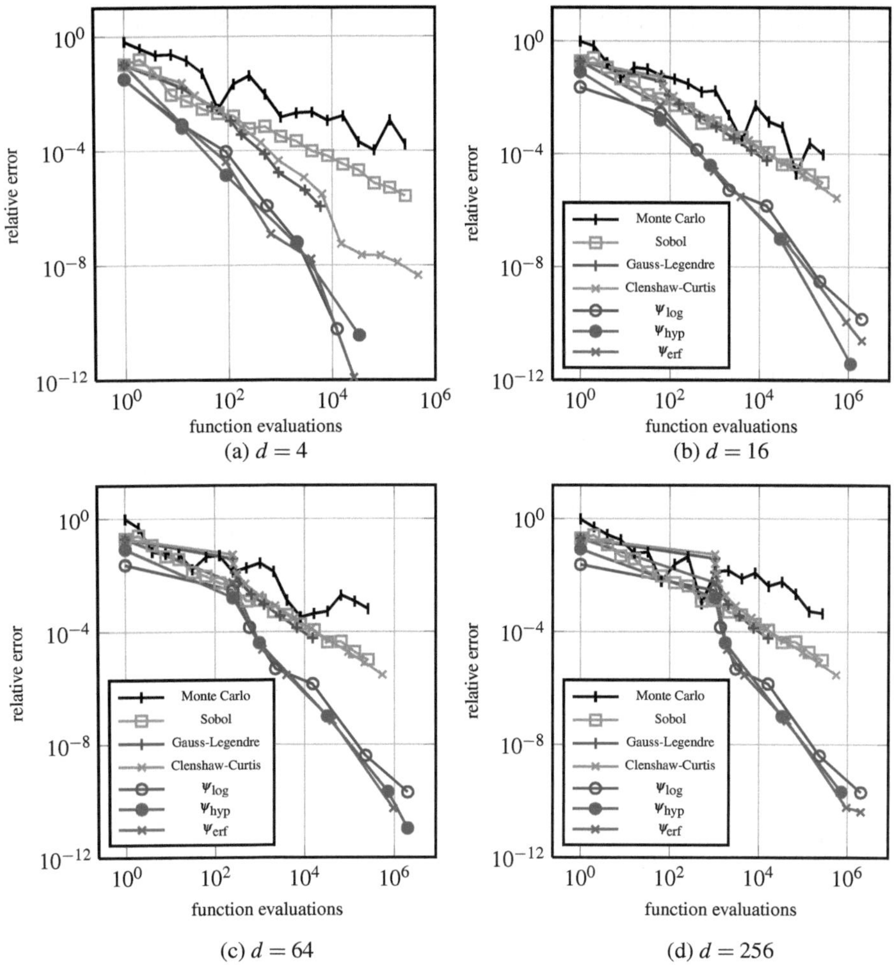

Fig. 8 Convergence behaviour of the Genz-integrand with $\Sigma_{i,j} = 2^{-(i+j)}$ and $b_i = -1/2 + \frac{i}{10}$

nominal but low intrinsic dimensions, where measured data are involved, like for example in the panel probit models in econometrics [24], in density estimation in statistics [13] or in the pricing of various instruments in financial engineering [15].

It can be observed in Fig. 8 that the dimension-adaptive sparse grid algorithm has now no trouble with nominally high dimensions and is able to correctly detect the important intrinsic coordinate directions. Moreover our new approach with $\psi_{\log}$, ψ_{hyp} and ψ_{erf} clearly outperforms all other methods in the case $d = 256$, even though exponential convergence, as in Fig. 6, is not apparent yet. Still, to achieve a relative accuracy of 10^{-7} it needs less than 10^5 function evaluations whereas QMC and the Clenshaw-Curtis or Gauss-Legendre based sparse grids would need about 10^8 function evaluations and plain Monte Carlo would even involve about 10^{13} function values.

5 Conclusion

In the present paper we constructed a new univariate generalized Gaussian quadrature rule for bounded domains, which is related to classical Gaussian quadrature on unbounded domains with respect to a certain weight function. Special cases involve the Gauss-Laguerre and Gauss-Hermite rules and thus allow for an easy construction of the new generalized Gaussian quadrature by building onto existing implementations. Another example, which is related to the double exponential quadrature approach, was also presented.

Moreover, we used sparse tensor-products of this new univariate approach to cover multivariate problems. As application we considered a variant of the Genz-algorithm in which the multivariate integrals are evaluated by a dimension-adaptive sparse grid approach that was based on the new generalized Gaussian quadrature formulae. We demonstrated that our new method is able to significantly outperform dimension-adaptive sparse grid methods based on Gauss-Legendre or Clenshaw-Curtis quadrature as well as Monte Carlo and Quasi-Monte Carlo methods in moderate dimensions up to 16 and for special cases also in a truly high-dimensional setting with dimensions of $d = 256$.

References

1. I. Bertschek, M. Lechner, Convenient estimators for the panel probit model. J. Econom. **87**(2), 329–371 (1998)
2. T. Bonk, A new algorithm for multi-dimensional adaptive numerical quadrature, in *Adaptive Methods – Algorithms, Theory, and Applications*, ed. by W. Hackbusch, G. Wittum. NNFM, vol. 46 (Vieweg, Braunschweig, 1994), pp. 54–68
3. A. Borsch-Supan, V. Hajivassiliou, Smooth unbiased multivariate probability simulators for maximum likelihood estimation of limited dependent variable models. J. Econom. **58**(3), 347–368 (1993)
4. H.-J. Bungartz, S. Dirnstorfer, Multivariate quadrature on adaptive sparse grids. Computing **71**(1), 89–114 (2003)
5. H.-J. Bungartz, M. Griebel, A note on the complexity of solving Poisson's equation for spaces of bounded mixed derivatives. J. Complex. **15**, 167–199 (1999)
6. H.-J. Bungartz, M. Griebel, Sparse grids. Acta Numer. **13**, 1–123 (2004)
7. P. Davis, P. Rabinowitz, *Methods of Numerical Integration* (Academic, London, 1984)
8. R. DeVore, Nonlinear approximation. Acta Numer. **7**, 51–150 (1998)
9. R. DeVore, L. Scott, Error bounds for Gaussian quadrature and weighted L_1 polynomial approximation. SIAM J. Numer. Anal. **21**(2), 400–412 (1984)
10. C. Dunnett, M. Sobel, Approximations to the probability integral and certain percentage points of a multivariate analogue of Student's t-distribution. Biometrika **42**, 258–260 (1955)

11. W. Gautschi, Construction of Gauss-Christoffel quadrature formulas. Math. Comput. **22**(102), 251–270 (1968)
12. A. Genz, Numerical computation of multivariate normal probabilities. J. Comput. Graph. Stat. **1**, 141–149 (1992)
13. A. Genz, F. Bretz, Comparison of methods for the computation of multivariate t probabilities. J. Comput. Graph. Stat. **11**(4), 950–971 (2002)
14. A. Genz, F. Bretz, *Computation of Multivariate Normal and t Probabilities* (Springer, Berlin, 2009)
15. T. Gerstner, Sparse grid quadrature methods for computational finance. Habilitation, Institut für Numerische Simulation, University of Bonn, 2007
16. T. Gerstner, M. Griebel, Numerical integration using sparse grids. Numer. Algorithms **18**, 209–232 (1998)
17. T. Gerstner, M. Griebel, Dimension–adaptive tensor–product quadrature. Computing **71**(1), 65–87 (2003)
18. J. Geweke, Bayesian inference in econometric models using Monte Carlo integration. Econometrica **57**(6), 1317–1339 (1989)
19. A. Ghizzetti, A. Ossicini, Sull'esistenza e unicitat delle formule di quadrature Gaussiane. Rend. Mat. **8**, 1–15 (1975)
20. M. Griebel, Sparse grids and related approximation schemes for higher dimensional problems, in *Proceedings of the Conference on Foundations of Computational Mathematics (FoCM05)*, Santander, 2005, pp. 106–161
21. M. Griebel, M. Holtz, Dimension-wise integration of high-dimensional functions with applications to finance. J. Complex. **26**(5), 455–489 (2010)
22. M. Griebel, S. Knapek, Optimized tensor-product approximation spaces. Constr. Approx. **16**(4), 525–540 (2000)
23. M. Hegland, Adaptive sparse grids, in *Proceedings of CTAC*, Brisbane, ed. by K. Burrage, R.B. Sidje, 2001, pp. 335–353
24. F. Heiss, The panel probit model: adaptive integration on sparse grids. Adv. Econom. **26**, 41–64 (2010)
25. S. Karlin, W.J. Studden, *Tchebycheff Systems: With Applications in Analysis and Statistics (Pure and Applied Mathematics)* (Interscience, New York, 1966)
26. M.P. Keane, A computationally practical simulation estimator for panel data. Econometrica **62**(1), 95–116 (1994)
27. F. Kuo, I. Sloan, H. Woźniakowski, Periodization strategy may fail in high dimensions. Numer. Algorithms **46**(4), 369–391 (2007)
28. D. Lubinsky, P. Rabinowitz, Rates of convergence of Gaussian quadrature. Math. Comput. **43**, 219–242 (1984)
29. J. Ma, V. Rokhlin, S. Wandzurra, Generalized Gaussian quadrature rules for systems of arbitrary functions. SIAM J. Numer. Anal. **33**(3), 971–996 (1996)
30. G. Mastroianni, G. Monegato, Error estimates for Gauss-Laguerre and Gauss-Hermite quadrature formulas, in *Approximation and Computation: A Festschrift in Honor of Walter Gautschi*, ed. by R. Zahar. ISNM International Series of Numerical Mathematics, vol. 119 (Birkhäuser, Boston, 1994), pp. 421–434
31. M. Masujima, *Path Integral Quantization and Stochastic Quantization* (Springer, Berlin, 2008)
32. M. Mori, Discovery of the double exponential transformation and its developments. Publ. Res. Inst. Math. Sci. **41**(4), 897–935 (2005)
33. T. Nahm, Error estimation and index refinement for dimension adaptive sparse grid quadrature with applications to the computation of path integrals. Master's thesis, Institut für Numerische Simulation, Universität Bonn, 2005
34. F. Nobile, R. Tempone, C. Webster, An anisotropic sparse grid stochastic collocation method for partial differential equations with random input data. SIAM J. Numer. Anal. **46**(5), 2411–2442 (2008)
35. F. Nobile, R. Tempone, C. Webster, A sparse grid stochastic collocation method for partial differential equations with random input data. SIAM J. Numer. Anal. **46**(5), 2309–2345 (2008)

36. E. Novak, K. Ritter, High dimensional integration of smooth functions over cubes. Numer. Math. **75**, 79–97 (1996)
37. D. Nuyens, R. Cools, Higher order quasi-Monte Carlo methods: a comparison, in *AIP Conference Proceedings*, ed. by T.E. Simos, G. Psihoyios, C. Tsitouras (Springer, New York, 2010), pp. 553–557
38. K. Petras, Smolyak cubature of given polynomial degree with few nodes for increasing dimension. Numer. Math. **93**(4), 729–753 (2003)
39. A. Sidi, Extension of a class of periodizing variable transformations for numerical integration. Math. Comput. **75**(253), 327–343 (2005)
40. I. Sloan, S. Joe, *Lattice Methods for Multiple Integration*. Oxford Science Publications (Clarendon Press, Oxford, 1994)
41. I. Sloan, X. Wang, H. Woźniakowski, Finite-order weights imply tractability of multivariate integration. J. Complex. **20**(1), 46–74 (2004)
42. I. Sloan, H. Wozniakowski, When are quasi-Monte Carlo algorithms efficient for high-dimensional integrals? J. Complex. **14**, 1–33 (1998)
43. S. Smolyak, Quadrature and interpolation formulas for tensor products of certain classes of functions. Dokl. Akad. Nauk SSSR **4**, 240–243 (1963)
44. F. Stenger, Integration formulae based on the trapezoidal formula. IMA J. Appl. Math. **12**(1), 103–114 (1973)
45. H. Takahasi, M. Mori, Double exponential formulas for numerical integration. Publ. Res. Inst. Math. Sci. **9**(3), 721–741 (1973)
46. G. Wasilkowski, H. Wozniakowski, Weighted tensor product algorithms for linear multivariate problems. J. Complex. **15**, 402–447 (1999)
47. B. Zwicknagl, R. Schaback, Interpolation and approximation in Taylor spaces. J. Approx. Theory **171**, 65–83 (2013)

An Adaptive Wavelet Stochastic Collocation Method for Irregular Solutions of Partial Differential Equations with Random Input Data

Max Gunzburger, Clayton G. Webster, and Guannan Zhang

Abstract A novel multi-dimensional multi-resolution adaptive wavelet stochastic collocation method (AWSCM) for solving partial differential equations with random input data is proposed. The uncertainty in the input data is assumed to depend on a finite number of random variables. In case the dimension of this stochastic domain becomes moderately large, we show that utilizing a hierarchical sparse-grid AWSCM (sg-AWSCM) not only combats the curse of dimensionality but, in contrast to the standard sg-SCMs built from global Lagrange-type interpolating polynomials, maintains fast convergence without requiring sufficiently regular stochastic solutions. Instead, our non-intrusive approach extends the sparse-grid adaptive linear stochastic collocation method (sg-ALSCM) by employing a compactly supported wavelet approximation, with the desirable multi-scale stability of the hierarchical coefficients guaranteed as a result of the wavelet basis having the Riesz property. This property provides an additional lower bound estimate for the wavelet coefficients that are used to guide the adaptive grid refinement, resulting in the sg-AWSCM requiring a significantly reduced number of deterministic simulations for both smooth and irregular stochastic solutions. Second-generation wavelets constructed from a lifting scheme allows us to preserve the framework of the multi-resolution analysis, compact support, as well as the necessary interpolatory and Riesz property of the hierarchical basis. Several numerical examples are given

M. Gunzburger (✉)
Department of Scientific Computing, 400 Dirac Science Library, Florida State University, Tallahassee, FL 32306-4120, USA
e-mail: gunzburg@fsu.edu

C.G. Webster • G. Zhang
Computer Science and Mathematics Division, Oak Ridge National Laboratory, One Bethel Valley Road, P.O. Box 2008, MS-6164, Oak Ridge, TN 37831-6164, USA
e-mail: webstercg@ornl.gov; zhangg@ornl.gov

J. Garcke and D. Pflüger (eds.), *Sparse Grids and Applications - Munich 2012*, Lecture Notes in Computational Science and Engineering 97,
DOI 10.1007/978-3-319-04537-5_6,

to demonstrate the improved convergence of our numerical scheme and show the increased efficiency when compared to the sg-ALSCM method.

1 Introduction

Many applications in engineering and science are affected by uncertainty in input data, including model coefficients, forcing terms, boundary condition data, media properties, source and interaction terms, as well as geometry. For example, highly heterogeneous materials may have properties that vary over small length scales so that these properties have to be often determined, e.g., by interpolating or extrapolating measurements obtained at a few locations. These types of uncertainties are known as *epistemic* because they are related to incomplete knowledge. In other situations, referred to as *aleatoric*, uncertainty is due to intrinsic variability in the system, e.g., fluctuations in turbulent flow fields. In practice, it is necessary to quantify both types of uncertainties, a process which is naturally referred to as *uncertainty quantification* (UQ).

The presence of random input uncertainties can be incorporated into a system of partial differential equations (PDEs) by formulating the governing equations as PDEs with random inputs. In practice, such PDEs may depend on a set of distinct random parameters with the uncertainties represented by a given joint probability distribution. In other situations, the input data varies randomly from one point of the physical domain to another and/or from one time instant to another; in these cases, uncertainties in the inputs are instead described in terms of *random fields* that can be expressed as an expansion containing an infinite number of random variables. For example, for correlated random fields, one has Karhunen-Loève (KL) expansions [37, 38], Fourier-Karhunen-Loève expansions [36], or expansions in terms of global orthogonal polynomials [26, 57, 59]. However, in a large number of applications, it is reasonable to limit the analysis to just a finite number of random variables, either because the problem input itself can be described in that way (e.g., the random parameter case) or because the input random field can be approximated by truncating an infinite expansion [24] (e.g., the correlated random field case).

Currently, there are several numerical methods available for solving PDEs with random input data. Monte Carlo methods (MCMs) (see, e.g., [22]) are the classical and most popular approaches for approximating expected values and other statistical moments of quantities of interest that depend on the solution of PDEs with random inputs. MCMs are very flexible and trivial to implement and parallelize using existing deterministic PDE solvers, but they feature very slow convergence because they do not exploit any regularity the solution may possess with respect to the input stochastic parameters. On the other hand, the convergence rates of MCMs have mild dependence on the number of random variables so that for problems involving a large number of random parameters, MCMs remain the methods of choice.

Several numerical approaches have been proposed that often feature much faster convergence rates. These include quasi MCMs [31, 32], multilevel MCMs

[4–6, 10] stochastic Galerkin methods (SGMs) [1, 2, 26, 42] and stochastic collocation methods (SCMs) [3, 41, 45, 46, 58]. All approaches transform the original stochastic problem into a deterministic one with a large number of parameters, however, the SGMs and the SCMs are stochastic polynomial methods and differ in the choice of the polynomial bases used and the resulting approximating spaces. To achieve increased rates of convergence relative to MCMs, both approaches are typically based on global polynomial approximations that take advantage of smooth behavior of the solution in the multi-dimensional parameter space. SGMs are based on orthogonal polynomials which lead to a coupling of the probabilistic and space/time degrees of freedom; for this reason, SGMs are referred to as being *intrusive*. On the other hand, SCMs are based on interpolatory polynomials so that, when implemented, they result in ensemble-based *non-intrusive* approaches for which the probabilistic and space/time degrees of freedom are uncoupled.

We emphasize that the better convergence behavior of SGMs and SCMs relative to MCMs requires high regularity with respect to the random variables. However, often in scientific and engineering problems there are irregular dependences, e.g., steep gradients, sharp transitions, bifurcations, or finite jump discontinuities, of a quantity of interest (QoI) with respect to the random variables. In such cases, global polynomial-based approximations such as SGMs and SCMs seriously deteriorate to the point that they converge very slowly or may even fail to converge. Indeed, for such applications, the use of SGMs and SCMs often result in no improvements over MCMs. As a result, one turns to local approximation methods. To be effective, such approximations have to be implemented using refinement strategies that focus on regions of irregular behavior; otherwise, there would be an explosion in the required computational effort as the number of random variables increases, a phenomenon commonly referred to as *the curse of dimensionality*.

Not surprisingly, there have been many proposed methods that attempt to control the curse, i.e., to put off its inevitable fatal effect to higher dimensions. Several techniques involve domain decomposition approaches using h-type finite element basis functions, similar to those constructed in the physical spatial domain. A *multi-element* approach utilized in [23] decomposes each parameter dimension into sub-domains and then uses tensor products to reconstruct the entire parameter space. This method has successfully been applied to moderate dimension problems, but the tensor-product decomposition inevitably re-introduces the curse of dimensionality. Similarly, [34, 35] presents a tensor product-based multi-resolution approximation based on a Galerkin projection onto a Wiener-Haar basis. This approach provides significant improvements over global orthogonal approximation approaches. However, in terms of robustness, dimension scaling is not possible due to the resulting dense coupled system and the lack of any rigorous criteria for triggering refinement.

It is recognized that any refinement strategy employed must be guided by an accurate estimation of both local and global errors. In [39, 40], an adaptive sparse-grid stochastic collocation strategy is applied that uses piecewise multi-linear hierarchical basis functions developed in [25, 27, 30]. This approach utilizes the hierarchical surplus as an error indicator to automatically detect the regions of importance (e.g., discontinuities) in stochastic parameter space and to adaptively refine the collocation points in this region. The adaptation process is continued until a prescribed global error tolerance is achieved. This goal, however, might be reached by using more collocation points than necessary due to the instability of the multi-scale basis used; see Sect. 3.3 for a complete description of the properties of such multi-scale sparse grid approximations using hierarchical subspace splitting and see Sect. 4 for the additional properties required to construct a stable and efficient multi-dimensional multi-resolution approximation.

In Sect. 4, we propose an adaptive wavelet stochastic collocation method that possesses the additional properties. The intent of our approach is to combat the curse of dimensionality while maintaining the increased convergence rates of standard SCM approaches by utilizing compactly supported wavelet basis functions. The construction principles of such functions are highly developed; see, e.g., [11, 12, 17, 18, 28] and the references therein. Such bases are in ubiquitous use in signal processing and other applications. They have also been *rigorously* shown to result in optimal approximations of PDEs (see, e.g., [13–15, 19, 20, 43, 49, 50]) and of PDE constrained optimal control problems (see, e.g., [16, 28]), when compared with traditional finite element approximations. In this paper, due to their general applicability to arbitrary domains, we consider second-generation wavelets constructed from a lifting scheme [53–55]. Moreover, in addition to maintaining compact support and the interpolatory properties of nodal bases, the beauty of the second-generation wavelets is that they also form a Riesz basis, a property that guarantees the stability of the hierarchical basis and allows one to construct a multi-resolution approximation that utilizes significantly fewer collocation points.

The outline of the paper is as follows. In Sect. 2, we introduce the mathematical description of a general stochastic initial-boundary problem and the main notation used throughout the paper. In Sect. 3, we briefly recall the stochastic collocation method and adaptive strategies using both global as well as piecewise linear hierarchical polynomials. In Sect. 4, we propose our novel adaptive wavelet stochastic collocation method and the properties of the second-generation wavelets we employ. In Sect. 5, several numerical examples are given to demonstrate the effectiveness and efficiency of our method compared with classic approaches.

2 Problem Setting

We follow the notation in [3,45,46] and begin by letting D denote a bounded domain in $\mathbb{R}^d$, $d \in \{1,2,3\}$, and $(\Omega, \mathcal{F}, P)$ denote a complete probability space. Here, Ω denotes the set of outcomes, $\mathcal{F} \subset 2^{\Omega}$ the σ-algebra of events, and $P : \mathcal{F} \to [0,1]$ a probability measure. We are interested in the following stochastic initial-boundary value problem: find $u\colon \Omega \times \overline{D} \times [0,T] \to \mathbb{R}^m$ such that P-almost everywhere in Ω

$$\mathcal{L}(a)(u) = f \quad \text{in } D \times [0,T] \tag{1}$$

subject to the boundary and initial conditions

$$\begin{aligned} \mathcal{B}(b)(u) &= g \quad \text{on } \partial D \times [0,T] \\ u &= u_0 \quad \text{on } D \times \{t = 0\}. \end{aligned} \tag{2}$$

Here, $\mathcal{L}$ denotes a differential operator (linear or non-linear) depending on the coefficient(s) $a(\omega, x, t)$ with $(\omega, x, t) \in \Omega \times D \times [0, T]$; $\mathcal{B}$ denotes a boundary operator depending on the coefficient(s) $b(\omega, x, t)$ with $(\omega, x, t) \in \Omega \times \partial D \times [0, T]$. Similarly, the right-hand sides $f(\omega, x, t)$, $g(\omega, x, t)$, and $u_0(\omega, x)$ can be assumed to be random fields as well. Note that, in general, a, b, f, g, and u_0 belong to different probability spaces, but for economy of notation, we simply denote the stochastic dependences of these random data as if all belong to the same probability space. We denote by $W(D)$ a Banach space and assume the underlying stochastic input data are chosen so that the corresponding stochastic system (1)–(2) is well-posed so that it has an unique solution $u(\omega, x, t) \in L^2_P(\Omega) \otimes L^2(W(D); 0, T)$, where the space

$$L^2_P(\Omega) \otimes L^2(W(D); 0, T) := \left\{ u \,:\, \Omega \times \overline{D} \times [0, T] \to \mathbb{R}^m \;\middle|\; u \text{ is strongly measurable} \right.$$
$$\left. \text{and} \int_0^T \int_\Omega \|u\|^2_{W(D)} \, dP(\omega) dt < +\infty \right\} \tag{3}$$

consists of Banach-space valued functions that have finite second moments. Finally, we note that in this setting the solution u can either be a scalar or vector-valued function depending on the system of interest.

An example problem posed in this setting is given as follows.

Example 2.1 (Linear Parabolic PDE with Random Inputs). Consider the initial-boundary value problem [60]: find a random field $u : \Omega \times \overline{D} \times [0, T] \to \mathbb{R}$ such that P-almost surely

$$\begin{aligned} \partial_t u(\omega, x, t) - \nabla \cdot [a(\omega, x) \nabla u(\omega, x, t)] &= f(\omega, x, t) \quad && \text{in} \quad \Omega \times D \times [0, T] \\ u(\omega, x, t) &= 0 && \text{on} \quad \Omega \times \partial D \times [0, T] \\ u(\omega, x, 0) &= u_0(\omega, x) && \text{on} \quad \Omega \times D, \end{aligned} \tag{4}$$

where ∇ denotes the gradient operator with respect to the spatial variable $x \in D$. To guarantee the well-posedness of the solution of (4) in $L^2_P(\Omega) \otimes L^2(H^1(D); 0, T)$, one assumes that almost surely the coefficient $a(x, \omega)$ is positive and uniformly bounded, i.e.,

$$P\big(\omega \in \Omega : a_{\min} \le a(\omega, x) \le a_{\max} \; \forall\, x \in \overline{D}\big) = 1 \;\text{ with }\; a_{\min}, a_{\max} \in (0, \infty) \tag{5}$$

and that the right-hand side satisfies

$$\int_0^T \int_D \mathbb{E}[f^2]dx\,dt := \int_0^T \int_D \int_\Omega f^2(\omega, x, t)\,dP(\omega)dx\,dt < +\infty.$$

2.1 Finite Dimensional Noise

In many applications, the source of randomness can be approximated using just a finite number of uncorrelated, or even independent, random variables. As such, similar to [3, 45, 46], we make the following assumptions regarding the stochastic input data, i.e., the random coefficients a and b in $\mathcal{L}$ and $\mathcal{B}$ and the right-hand sides f, g, and u_0 in (1)–(2).

(A_1) *The stochastic input coefficient a satisfies* (5) *and the other stochastic input data are bounded from above and below with probability* 1*, e.g., for the right-hand side $f(\omega, x, t)$, there exists $f_{\min} > -\infty$ and $f_{\max} < \infty$ such that*

$$P\big(\omega \in \Omega \ : \ f_{\min} \le f(\omega, x, t) \le f_{\max} \quad \forall x \in \overline{D}, \ \forall t \in [0, T]\big) = 1 \quad (6)$$

and similarly for all remaining inputs.

(A_2) *The stochastic input data have the form*

$$\begin{aligned} &a(\omega, x, t) = a(\mathbf{y}_a(\omega), x, t), \ f(\omega, x, t) = f(\mathbf{y}_f(\omega), x, t) \ \text{ on } \ \Omega \times D \times [0, T] \\ &b(\omega, x, t) = b(\mathbf{y}_b(\omega), x, t), \ g(\omega, x, t) = g(\mathbf{y}_g(\omega), x, t) \ \text{ on } \ \Omega \times \partial D \times [0, T] \\ &u_0(\omega, x) = u_0(\mathbf{y}_{u_0}, x) \ \text{ on } \ \Omega \times D. \end{aligned} \quad (7)$$

where $N_a \in \mathbb{N}_+$ and $\mathbf{y}_a(\omega) = (y_{a,1}(\omega), \dots y_{a,N_a}(\omega))$ is a vector of uncorrelated real-valued random variables and likewise for N_b, N_f, N_g, $N_{u_0} \in \mathbb{N}_+$ and $\mathbf{y}_b = (y_{b,1}(\omega), \dots y_{b,N_b}(\omega))$, $\mathbf{y}_f = (y_{f,1}(\omega), \dots y_{f,N_f}(\omega))$, $\mathbf{y}_g = (y_{g,1}(\omega), \dots y_{g,N_g}(\omega))$ and $\mathbf{y}_{u_0} = (y_{u_0,1}(\omega), \dots y_{u_0,N_{u_0}}(\omega))$, respectively.

(A_3) *The random functions $a(\mathbf{y}_a(\omega), x, t), b(\mathbf{y}_b(\omega), x, t), f(\mathbf{y}_f(\omega), x, t), g(\mathbf{y}_g(\omega), x, t)$ and $u_0(\mathbf{y}_{u_0}(\omega), x, t)$ are assumed to be σ-measurable with respect to $\mathbf{y}_a$, $\mathbf{y}_b$, $\mathbf{y}_f$, $\mathbf{y}_g$ and $\mathbf{y}_{u_0}$, respectively.*

In many applications, the stochastic input data may have a simple piecewise random representation whereas, in other applications, the coefficients a and b in (1) and the right-hand sides f, g, and u_0 in (2) may have spatial variation that can be modeled as a correlated random field, making them amenable to description by a Karhunen-Loève (KL) expansion [37, 38]. In practice, one has to truncate such expansions according to the degree of correlation and the desired accuracy of the simulation. Examples of both types of random input data, each satisfying assumptions A_1–A_3, are given next. Without lost of generality, we only consider the coefficient $a(\omega, x, t)$ in the examples.

Example 2.2 (Piecewise Constant Random Field). We assume the spatial domain D is the union of non-overlapping subdomains D_j, $j = 1, \dots, J$, and the time interval $(0, T)$ is the union of disjoint subintervals (T_{k-1}, T_k), $k = 1, \dots, K$. Then, we consider the coefficient $a(\omega, x, t)$ is piecewise constant and random on each space-time subdomain $D_j \times (T_{k-1}, T_k)$, i.e.,

$$a(\omega, x, t) = \sigma_0 + \sum_{n=1}^{N_a} y_{a,n}(\omega)\sigma_n 1_{D_j \times (T_{k-1}, T_k)}(x, t) \quad \text{for} \quad n = j + (k-1)J,$$

where σ_n, $n = 0, \dots, N$, denote constants, $1_{D_j \times (T_{k-1}, T_k)}$ denotes the indicator function of the set $D_j \times (T_{k-1}, T_k) \subset D \times [0, T]$, and the random variables $y_{a,n}(\omega)$ are bounded and independent. Note that assumption A_1 requires restrictions on the constants σ_n and the bounds on the random variables $y_{a,n}(\omega)$; in practice, such restrictions would be deduced from the physics of the problem.

Example 2.3 (Karhunen-Loève Expansion). Any second-order correlated random field $a(\omega, x, t)$ with continuous covariance function $Cov[a](\tilde{x}_1, \tilde{x}_2)$, where $\tilde{x}_1 = (x_1, t_1)$ and $\tilde{x}_2 = (x_2, t_2)$ are space-time coordinates, can be represented as an infinite sum of random variables by means of, e.g., a KL expansion. For $\tilde{x} = (x, t)$, we define the operator $F : L^2(D) \times L^2(0, T) \to L^2(D) \times L^2(0, T)$ by

$$Fv(\tilde{x}) := \int_0^T \int_D Cov[a](\tilde{x}_1, \tilde{x})v(\tilde{x}_1)\, d\tilde{x}_1 \quad \forall\, v \in L^2(D) \times L^2(0, T). \tag{8}$$

Because of the symmetry and non-negativity properties of covariance functions, the operator F has real, non-negative eigenvalues $\{\lambda_n\}_{n=1}^{\infty}$ that may be arranged in non-increasing order and corresponding real orthonormal eigenfunctions $\{a_n(\tilde{x})\}_{n=1}^{\infty}$. For simplicity of the exposition, we assume that the eigenvalues are positive. Furthermore, if we define mutually uncorrelated real random variables by

$$y_n(\omega) := \frac{1}{\sqrt{\lambda_n}} \int_0^T \int_D (a(\omega, \tilde{x}) - \mathrm{e}[a](\tilde{x}))\, a_n(\tilde{x})\, d\tilde{x}, \qquad n = 1, 2, \dots$$

with zero mean and variance $Var[y_n] = \sqrt{\lambda_n}$, then $a(\omega, x, t)$ can be represented by the truncated N-term KL expansion, i.e.

$$a(\omega, x, t) = \mathbb{E}[a](x, t) + \sum_{n=1}^{N_a} \sqrt{\lambda_n} a_n(x, t) y_{a,n}(\omega).$$

Finally, note that if the process is Gaussian, then the random variables $\{y_{a,n}\}_{n=1}^{\infty}$ are standard independent identically distributed random variables.

We remark that assumption A_2 and the Doob-Dynkin lemma guarantee that $a(\mathbf{y}_a(\omega), x, t)$ is a Borel-measurable function of the random vector $\mathbf{y}_a$ and likewise for b, f, g, u_0, with respect to $\mathbf{y}_a$, $\mathbf{y}_b$, $\mathbf{y}_f$, $\mathbf{y}_g$ and $\mathbf{y}_{u_0}$. As discussed in [60], the random fields a, b, f, g, u_0 are independent because of their physical properties, so that $\mathbf{y}_a$, $\mathbf{y}_b$, $\mathbf{y}_f$, $\mathbf{y}_g$, $\mathbf{y}_{u_0}$ are independent random vectors. Thus, we relabel the elements of the five random vectors and define $\mathbf{y} = (y_1, \ldots, y_N) = (\mathbf{y}_a, \mathbf{y}_b, \mathbf{y}_f, \mathbf{y}_g, \mathbf{y}_{u_0})$ where $N = N_a + N_b + N_f + N_g + N_{u_0}$. By definition, the random variables $\{y_n\}_{n=1}^N$ are mappings from the sample space Ω to the real space $\mathbb{R}^N$, so we denote by $\Gamma_n = y_n(\Omega) \subset \mathbb{R}$ the image of the random variable y_n, and set $\Gamma = \prod_{n=1}^N \Gamma_n$, where $N \in \mathbb{N}_+$. If the distribution measure of $\mathbf{y}(\omega)$ is absolutely continuous with respect to Lebesgue measure, there exists a joint probability density function (PDF) for $\{y_n\}_{n=1}^N$ denoted by

$$\rho(\mathbf{y}) : \Gamma \to \mathbb{R}_+ \quad \text{with} \quad \rho(\mathbf{y}) \in L^\infty(\Gamma).$$

Thus, based on the assumption A_2, the probability space $(\Omega, \mathcal{F}, P)$ is mapped to $(\Gamma, \mathcal{B}(\Gamma), \rho(\mathbf{y})d\mathbf{y})$, where $\mathcal{B}(\Gamma)$ is the Borel σ-algebra on Γ and $\rho(\mathbf{y})d\mathbf{y}$ is the finite measure.

Finally, we are in position to restate the random input data in terms of $\mathbf{y}$ as follows:

$$\begin{aligned}
&a(\omega,x,t) = a(\mathbf{y}(\omega),x,t),\; f(\omega,x,t) = f(\mathbf{y}(\omega),x,t) \;\text{ for } (\mathbf{y}(\omega),x,t) \in \Gamma \times D \times [0,T]\\
&b(\omega,x,t) = b(\mathbf{y}(\omega),x,t),\; g(\omega,x,t) = g(\mathbf{y}(\omega),x,t) \;\text{ for } (\mathbf{y}(\omega),x,t) \in \Gamma \times \partial D \times [0,T]\\
&u_0(\omega,x) = u_0(\mathbf{y}(\omega),x) \;\text{ for } (\mathbf{y}(\omega),x) \in \Gamma \times D.
\end{aligned} \tag{9}$$

As a result, the problem (1)–(2) can be restated as follows: find a function $u(\mathbf{y}, x, t) : \Gamma \times \overline{D} \times [0, T] \to \mathbb{R}^m$ such that ρ-almost every $\mathbf{y} \in \Gamma$, we have that

$$\mathcal{L}(a(\mathbf{y}, x, t))(u) = f(\mathbf{y}, x, t) \quad \text{in } D \times [0, T] \tag{10}$$

subject to the boundary and initial conditions

$$\begin{aligned}
\mathcal{B}(b(\mathbf{y}, x, t))(u) &= g(\mathbf{y}, x, t) \;\text{ on } \partial D \times [0, T]\\
u &= u_0(\mathbf{y}, x) \quad \text{on } D \times \{t = 0\}.
\end{aligned} \tag{11}$$

By assuming the solution u of (1) and (2) is σ-measurable with respect to a, b, f, g and u_0, then, by the Doob-Dynkin lemma [47], $u(\omega, \cdot, \cdot)$ can also be characterized by the same random vector $\mathbf{y}$, i.e.,

$$u(\omega, x, t) = u(y_1(\omega), \ldots, y_N(\omega), x, t) \in L^2_\rho(\Gamma) \otimes L^2(W(D); 0, T), \tag{12}$$

where $L^2_\rho(\Gamma)$ is the space of square integrable functions of Γ with respect to the measure $\rho(\mathbf{y})d\mathbf{y}$. As indicated in (12), the solution $u(\mathbf{y}, x, t)$ belongs to the function space $L^2_\rho(\Gamma) \otimes L^2(W(D); 0, T)$ that is defined by

$$L^2_\rho(\Gamma) \otimes L^2(W(D); 0, T) := \Big\{ u : \Gamma \times \overline{D} \times [0, T] \to \mathbb{R}^m \ \Big|\ u \text{ is strongly measurable}$$
$$\text{and} \int_\Gamma \int_0^T \|u\|^2_{W(D)} \rho(\mathbf{y}) dt d\mathbf{y} < \infty \Big\}. \qquad (13)$$

We once again note that, in general, each appearance of $\mathbf{y}$ in (10)–(11) can be a different vector of random variables each belonging to a different probability space and that in the end, the solution u depends on all the different $\mathbf{y}$'s which collectively belong to the product space of the individual probability spaces. However, again to economize notation, we do not explicitly differentiate between the different vectors of random variables.

Thus far we have turned the possibly infinite-dimensional stochastic initial-boundary value problem given by (1)–(2) into a finite-dimensional parametric problem (10)–(11). Without loss of generality, we will assume the support of the random variables y_n is $\Gamma_n = [0, 1]$ for $n = 1, \dots, N$ and therefore the bounded stochastic (or parameter) space is the N-dimensional unit hypercube $\Gamma = [0, 1]^N$. At this point, we can apply any stochastic approximation technique, e.g., spectral-Galerkin, locally adaptive, etc. However, the focus of our work involves *non-intrusive* approximations (such as Monte Carlo sampling or stochastic collocation methods) in probability space for which, for any realization of the random parameters $\mathbf{y}(\omega_k)$, solutions can be constructed using standard deterministic approximation techniques in space-time, e.g., finite difference methods, finite element methods, finite volume methods, etc. for spatial discretization and backward Euler or Crank-Nicolson schemes for temporal discretization [44, 60].

3 Adaptive Stochastic Collocation Methods

To provide context and background for the new method we present in Sect. 4, in this section we discuss, in general terms, adaptive stochastic collocation methods (SCMs). These approximations are computed via Lagrange interpolation of the random parameter dependence of solutions of (10)–(11), described in Sect. 3.1, constructed from either globally or locally supported basis functions, described in Sects. 3.2 and 3.3 respectively. In Sect. 3.3, we discuss in somewhat more detail the special case of hierarchical piecewise polynomial basis functions, leading to the hierarchical, locally adaptive, piecewise linear approximations. The latter is the closest precursor to our new method and, naturally, we use it for comparison purposes.

We note that the use of polynomials having the property that the interpolation matrix is diagonal, i.e. the "delta property" (see Remark 3.1), leads to approximations that some authors refer to as *stochastic collocation methods* (SCMs). Others use that terminology to refer to any method for which the parameter and spatial/temporal degrees of freedom uncouple; with this view, which is the one we adopt, all methods discussed below and in this section and in Sect. 4 would be referred to as being SCMs.

3.1 Lagrange Interpolation in the Probabilistic Domain

The goal is to construct a numerical approximation of the solution of (10)–(11) in a finite-dimensional subspace $\mathcal{P}(\Gamma) \otimes L^2(W_h(D); 0, T)$. Here, $W_h(D) \subset W(D)$ is a standard finite element space of dimension $\dim(W_h) = M_h$, used for spatial discretization and $\mathcal{P}(\Gamma) \subset L^2_\rho(\Gamma)$ is a finite-dimensional space of dimension $\dim(\mathcal{P}(\Gamma)) = M$, used for approximation in parameter space. Of course, a temporal discretization, usually via a finite difference method, is implied as well. Interpolatory approximations in parameter space start with the selection of a set of distinct points $\{\mathbf{y}_k\}_{k=1}^M \in \Gamma$, in parameter space and a set of basis functions[1] $\{\psi_k(\mathbf{y})\}_{k=1}^M \in \mathcal{P}(\Gamma)$. Then, we seek an approximation $u_{M_h,M} \in \mathcal{P}(\Gamma) \otimes L^2(W(D); 0, T)$ of the solution u of the problem (10)–(11) of the form

$$u_{M_h,M}(\mathbf{y}, x, t) = \sum_{k=1}^{M} c_k(x,t)\psi_k(\mathbf{y}). \tag{14}$$

The Lagrange interpolant is defined by first obtaining M realizations $u_{M_h}(\mathbf{y}_k, x, t)$ of the finite element approximation[2] of the solution $u(\mathbf{y}_k, x, t)$ of the problem (10)–(11), i.e., one solves for the finite element approximation for each of the interpolation points in the set $\{\mathbf{y}_k\}_{k=1}^M$. Then, the coefficient functions $\{c_k(x,t)\}_{k=1}^M$ are determined by imposing the interpolation conditions

$$\sum_{\ell=1}^{M} c_\ell(x,t)\psi_\ell(\mathbf{y_k}) = u_{M_h}(\mathbf{y}_k, x, t) \qquad \text{for } k = 1, \ldots, M. \tag{15}$$

Thus, the coefficient functions $\{c_\ell(x,t)\}_{\ell=1}^M$ are each a linear combination of the data functions $\{u_{M_h}(\mathbf{y}_k, x, t)\}_{k=1}^M$; the specific linear combinations are determined in the usual manner from the entries of the inverse of the $M \times M$ interpolation matrix $\mathbf{K}$ having entries $K_{k\ell} = \psi_\ell(\mathbf{y}_k)$, $k, \ell = 1, \ldots, M$. The sparsity of $\mathbf{K}$ heavily depends on the choice of basis; that choice could result in matrices that range from fully dense to diagonal.

---

[1] In general, the number of points and number of basis functions do not have to be the same, e.g., for Hermite interpolation. However, because here we only consider Lagrange interpolation, we let M denote both the number of points and the dimension of the basis.

[2] Extensions to other methods, e.g., finite difference, finite volume, spectral or h-p, etc. are straightforward.

A main attraction of interpolatory approximations of parameter dependences is that it effects a complete decoupling of the spatial/temporal and probabilistic degrees of freedom. Clearly, once the interpolation points $\{\mathbf{y}_k\}_{k=1}^M$ are chosen, one can solve M deterministic problems [i.e., the spatial/temporal discretization of (10)–(11)], one for each parameter point $\mathbf{y}_k$, with total disregard to what basis $\{\psi_k(\mathbf{y})\}_{k=1}^M$ one choose to use. Then, the coefficients $\{c_k(x,t)\}_{k=1}^M$ defining the approximation (14) of the solution of (10)–(11) are found from the interpolation matrix $\mathbf{K}$ as discussed above; its only in this last step that the choice of basis enters into the picture. Note that this decoupling property makes the implementation of Lagrange interpolatory approximations of parameter dependences as trivial as it is for Monte Carlo sampling. However, if that dependence is smooth, because of the higher accuracy of, e.g., polynomial approximations in the space $\mathcal{P}(\Gamma)$, interpolatory approximations require substantially fewer sampling points to achieve a desired tolerance.

Remark 3.1 (The "Delta Property"). Given a set of interpolation points, to complete the setup of a Lagrange interpolation problem, one has to then choose a basis. The simplest and most popular choice are the Lagrange polynomials, i.e., polynomials that have the "delta property" $\psi_\ell(\mathbf{y}_k) = \delta_{k\ell}$, where $\delta_{k\ell}$ denotes the Kroeneker delta. In this case, the interpolating conditions (15) reduce to $c_k(x,t) = u_h(\mathbf{y}_k, x, t)$ for $k = 1, \ldots, M$, i.e., the interpolation matrix $\mathbf{K}$ is simply the $M \times M$ identity matrix. In this sense, the use of Lagrange polynomial bases can be viewed as resulting in pure sampling methods, much the same as Monte Carlo methods, but instead of randomly sampling in the parameter space Γ, the sample points are deterministically structured as, e.g., tensor product or sparse grid points.

3.2 Adaptive Global Sparse-Grid Lagrange Interpolation

When the solution is analytic with respect to the noise parameterization, the most widely used approaches for constructing approximations of the form (14), involves building global Lagrange interpolants [3,45,46]. This is accomplished by replacing the polynomial space $\mathcal{P}(\Gamma)$ by $\mathcal{P}_p(\Gamma)$, defined as the span of product polynomials, i.e.,

$$\mathcal{P}_p(\Gamma) = \text{span}\left\{ \prod_{n=1}^{N} y_n^{p_n} \quad \text{with } \mathbf{p} = (p_1, \ldots, p_N) \in \mathcal{J}(p) \right\},$$

where the index set $\mathcal{J}(p)$ determines the type of polynomial space used. Thus, the dimension M of $\mathcal{P}_p(\Gamma)$ is the cardinality of the index set $\mathcal{J}(p)$. Two obvious choices are tensor product spaces of one-dimensional polynomials of degree p for which $\mathcal{J}(p) = \{\mathbf{p} \in \mathbb{N}^N : \max_{1\le n\le N} p_n \le p\}$ and total degree p spaces for which $\mathcal{J}(p) = \{\mathbf{p} \in \mathbb{N}^N : \sum_{n=1}^N p_n \le p\}$.

Both these choices are problematical even for problems having moderately large parameter dimension N. The first choice results in $M = (p+1)^N$ interpolation points, a number which grows explosively with increasing N; this is perhaps the most egregious instance of the curse of dimensionality. For the second, we have $M = (N+p)!/(N!p!)$ interpolation points, i.e., much slower growth than for the tensor product case, so that the inevitable fatal effects of the curse are postponed to higher dimensions. However, this choice requires a judicious choice of the location of the interpolation points because arbitrary choices can result in large Lebesgue constants which can lead to serious deterioration in accuracy. Unfortunately, good choices of total degree interpolation points in N-dimensional cubes are not known, even for moderate values of N.

A third choice for the interpolation abscissas are sparse-grid points, constructed from the roots of either the nested Chebyshev (Clenshaw-Curtis) or the Gaussian polynomials [3,45,46]. Typically, in these approaches the index set is defined using the Smolyak method [9,52] where

$$\mathcal{J}(p) = \left\{ \mathbf{p} \in \mathbb{N}^N : \sum_{n=1}^{N} f(p_n) \leq f(p) \right\} \text{ with } f(p) = \begin{cases} 0, & p = 0 \\ 1, & p = 1 . \\ \lceil \log_2(p) \rceil, & p \geq 2 \end{cases}$$

Other polynomial spaces have been described and considered in, e.g., [7,56].

For any choice of interpolation points, a reduction in the number of interpolation points can be effected by using *dimension-adaptive* global polynomials. For example, for the tensor product, total degree and Smolyak cases, one can use the index sets $\mathcal{J}(p) = \{\mathbf{p} \in \mathbb{N}^N : \max_{1 \leq n \leq N} \alpha_n p_n \leq \alpha_{min} p\}$, $\mathcal{J}(p) = \{\mathbf{p} \in \mathbb{N}^N : \sum_{n=1}^{N} \alpha_n p_n \leq \alpha_{min} p\}$, and $\mathcal{J}(p) = \left\{\mathbf{p} \in \mathbb{N}^N : \sum_{n=1}^{N} \alpha_n f(p_n) \leq \alpha_{min} f(p)\right\}$, respectively, where the weights $\alpha_n > 0$, $n = 1, \ldots, N$, can be computed either a priori or a posteriori; see [46].

3.3 *Adaptive Hierarchical Sparse-Grid Lagrange Interpolation*

None of the approaches discussed above, as well as those commonly refereed to as polynomial chaos methods, that use global orthogonal polynomials in the parameter space, are effective in approximating solutions $u(\mathbf{y}, x, t)$ of (10)–(11) that have irregular dependence with respect to the random parameters. What is required for the effective approximation of solutions having irregular dependence with respect to the random parameters is an approximating space that allows for, through a judicious choice of basis, a multi-level, multi-scale decomposition. Such an approach can be constructed using piecewise polynomial approximations in the parameter space with multi-level, multi-scale hierarchical bases. A step in this direction was the development of an adaptive piecewise linear hierarchical sparse-grid approximation

[9,27] and their utilization for solving problems with random inputs [39,40]. In this section we discuss hierarchical sparse-grid Lagrange interpolation approaches and also specialize to the approach [9, 27, 39, 40]. In Sect. 4, we extend this technique by developing a multi-dimensional multi-resolution interpolating wavelet-based approximation.

Instead of using global polynomial interpolating spaces that attempt to achieve greater accuracy by increasing the degree p of the polynomial space, piecewise polynomial interpolation spaces attempt to achieve greater accuracy with a fixed polynomial degree by refining the grid that is the underpinning of the definition of the space. Problems having solutions with irregular behavior cannot take advantage of increases in the degree of the polynomials used; however, through local grid refinement in regions where the solution exhibits irregular behavior, piecewise polynomial spaces have the potential to be effective for such problems. However, realizing that potential for problems with even moderate parameter dimension N is not a straightforward matter.

Piecewise polynomial spaces for Lagrange interpolation are most often implemented in a standard "finite element" manner using locally supported nodal basis functions. One advantage of this approach is that the basis functions have the "delta property" (see Remark 3.1). However, such choices do not result in a multi-scale basis so that defining reliable error indicators for adaptive refinement is a difficult matter and, in fact, obtaining approximations that are efficient with respect to the number of degrees of freedom used to achieve a desired accuracy is not possible. We also focus the discussion on sparse-grid *hierarchical* polynomial interpolation because multi-dimensional approximations based on tensor product grids are not viable for high-dimensional parameter spaces, even for polynomial degree $p = 1$, because of the large number of degrees of freedom involved.

That is, for each parameter dimension $n = 1, \ldots, N$, we define $V_n := L^2_\rho(\Gamma_n)$. Then, the desired approximation is based on a sequence of subspaces $\{V_{i_n}\}_{i_n=0}^{\infty}$ of V_n of increasing dimension M_{i_n},which is dense in V_n, i.e., $\cup_{i_n=0}^{\infty} V_{i_n} = V_n$. The sequence of spaces is also required to be nested, i.e.,

$$V_0 \subset V_1 \subset V_2 \subset \cdots \subset V_{i_n} \subset V_{i_n+1} \subset \cdots \subset V_n. \tag{16}$$

A set of subspaces satisfying these requirements are defined as the span of a nodal piecewise polynomial basis of order p, i.e.,

$$V_{i_n} = \text{span}\{\phi_{j_n}^{i_n}(y_n) \mid 0 \leq j_n \leq 2^{i_n}\}, \tag{17}$$

where i_n denotes the scaling level of all the basis functions $\phi_{j_n}^{i_n}$ with compact support, i.e., $\text{supp}(\phi_{j_n}^{i_n}) = O(2^{-i_n})$, and $\phi_{j_n}^{i_n}(y_n)$ is a polynomial of degree p. For example, suppose M_{i_n} distinct points are selected in the interval $\overline{\Gamma}_n$ such that the maximum distance between any two neighboring points is of order $O(2^{-i_n})$. Then, the simplest choice for the set $\{\phi_{j_n}^{i_n}\}_{j_n=1}^{M_{i_n}}$ are the linear "hat" functions corresponding to the selected points in $\overline{\Gamma}_n$; in this case, we indeed have that the support of each $\phi_{j_n}^{i_n}(y_n)$ is of order $O(2^{-i_n})$.

Similarly, for an N-dimensional problem, we define $\mathcal{V}^N := L^2_\rho(\Gamma)$. Then, a sequence of subspaces $\{\mathcal{V}_l^N\}_{l=0}^{\infty}$ of $\mathcal{V}^N$ can be constructed using a sparse-grid framework, i.e.,

$$\mathcal{V}_l^N = \bigcup_{|\mathbf{i}|\leq l} \bigotimes_{n=1}^{N} V_{i_n} = \bigcup_{|\mathbf{i}|\leq l} \text{span}\left\{ \prod_{n=1}^{N} \phi_{j_n}^{i_n}(y_n) \;\middle|\; 0 \leq j_n \leq 2^{i_n} \right\}, \tag{18}$$

where $\mathbf{i} = (i_1, \ldots, i_N) \in \mathbb{N}_+^N$ is a multi-index and $|\mathbf{i}| \equiv i_1 + \cdots + i_N \leq l$ defines the resolution of the sparse-grid approximation in $\mathcal{V}_L^N$. Note that full tensor-product resolution is defined by simply replacing the index set by $\max_{n=1,\ldots,N} i_n \leq l$.

Instead of using locally supported nodal bases, we construct a hierarchical approximation at level L using a truncation $\mathcal{V}_L^N$ of the infinite expansion $\mathcal{V}^N$. We begin with a basis for $\mathcal{V}_0^N$ and then, due to the nested property of $\{\mathcal{V}_l^N\}_{l=0}^{\infty}$, we express the finer subspaces of $\mathcal{V}_l^N$ as a direct sum $\mathcal{V}_l^N = \mathcal{V}_{l-1}^N \oplus \mathcal{W}_l^N$, where $\mathcal{W}_l^N = \mathcal{V}_l^N / \oplus_{m=0}^{l-1} \mathcal{V}_m^N$. Therefore, we have that

$$\mathcal{V}_L^N = \mathcal{V}_0^N \oplus \mathcal{W}_1^N \oplus \cdots \oplus \mathcal{W}_L^N. \tag{19}$$

Then, the *hierarchical sparse-grid approximation* of $u_{M_h,M}(\mathbf{y}, x, t) \in \mathcal{V}_L^N \otimes L^2(W(D); 0, T)$ in (14) is defined by

$$u_{M_h,M}(\mathbf{y}, x, t) \equiv \mathcal{I}_L^N(u)(\mathbf{y}, x, t) = \sum_{l=0}^{L} \sum_{|\mathbf{i}|=l} \sum_{\mathbf{j}\in B_{\mathbf{i}}} c_{\mathbf{j}}^{\mathbf{i}}(x,t)\psi_{\mathbf{j}}^{\mathbf{i}}(\mathbf{y}), \tag{20}$$

where $\mathcal{I}_L^N : \mathcal{V}^N \to \mathcal{V}_L^N$ denotes the approximation operator, $\psi_{\mathbf{j}}^{\mathbf{i}} = \prod_{n=1}^{N} \phi_{j_n}^{i_n}$ denotes a multi-dimensional hierarchical polynomial, and $B_{\mathbf{i}}$ a multi-index set defined by

$$B_{\mathbf{i}} \equiv \left\{ \mathbf{j} \in \mathbb{N}_+^N \;\middle|\; \begin{array}{ll} 0 \leq j_n \leq 2^{i_n},\ j_n \text{ odd}, & 1 \leq n \leq N, \text{ if } i_n > 0 \\ j_n = 0, 1, & 1 \leq n \leq N, \text{ if } i_n = 0 \end{array} \right\}. \tag{21}$$

The approximation space $\mathcal{P}_p(\Gamma) = \mathcal{V}_L^N$ and the particular basis chosen are required to possess the following properties.

(P_1) **Nested hierarchical subspaces**: $\mathcal{V}_0^N \subset \mathcal{V}_1^N \subset \cdots \subset \mathcal{V}_\infty^N$.

(P_2) **Small compact support**: $\text{supp}\left(\prod_{n=1}^{N} \phi_{j_n}^{i_n}\right) = O\left(2^{-\sum_{n=1}^{N} i_n}\right)$.

(P_3) **Interpolatory basis**: $\{\phi_j^i\}$ in (17) is an interpolating basis for V_i, e.g., the "hat" functions, so that the approximation operator $\mathcal{I}_L^N$ in (20) is a multi-dimensional interpolation operator.

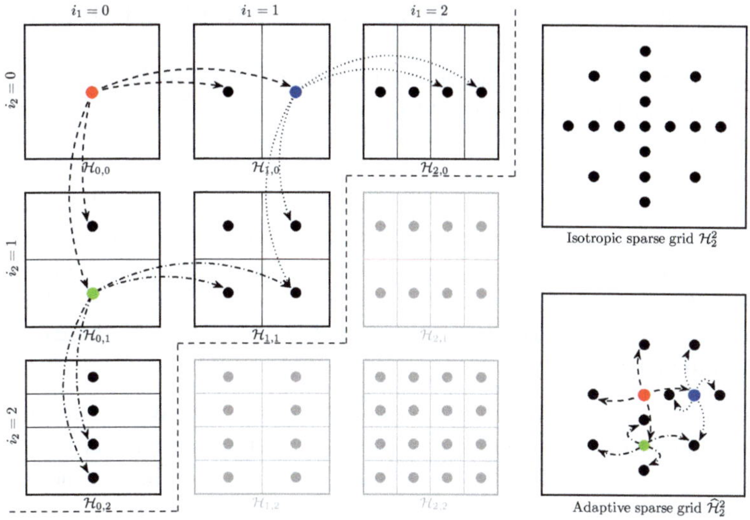

Fig. 1 Nine tensor-product sub-grids (*left*) for level $L = 0, 1, 2$ of which only the six sub-grids with $i_1 + i_2 \leq 2$ are chosen to appear in the level $L = 2$ isotropic sparse grid $\mathcal{H}_2^2$ (*right-top*) containing 17 points. With adaptivity, each point that corresponds to a large surplus, e.g., the points in *red*, *blue*, or *green*, lead to 2 children points added in each direction resulting in the adaptive sparse grid $\widehat{\mathcal{H}}_2^2$ (*right-bottom*) containing 12 points

(P_4) **Decay of the coefficients for smooth functions in** $L^2_\rho(\Gamma)$: there exits a constant C, independent of the level L, such that for every $u(\mathbf{y}, \cdot, \cdot) \in L^2_\rho(\Gamma)$ the following holds:

$$\sum_{l=0}^{L} \sum_{|\mathbf{i}|=l} \sum_{\mathbf{j} \in B_\mathbf{i}} \left|c_\mathbf{j}^\mathbf{i}\right|^2 2^{2l} \leq C L \|u\|^2_{L^2_\rho(\Gamma)}. \tag{22}$$

Denote by $\mathcal{H}_\mathbf{i} = \{\mathbf{y}_\mathbf{j}^\mathbf{i} \mid \mathbf{j} \in B_\mathbf{i}\}$ the set of points corresponding to the basis functions $\psi_\mathbf{j}^\mathbf{i}(\mathbf{y})$ with $\mathbf{j} \in B_\mathbf{i}$; then, the set of points corresponding to the subspace $\mathcal{W}_l^N$ is given by $\cup_{|\mathbf{i}|=l} \mathcal{H}_\mathbf{i}$ and the set of points used by $\mathcal{I}_L^N(u)$ is defined by

$$\mathcal{H}_L^N = \bigcup_{|\mathbf{i}| \leq L} \mathcal{H}_\mathbf{i} \tag{23}$$

which is the sparse grid corresponding to $\mathcal{V}_L^N$. Note that due to property P_1, the sparse grid $\mathcal{H}_L^N$ is also nested, i.e., $\mathcal{H}_{L-1}^N \subset \mathcal{H}_L^N$. In Fig. 1, we plot the structure of a level $L = 2$ sparse grid in $N = 2$ dimensions, without considering boundary points. The left nine sub-grids $\mathcal{H}_\mathbf{i}$ correspond to the nine multi-index sets $B_\mathbf{i}$, where

$$\mathbf{i} \in \{(0,0),(0,1),(0,2),(1,0),(1,1),(1,2),(2,0),(2,1),(2,2)\}.$$

The level $L = 2$ sparse grid $\mathcal{H}_2^2$ shown on the right (top) includes only six of the nine sub-grids in black according to the criterion $|\mathbf{i}| \leq 2$. Moreover, due to the nested property of the hierarchical basis, $\mathcal{H}_2^2$ has only 17 points, as opposed to the 49 points of the full tensor-product grid.

Next, we explain how to compute the coefficients $c_{\mathbf{j}}^{\mathbf{i}}(x,t)$. In general, this requires the solution of a linear system whose right-hand-side depends only on the value of the finite element approximation of the solution u at each collocation point. Moreover, the structure of the coefficient matrix depends on the type of hierarchical polynomials used in (20). However, for some choices of the basis, these coefficients can be computed explicitly.

Example 3.2 (Linear Hierarchical Piecewise Polynomials). We can take the hierarchical one-dimensional functions to be the standard piecewise linear finite element basis, i.e., the basis function ϕ_j^i in (17) are obtained by the dilation and translation of the function

$$\phi(y) = \begin{cases} 1 - |y| & \text{if } y \in [-1,1] \\ 0 & \text{otherwise.} \end{cases} \tag{24}$$

This basis possesses properties P_1–P_4. Examples of higher-degree bases are given in [9]. Then, $\mathcal{I}_L^N(u)$ in (20) can be rewritten as

$$\mathcal{I}_L^N(u)(\mathbf{y},x,t) = \mathcal{I}_{L-1}^N(u)(\mathbf{y},x,t) + \Delta\mathcal{I}_L^N(u)(\mathbf{y},x,t), \tag{25}$$

where $\mathcal{I}_{L-1}^N(u)$ is the sparse-grid approximation in $\mathcal{V}_{L-1}^N$ and $\Delta\mathcal{I}_L^N(u)$ is the hierarchical difference interpolant corresponding to $\mathcal{W}_L^N$. Due to property P_1, the set of grid points used by $\Delta\mathcal{I}_L^N(u)$ can be denoted by $\Delta\mathcal{H}_L^N = \mathcal{H}_L^N \backslash \mathcal{H}_{L-1}^N$. Then, due to the interpolatory property P_3 and the choice of the basis function (24), by substituting $\mathbf{y}_{\mathbf{j}}^{\mathbf{i}} \in \Delta\mathcal{H}_L^N$ in (25), we obtain that

$$\begin{aligned} c_{\mathbf{j}}^{\mathbf{i}}(x,t) &= \mathcal{I}_L^N(u)(\mathbf{y}_{\mathbf{j}}^{\mathbf{i}},x,t) - \mathcal{I}_{L-1}^N(u)(\mathbf{y}_{\mathbf{j}}^{\mathbf{i}},x,t) \\ &= u(\mathbf{y}_{\mathbf{j}}^{\mathbf{i}},x,t) - \mathcal{I}_{L-1}^N(u)(\mathbf{y}_{\mathbf{j}}^{\mathbf{i}},x,t) \end{aligned} \tag{26}$$

as the hierarchical surplus. This is simply the difference between the solution u at a point $\mathbf{y}_{\mathbf{j}}^{\mathbf{i}}$ on the current level of interpolation and the interpolated value of the previous level [30] at that point. Therefore, using the recursive formula (25), we can compute all the coefficients $c_{\mathbf{j}}^{\mathbf{i}}$ in (20) by calculating the coefficients of $\Delta\mathcal{I}_L^N(u)$ for $l = 1,\ldots,L$.

According to the analysis in [30], for smooth functions described by property P_4, the hierarchical surpluses tend to zero as the interpolation level goes to infinity. On the other hand, for irregular functions having, e.g., steep slopes or jump discontinuities, the magnitude of the surplus is an indicator of the interpolation error and, as such, can be used to control the error adaptively. That is, for the sparse grid $\mathcal{H}_L^N$, abscissas involved in each direction can be considered as a tree-like data structure as shown in Fig. 1.

For example, on the left, we show that the red point in $\mathcal{H}_{0,0}$ has two children points at level $L = 1$ in each of the horizontal and vertical directions; the four children are indicated by the arrows emanating from the red point. Each of its four children also have four children of their own at level $L = 2$, and so on for subsequent levels. Suppose the magnitude of the coefficients (the surplus) associated with the blue and green children are larger than a prescribed tolerance, but those for the two black children of the red point are smaller than the tolerance. In this case, refinement is effected only from the blue and green children; no refinement is done of the black children. This is indicated by having, at level $L = 2$, four arrows emanate from the blue and green points, but none from the black points. We arrive at the adaptive space grid $\widehat{\mathcal{H}}_{2,2}$ that has 12 total collocation points, shown on the right (bottom) of Fig. 1. The analagous (non-adaptive) isotropic sparse grid, which has 17 collocation points, is also shown on the right (top).

In general, a grid point in a N-dimensional space has $2N$ children which are also the neighbor points of the parent node. However, note that the children of a parent point correspond to hierarchical basis functions on the next interpolation level, so that we can build the interpolant $\mathcal{I}_L^N(u)$ in (20) from level $L-1$ to L by adding those points on level L whose parent has a surplus greater than our prescribed tolerance. In this way, we can refine the sparse grid locally and end up with an adaptive sparse grid which is a sub-grid of the corresponding isotropic sparse grid.

A *sparse grid adaptive linear stochastic collocation method* (sg-ALSCM) that utilizes a locally supported linear hierarchical basis, given by (24), to approximate random functions in the multi-dimensional hypercube $\Gamma \subset \mathbb{R}^N$ were considered in [21,29,39,40]. As mentioned in Example 3.2, the expansion coefficients $c_{\mathbf{j}}^{\mathbf{i}}(x,t)$ in (20) are simply the hierarchical surpluses and adaptive refinement is guided by the magnitude $|c_{\mathbf{j}}^{\mathbf{i}}|$ of those coefficients. However, this approach has a major drawback: one cannot estimate the error from *below* with constants independent of the number of hierarchical levels involved. Thus, the linear hierarchical basis does not form a stable multi-scale splitting of the approximation space [48] and the absolute value of a hierarchical coefficient is just a local error indicator and not a true error estimator. As a result, one obtains sufficiently refined sparse approximations for which the error is behaving as predicted, but in doing so, one may have used many more grid points than needed to achieve a prescribed error tolerance for the adaptive procedure. This scheme has no guarantee of efficiency so that some previous claims of optimality with respect to complexity for this approach are heuristic, not provable and, in general, not valid.

Our approach is generally similar, but uses multi-resolution *wavelet* approximations that possess all the properties (P_1–P_4) of the linear basis functions, but also possess an additional property that guarantee stability and efficiency. We will introduce this essential criteria in Sect. 4 and more importantly, also explain the advantages of our novel adaptive wavelet stochastic collocation method (AWSCM).

4 Adaptive Wavelet Stochastic Collocation Method

As discussed several times in the paper, UQ for complex stochastic systems that require the approximation and/or resolution of statistical QoIs involving, e.g., steep gradients, sharp transitions, bifurcations, or finite jump discontinuities, in possibly high-dimensional probabilistic domains, require sophisticated multi-dimensional multi-resolution adaptive algorithms. To be effective, however, refinement strategies must be guided by accurate estimates of errors (both local and global) while not expending unnecessary computational effort approximating the QoI with respect to any random dimension. In the sg-ALSCM described in Sect. 3.3, given the hierarchical sparse-grid approximation (20) that satisfies properties P_1–P_4, stable and efficient approximations of such irregular problems cannot be guaranteed. Here, by efficiency we mean achieving a prescribed error tolerance with a reduced number of grid points. This can result in an explosion in computational effort for high-dimensional problems. Towards alleviating this effect, we require the following additional property of the basis functions $\psi_{\mathbf{j}}^{\mathbf{i}}$ (20), namely

(P_5) **Riesz property**: the basis $\{\psi_{\mathbf{j}}^{\mathbf{i}}\}$ in (20) is a Riesz basis so that there exists a constant $C_R > 0$, independent of the level L, such that for all $\mathcal{I}^L(u)$ given by (20) the following holds:

$$C_R^{-1} \sum_{l=0}^{L} \sum_{|\mathbf{i}|=l} \sum_{\mathbf{j}\in B_{\mathbf{i}}} |c_{\mathbf{j}}^{\mathbf{i}}|^2 \leq \left\| \mathcal{I}_N^L(u) \right\|_{\mathcal{V}_N}^2 \leq C_R \sum_{l=0}^{L} \sum_{|\mathbf{i}|=l} \sum_{\mathbf{j}\in B_{\mathbf{i}}} |c_{\mathbf{j}}^{\mathbf{i}}|^2 , \qquad (27)$$

where the set of multi-indices $B_{\mathbf{i}}$ is defined as in (21) and $\mathcal{I}_N^L(u) \equiv \mathcal{I}_N^L(u)(\mathbf{y},\cdot,\cdot)$.

Unfortunately, finite element bases such as the linear hierarchical polynomials used in the sg-ALSCM of [39, 40] are not Riesz bases, so norms of such approximations can only be bounded from above (but not from below) by sequence norms of the corresponding coefficients. In other words, they are not L_ρ^2-stable, as implied by property P_5. The same can be said for the high-order hierarchical polynomial basis in [9], the Lagrangian interpolation polynomials used in [45, 46], as well as the orthogonal polynomials in [33, 34].

On the other hand, standard Riesz bases, e.g., Fourier and orthogonal polynomials, consist of functions that are globally supported. In the numerical PDE setting, this has the disadvantage of leading to dense stiffness matrices and, in the UQ setting, to intrusive methods. However, *certain classes of hierarchical wavelet and pre-wavelet bases are not only Riesz bases, but consist of compactly supported basis functions.* Thus, we have the best of both worlds: the compact support property of standard finite element bases and the Riesz basis property of spectral bases. Moreover, an interpolating wavelet basis can be utilized for the approximation given by (20), satisfies all the properties P_1–P_5, and forms a stable multi-resolution analysis of the stochastic space L_ρ^2 as defined in [18]. Hence, for the interpolating wavelet basis, we obtain the two-sided estimates given by P_5. Therefore, the magnitude of the wavelet coefficients $|c_{\mathbf{j}}^{\mathbf{i}}|$ in (20) can serve as *true local error estimators* and the lower bounds provided by the Riesz basis property gives us *a rigorous indicator of the efficiency of adaptive schemes.* This means a prescribed error tolerance is reached at a significantly reduced number of points in a sparse grid adaptation process. This results in a superior convergence rate when compared to methods using other hierarchical multi-scale basis functions. We choose one particular class of second-generation wavelets, namely lifted interpolating wavelets on the bounded interval, to achieve this goal. We next provide details about such wavelets.

4.1 Second-Generation Wavelets and the Lifting Scheme

Second-generation wavelets are a generalization of biorthogonal wavelets that are more easily applied to functions defined on bounded domains. Second-generation wavelets form a Riesz basis for some function space, with the wavelets being local in both "spatial" domain (in our context, the parameter domain) and the frequency domain and often having many vanishing polynomial moments, but they do not possess the translation and dilation invariance of their biorthogonal cousins. The lifting scheme [53,54] is a tool for constructing second-generation wavelets that are no longer dilates and translates of one single scaling function. In contrast to first-generation wavelets, which use the Fourier transform to build the wavelet basis, a construction using lifting is performed exclusively in the "spatial" domain so that wavelets can be custom designed for complex domains and irregularly sampled data.

The basic idea behind lifting is to start with simple multi-resolution analysis and gradually build a multi-resolution analysis with specific, a priori defined properties. The lifting scheme can be viewed as a process of taking an existing first-generation wavelet and modifying it by adding linear combinations of the scaling function at the coarse level. To explain the procedure in detail, we follow the notation in Sect. 3.3. The approximation space $V_i = \text{span}\{\phi_j^i \mid 0 \le j \le 2^i\}$ in (17) has a decomposition $V_i = V_{i-1} \oplus W_i$, where V_{i-1} and W_i are defined by

$$V_{i-1} = \text{span}\{\phi_j^{i-1} | 0 \le j \le 2^{i-1}\} \quad \text{and} \quad W_i = \text{span}\{\phi_j^i | 0 \le j \le 2^i,\ j \text{ odd}\}. \tag{28}$$

Here, W_i is viewed as the hierarchical subspace on level i, and $\phi_j^i \in W_i$ are the first-generation interpolating wavelets. Then, the corresponding second-generation wavelet $\hat{\phi}_j^i$ is constructed by "lifting" ϕ_j^i as

$$\hat{\phi}_j^i \equiv \phi_j^i + \sum_{\hat{j}=0}^{2^{i-1}} \alpha_{\hat{j},j}^{i-1} \phi_{\hat{j}}^{i-1}, \tag{29}$$

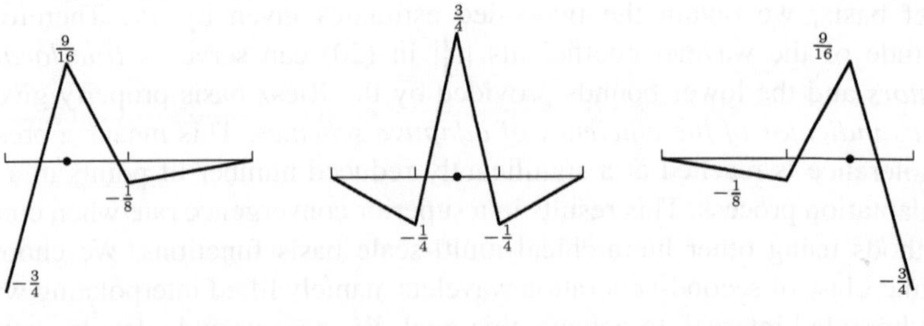

Fig. 2 Left-boundary wavelet (*left*), central wavelet (*middle*), right-boundary wavelet (*right*)

where the weights $\alpha_{\hat{j},i}^{i-1}$ in the linear combination are chosen in such a way that the new wavelet $\hat{\phi}_j^i$ has more vanishing moments than ϕ_j^i and thus provides a stabilization effect. If we apply this approach to the piecewise linear hierarchical basis, i.e., to the "hat" functions, in such a way that the lifting wavelet basis has two vanishing moments, we end up with

$$\begin{aligned} \hat{\phi}_j^i &= \phi_j^i - \frac{1}{4}\phi_{\frac{j-1}{2}}^{i-1} - \frac{1}{4}\phi_{\frac{j+1}{2}}^{i-1} && \text{for } 1 < j < 2^i - 1,\ j \text{ odd} \\ \hat{\phi}_j^i &= \phi_j^i - \frac{3}{4}\phi_{\frac{j-1}{2}}^{i-1} - \frac{1}{8}\phi_{\frac{j+1}{2}}^{i-1} && \text{for } j = 1 \\ \hat{\phi}_j^i &= \phi_j^i - \frac{1}{8}\phi_{\frac{j-1}{2}}^{i-1} - \frac{3}{4}\phi_{\frac{j+1}{2}}^{i-1} && \text{for } j = 2^i - 1, \end{aligned} \tag{30}$$

where the three equations define the central "mother" wavelet, the left-boundary wavelet, and the right-boundary wavelet, respectively. We illustrate the three lifting wavelets in Fig. 2. For additional details, see [55].

Due to the fact that our second-generation wavelets are lifted from the first-generation wavelets, properties P_1–P_4 are guaranteed. In addition, from the analysis provided in [53, 54], we know that they also constitute a Riesz basis so that property P_5 is satisfied. Therefore, introduction of the lifted wavelet basis into the hierarchical sparse-grid approximation framework results in a novel non-intrusive sparse grid adaptive wavelet stochastic collocation method (sg-AWSCM). This method allows us to encapsulate the advantages of the increased convergence rates of standard SCM and polynomial chaos approaches resulting from higher-order polynomial expansion (p-refinement) [46] with the robustness of efficient local decompositions (h-refinement) [17] for the approximation of irregular solutions and QoIs coming from PDEs with random inputs in high-dimensional stochastic domains.

Note that due to the interpolatory property of the wavelet basis, when computing the wavelet coefficients in (14), we only face an interpolation problem. That is, from the construction procedure of the lifted wavelets described above, we observe that neighboring wavelet basis function at the same level have overlapping support such that the resulting interpolation matrix for our sg-AWSCM is has greater bandwidth compared to that for sg-ALSCM. For the one-dimensional problem, the paper [55] proposed fast algorithms for computing wavelet coefficients. We are currently working on extending their algorithms to the multi-dimensional case, but in this paper, we just use mature numerical libraries, e.g., LINPACK, to solve the linear system for the interpolation coefficients.

5 Numerical Examples

This section illustrates the convergence properties of the sparse grid adaptive wavelet collocation method for solving three problems. In all examples, we use the linear hierarchical second generation lifted wavelets described in Sect. 4.1. The first example is used to compare our sg-AWSCM with the sg-ALSCM for approximating irregular (deterministic) functions in $N = 2$ parameter dimensions. In the second example, we apply our new approach to solve: (a) the Burgers equation with random boundary condition data and (b) the time-dependent Riemann problem for the Burgers equation with random initial conditions. Finally, in the third example, we investigate the ability of the sg-AWSCM to detect the important random dimensions in a elliptic problem having a moderately high number of random parameter inputs. As opposed to the previous dimension-adaptive approach of [45], our new sg-AWSCM does not require a priori nor a posterori estimates to guide adaptation. Instead, as described in Sect. 4, our multi-dimension multi-resolution adaptive approach uses only the sparse grid wavelet coefficient to guide refinement while maintaining increased convergence. We will also use this problem to compare the convergence of our sg-AWSCM with other ensemble-based methods such as the isotropic sparse grid method and the sg-ALSCM and to compare all these approaches to the best N-term sparse grid approximation.

5.1 Approximation of Irregular Deterministic Functions

Consider the two bivariate functions $f_1(y_1, y_2)$ on $[-1, 1]^2$ and $f_2(y_1, y_2)$ on $[0, 1]^2$ defined by

$$f_1(y_1, y_2) = \begin{cases} \exp(-2(y_1^2 + y_2^2)) & \text{if } y_1^2 + y_2^2 \geq 0.25 \\ 2\exp(-\frac{1}{2}) - \exp(-2(y_1^2 + y_2^2)) & \text{if } y_1^2 + y_2^2 < 0.25 \end{cases} \tag{31}$$

$$f_2(y_1, y_2) = \frac{1}{|0.15 - y_1^2 - y_2^2| + 0.1} \quad \text{on} \quad [0, 1] \times [0, 1]. \tag{32}$$

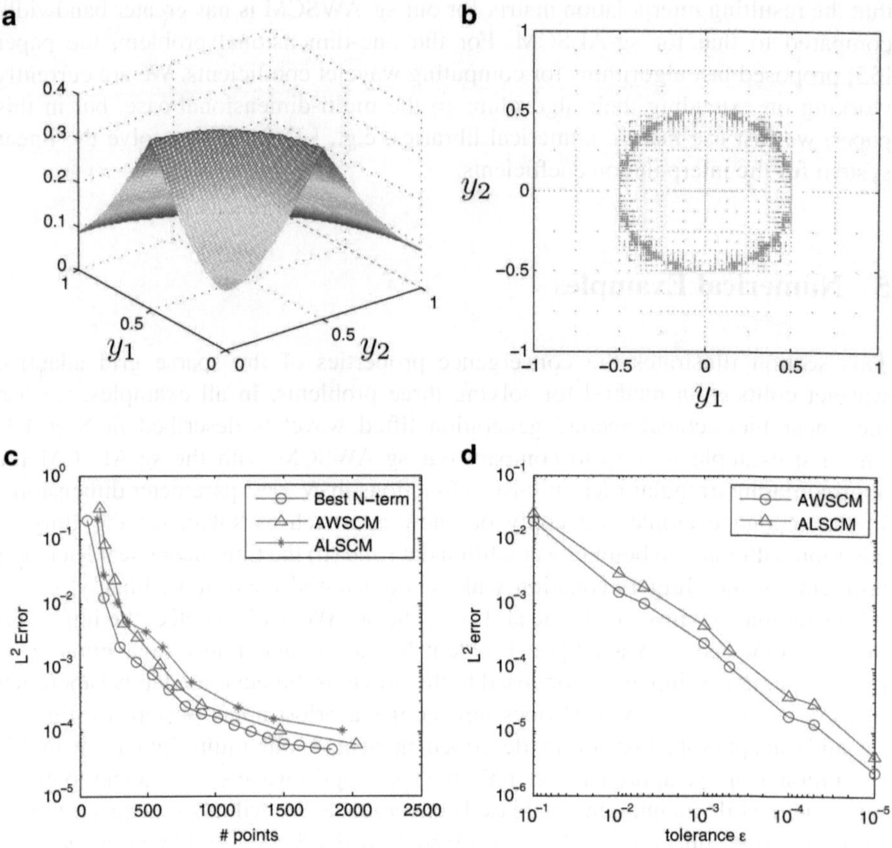

Fig. 3 Results for $f_1(y_1, y_2)$ in [31]: (**a**) the target function; (**b**) the points used by the sg-AWSCM for a tolerance $\varepsilon = 10^{-3}$; (**c**) error decay vs. number of points; (**d**) error decay vs. the tolerance ε

It is easy to see that $f_1(y_1, y_2)$ and $f_2(y_1, y_2)$ represent two types of irregular behavior. The function $f_1(y_1, y_2)$ has a jump in its first-order derivatives $\partial f_1/\partial y_1$ and $\partial f_1/\partial_2$ across the circle $y_1^2 + y_2^2 = 0.5$ whereas $f_2(y_1, y_2)$ has a steep gradient across the curve $y_1^2 + y_2^2 = 0.15$. To construct interpolants for both $f_1(y_1, y_2)$ and $f_2(y_1, y_2)$ using the sg-ALSCM and the sg-AWSCM, we first build a level $L = 3$ isotropic sparse grid as the initial grid, then add nodes adaptively guided by linear hierarchical surpluses or wavelet coefficients, respectively. The interpolation results for $f_1(y_1, y_2)$ are shown in Fig. 3. Figure 3a displays the function f_1; only the first quadrant is shown due to the symmetries of the function. Figure 3b reveals the resulting adaptive sparse interpolation grid constructed from the lifted wavelets for a tolerance $\varepsilon = 10^{-3}$. In Fig. 3c, we show the advantage of our approximation by plotting the convergence rates for the adaptive sg-ALSCM and sg-AWSCM approximations as well as for the best N-term approximations obtained by extracting the

Table 1 For $N = 2$ dimensions, we compare the number of grid points required by the sg-ALSCM, sg-AWSCM, the isotropic sparse grids (ISG) and the best N-term approximations to compute the interpolated approximation of $f_1(y_1, y_2)$ to an accuracy smaller than the prescribed error tolerance α, i.e., so that $\|\mathcal{I}_L^N(f_1)(y_1, y_2) - f_1(y_1, y_2)\| \leq \alpha$

Error α	sg-ALSCM	sg-AWSCM	Best N-term (linear)	Best N-term (wavelet)	ISG (linear)	ISG (wavelet)
5.0E-02	366	330	181	183	321	321
1.0E-02	366	330	240	233	849	849
5.0E-03	366	330	265	261	1,689	1,689
1.0E-03	774	623	479	482	7,169	7,169
5.0E-04	920	737	640	635	32,769	32,769
1.0E-04	1,927	1,548	1,261	1,254	147,497	147,497

N-terms with the N biggest coefficients from the non-adaptive sg-ALSCM and sg-AWSCM approximations, respectively. We observe that the convergence behavior of the sg-AWSCM more closely matches that of the best N-term approximation, compared to the sg-ALSCM, which results in a reduction in the number of function evaluations to achieve the desired accuracy $\varepsilon = 10^{-3}$. In Fig. 3d, we also plot the convergence behavior of both methods versus the tolerance ε. We see that for the same prescribed tolerance for the hierarchical surpluses, the sg-AWSCM can achieve higher accuracy than the sg-ALSCM. Similar conclusions can be made by examining Table 1, where we show the number of sparse grid points required by the various interpolants to achieve a desired accuracy. In all cases the sg-AWSCM outperforms the sg-ALSCM and more closely matches the best N-term approximation.

The same observations and conclusions can be reached for the function $f_2(y_1, y_2)$ by examining Fig. 4 and Table 2. Additionally, in Fig. 5, we show the condition number of the linear system used to construct the interpolation wavelet coefficients $f_1(y_1, y_2)$; we see that the interpolation matrix is well-conditioned. Therefore, as expected, due to the additional property P_5 and the well-conditioning of the interpolation matrix for the wavelet coefficients, when approximating functions with discontinuous derivatives, the sg-AWSCM substantially reduces the complexity of determining an accurate interpolant compared to the sg-ALSCM.

5.2 Burgers Equation with Random Inputs

We next apply our novel sg-AWSCM to construct approximations of solutions of two Burgers equation problems. First, we consider the steady viscous Burgers equation with random boundary condition data:

$$\begin{cases} \dfrac{1}{2}\dfrac{\partial u^2}{\partial x} - \nu\dfrac{\partial^2 u}{\partial x^2} = 0 \quad \text{in} \quad [-1, 1] \\ u(-1) = y(\omega), \qquad u(1) = 0, \end{cases} \tag{33}$$

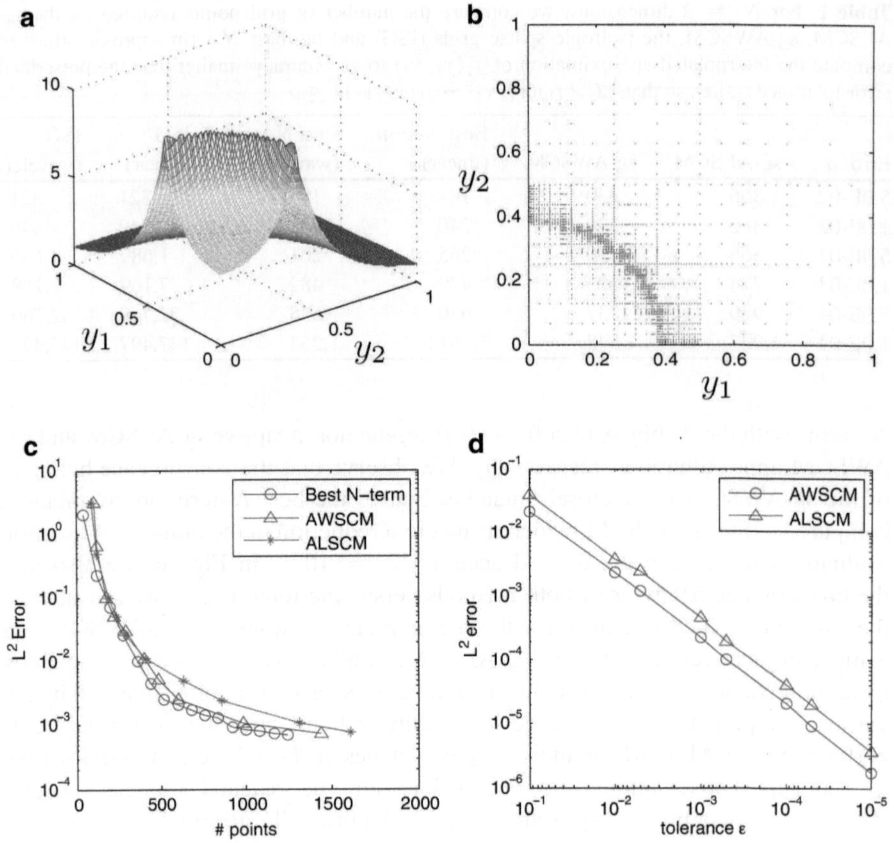

Fig. 4 Results for $f_2(y_1, y_2)$ in Example 1: (**a**) the target function; (**b**) the points used by the sg-AWSCM for a tolerance $\varepsilon = 10^{-2}$; (**c**) error decay vs. number of points; (**d**) error decay vs. the tolerance ε

Table 2 For $N = 2$ dimensions, we compare the number of function evaluations required by the sg-ALSCM, sg-AWSCM, the isotropic sparse grid (ISG), and the best N-term approximation to compute the interpolated approximation of $f_2(y_1, y_2)$ to an accuracy smaller than the prescribed error tolerance α, i.e., so that $\|\mathcal{I}_L^N(f_2)(y_1, y_2) - f_2(y_1, y_2)\| \le \alpha$

Error α	sg-ALSCM	sg-AWSCM	Best N-term (linear)	Best N-term (wavelet)	ISG (linear)	ISG (wavelet)
0.5E-01	120	124	74	81	145	145
1.0E-01	188	176	144	131	321	321
5.0E-02	243	238	237	222	1,689	1,689
1.0E-02	445	414	359	343	7,169	7,169
5.0E-03	638	491	431	419	32,769	32,769
1.0E-03	1,392	1,062	902	888	69,633	69,633

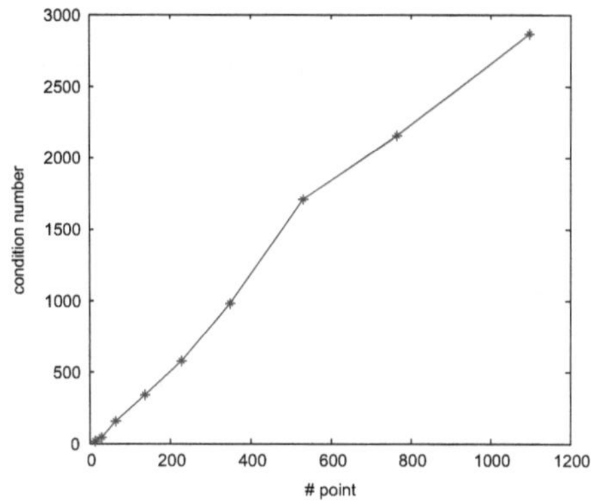

Fig. 5 Condition number vs. the number of function evaluations for sg-AWSCM for Example 1

where the left boundary condition data is a uniformly distributed random variable $y(\omega) \sim \mathcal{U}[0.95, 1.05]$, i.e., the parameter space $\Gamma = [0.95, 1.05]$ and the PDF $\rho(y) = 10$; the viscosity is set to $\nu = 0.1$.

The deterministic solver used for this problem is as a finite difference discretization of the conservative form of the equation followed by an application of Newton's method to solve the resulting nonlinear system. Figure 6 shows the computed realizations of the solution $u(y, \cdot)$ for several values of the left boundary value $y(\omega)$. Observe that perturbing $y(\omega)$ from 1 to 1.005 effects a startlingly large perturbation to the solution $u(y, \cdot)$. Thus, we conclude that the solution $u(y, \cdot)$ is very sensitive to $y(\omega)$ near $y(\omega) = 1$. In particular, this holds for the point x_0 at which the solution u changes sign. Thus, if we choose the quantity of interest to be the point x_0, we again have an instance of irregular behavior. Therefore, we focus on quantifying the uncertainty of x_0 propagated from $y(\omega)$ following the uniform distribution $\mathcal{U}[0.95, 1.05]$. To build the multi-scale interpolant using the AWSCM, we start with a 4-level uniform grid on $[0.95, 1.05]$ and then add points adaptively, guided by the size of the wavelet coefficients. The tolerance ε is set to 10^{-3}. The relation between x_0 and $y(\omega)$ and the corresponding adaptive grid are shown in Fig. 7. We can see that $x_0(y)$ has a steep slope around $y(\omega) = 1$ (which accounts for its high sensitivity near that value) and that the corresponding adaptive grid is refined around the point $y(\omega) = 1$. The convergence rate of $\mathbb{E}[x_0]$ is shown in Fig. 8 and compared to that of the best N-term approximation obtained by extracting N terms with N largest coefficients from an approximation on an non-adaptive, uniform grid.

Next, we solve a time-dependent Riemann problem for a Burgers equation with random initial shock location [51]:

$$\begin{cases} \dfrac{\partial u}{\partial t} + \dfrac{\partial}{\partial x}\left(\dfrac{1}{2}u^2\right) = 0, \quad (x,t) \in [-1,1] \times (0, +\infty) \\ u_0(x,\omega) = \begin{cases} 1 & \text{if} \quad x < y(\omega) \\ 0 & \text{if} \quad x \geq y(\omega). \end{cases} \end{cases} \tag{34}$$

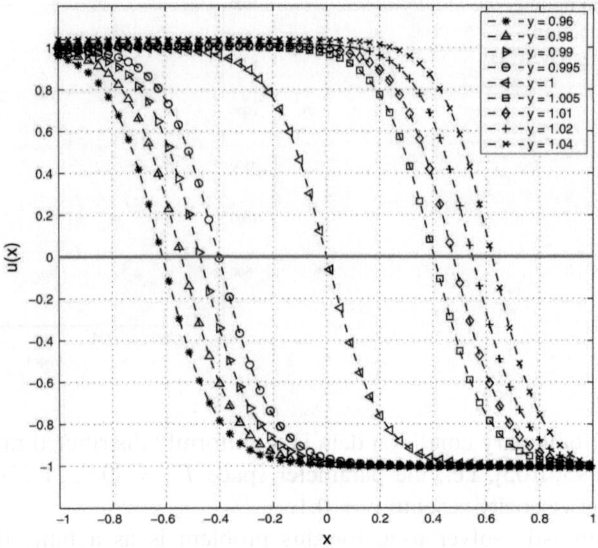

Fig. 6 Sensitivity of x_0 to the left boundary value $y(\omega)$ in example (33)

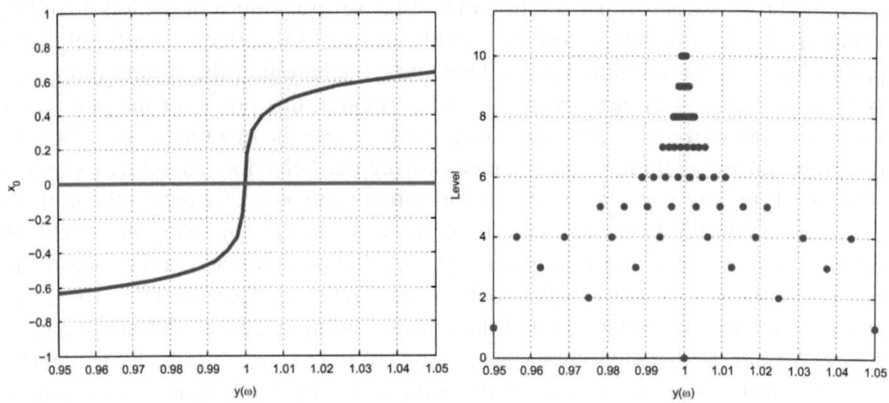

Fig. 7 The relation between x_0 and $y(\omega)$ (*left*), adaptive grid with tolerance $\varepsilon = 10^{-3}$ (*right*) in example (33)

The initial shock location depends on an uniform random variable $y(\omega) \sim \mathcal{U}[-0.1, 0.1]$, i.e., we have the parameter space $\Gamma = [-0.1, 0.1]$ and the PDF $\rho(y) = 5$. A formula for the expectation $\mathbb{E}[u]$ and variance $\mathrm{Var}[u]$ of the exact solution u can be found in [51].

The deterministic solver used for this problem is a weighted essentially non-oscillatory (WENO) scheme. Here we consider the solution at time $t = 0.2$. We compute the approximate deterministic solution on a uniform spatial grid with 1,025

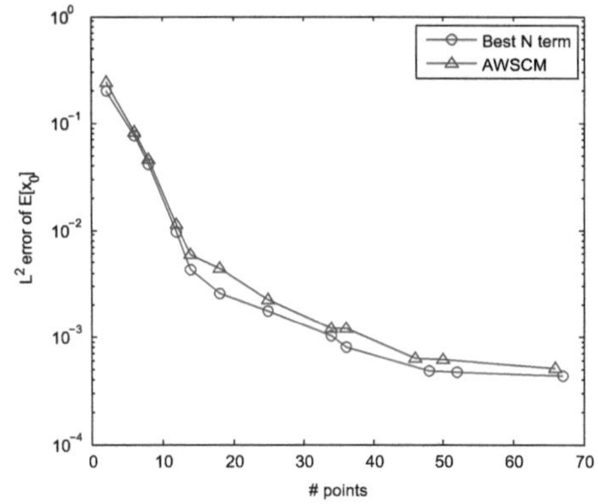

Fig. 8 The convergence rate of the sg-AWSCM with tolerance $\varepsilon = 10^{-3}$ in example (33)

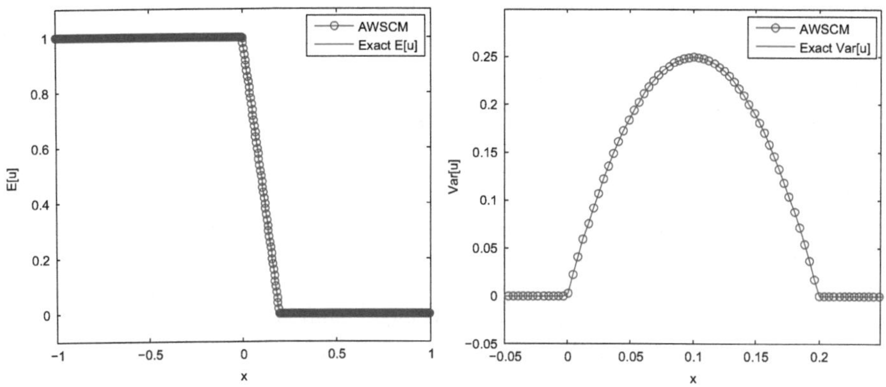

Fig. 9 Expectation (*left*) and variance (*middle*) of the probabilistic shock profile in example (34)

points in spatial domain $[-1, 1]$. In Fig. 9, we plot the expectation and variance of the approximate shock profile at $t = 0.2$, computed with the AWSCM; also plotted are the corresponding exact statistics. To test the adaptive wavelet procedure, we choose our quantities of interest to be the expectations of $u(y(\omega), x)$ at three locations, namely $\mathbb{E}[u](x = 0.036, t = 0.2)$, $\mathbb{E}[u](x = 0.127, t = 0.2)$, and $\mathbb{E}[u](x = 0.590, t = 0.2)$. We then build the grids using AWSCM with the tolerance $\varepsilon = 0.01$. At each location, we start with a two-level uniform grid on $[-0.1, 0.1]$ in the parameter space and then add points guided by the magnitudes of the wavelet coefficients. In Fig. 10, we plot the adaptive grids for the three cases. We can see that the singular point of $u(y(\omega), x, t = 0.2)$ with respect to $y(\omega)$ depends on the value of x. At the time instant $t = 0.2$: if $x \in [0, 0.2]$ such as $x = 0.036$ or $x = 0.127$, then $u(y(\omega), x, t = 0.2)$ has a singular point but its location is determined by the value of x; on the other hand, there is no singular point in $u(y(\omega), x, t = 0.2)$ for $x \in [-1, 0) \cup (0.1, 1]$, including for $x = 0.590$, so that grid refinement in parameter

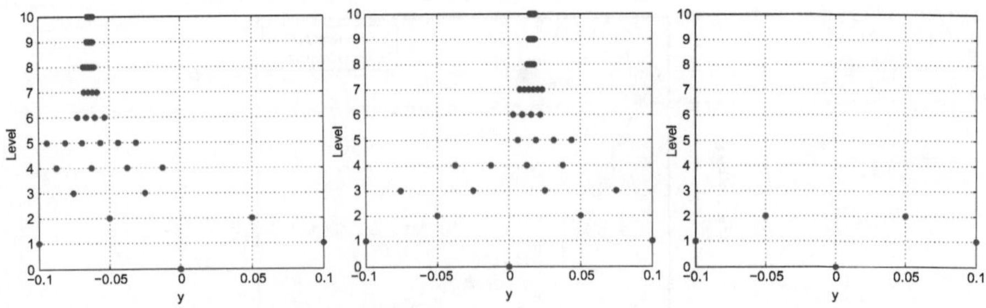

Fig. 10 Adaptive grids with $\varepsilon = 0.01$ for quantities of interest being $\mathbb{E}[u](x)$ at three spatial points: $x = 0.036$ (*left*), $x = 0.127$ (*middle*), $x = 0.590$ (*right*) in example (34)

spaces is not needed; the AWSCM method recognizes this so that the two-level initial grid is not changed by the adaptation procedure.

5.3 *Elliptic PDE with Random Inputs*

Similar to [45, 46], we consider an elliptic PDE in two spatial dimensions with random inputs. As shown in the previous examples, the AWSCM and the sg-AWSCM can accurately capture the irregular, even non-smooth regions in a low-dimensional stochastic parameter space. If the solution depends on a moderately large number of random variables with sufficient regularity (analytic in this case), the major challenge of the numerical approximation is anisotropic (dimension-adaptive) refinement. To this end, we note that global polynomial approaches, such as the a priori and a posteriori methods developed in [46] or the *quasi-optimal* techniques presented in [8], will obviously outperform our local multi-resolution approach. However, we use this example to demonstrate the ability of the sg-AWSCM method to also detect important dimensions when the random variables do not "weigh equally" in the stochastic solution. The specific problem we solve is given by

$$\begin{cases} -\nabla \cdot \big(a(\omega,x)\nabla u(\omega,x)\big) = \cos(x_1)\sin(x_2) & \text{in } \Omega \times D \\ u(\omega,x) = 0 & \text{on } \Omega \times \partial D, \end{cases} \tag{35}$$

where $D = [0,1]^2$, x_1 and x_2 denote the components of the spatial position vector x, and ∇ denotes the gradient operator with respect to x. The forcing term $f(\omega,x)$ is deterministic. The random diffusion coefficient $a(\omega,x)$ has a one-dimensional spatial dependence and is given by

$$\log(a(\omega,x) - 0.5) = 1 + y_1(\omega)\left(\frac{\sqrt{\pi}C}{2}\right)^{1/2} + \sum_{n=2}^{N} \zeta_n \varphi_n(x_1) y_n(\omega), \tag{36}$$

where, for $n \geq 2$,

$$\zeta_n := (\sqrt{\pi}C)^{1/2} \exp\left(-\frac{\left(\lfloor \frac{n}{2} \rfloor \pi C\right)^2}{8}\right) \tag{37}$$

and

$$\varphi_n(x) := \begin{cases} \sin\left(\frac{\lfloor \frac{n}{2} \rfloor \pi x_1}{C_p}\right) & \text{for } n \text{ even} \\ \cos\left(\frac{\lfloor \frac{n}{2} \rfloor \pi x_1}{C_p}\right) & \text{for } n \text{ odd.} \end{cases} \tag{38}$$

In this example, the random variables $\{y_n(\omega)\}_{n=1}^{\infty}$ are independent, have zero mean and unit variance, i.e., $\mathbb{E}[y_n] = 0$ and $\mathbb{E}[y_n y_m] = \delta_{nm}$ for $n, m \in \mathbb{N}_+$, and are uniformly distributed in the interval $[-\sqrt{3}, \sqrt{3}]$. For $x_1 \in [0, 1]$, let C_L be the physical correlation length of the stationary covariance function

$$\text{Cov}[\log(a - 0.5)](x_1, x_1') = \exp\left(-\frac{(x_1 - x_1')^2}{C_L^2}\right). \tag{39}$$

Then, the parameter C_p in (38) is $C_p = \max(1, 2C_L)$ and the parameter C in (36) and (37) is $C = C_L/C_p$. Also, ζ_n and φ_n, for $n = 1, \ldots, N$ are the eigenvalues and eigenfunctions [given by (37) and (38) respectively] of the covariance operator defined by substituting (39) into (8). The eigenvalues ζ_n in (36) decay with increasing n with large values of the correlation length C_L corresponding to fast decay. Thus, the parameter dimensions have decreasing influence on the solution as their index increases and, for large C_L, the influence decreases quickly. Therefore, an approximation requires less accurate resolution of the dimensions having small influence, compared with that for more influential dimensions so that, to achieve maximum efficiency (e.g., the fewest sample points in parameter space) for a given overall accuracy, one should use anisotropic, or dimension-adaptive set of sparse grid points.

For the numerical results, we set $C_L = \frac{1}{2}$ and retain seven terms of the expansion (36) and treat the truncated version as the exact diffusion field. In this case, the eigenvalues are $\zeta_1 = 0.665$, $\zeta_2 = \zeta_3 = 0.692$, $\zeta_4 = \zeta_5 = 0.274$, $\zeta_6 = \zeta_7 = 0.059$. In order to investigate convergence rates, we compare the expected value $\mathbb{E}[u]$ approximated by our sg-AWSCM method with a tolerance $\varepsilon = 10^{-5}$ to the "exact" solution determined from simulations based on 10^6 Monte Carlo (MC) samples of the seven-dimensional parameter. Specifically, in Fig. 11, for several values of the level L, we plot $\|\mathbb{E}[\mathcal{I}_L^N(u)] - \mathrm{e}_{MC}[u]\|$, i.e., the $L^2(D)$ norm of the "error" between the expected values obtained using the sg-AWSCM method and the densely sampled MC method. Also provided in that figure are the corresponding errors for the isotropic sparse grid, for the sg-ALSCM, and for the best N-term approximation defined by taking the N terms in the isotropic sparse

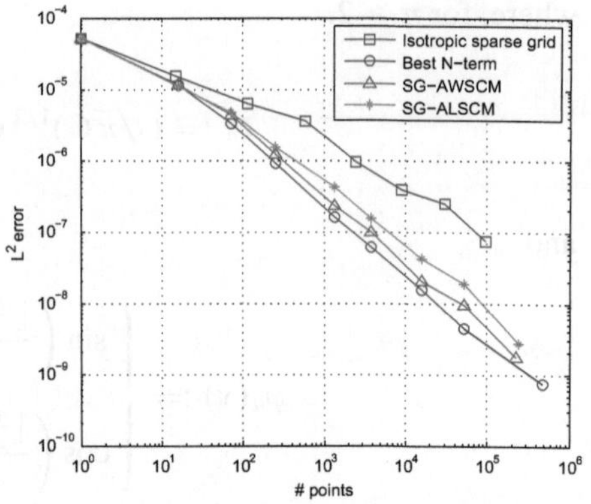

Fig. 11 The convergence rate of isotropic sparse grid, the sg-ALSCM, and the sg-AWSCM approximations with tolerance $\varepsilon = 10^{-5}$

Table 3 For $N = 7$ dimensions, we compare the number of function evaluations required by the isotropic sparse grid (ISG), the sg-ALSCM, the sg-AWSCM, and the best N-term approximation to compute the expected value of the solution to within a prescribed global error tolerance α, i.e., so that $\|\mathbb{E}[\mathcal{I}_L^N(u)] - \mathrm{e}_{MC}[u]\| \leq \alpha$

Error α	sg-ALSCM	sg-AWSCM	Best N-term (linear)	Best N-term (wavelet)	ISG (linear)	ISG (wavelet)
1.0E-05	30	25	34	35	73	73
5.0E-06	74	60	85	81	344	344
1.0E-06	476	248	772	763	2,435	2,435
5.0E-07	1,038	840	1,271	1,280	7,767	7,767
1.0E-07	7,333	3,824	2,812	2,732	85,861	85,861

grid solution with the largest coefficients. The errors are plotted against the number of points in parameter space each method requires to achieve the desired accuracy. As expected, due to the fast decay of the eigenvalues, the convergence, with respect to the number of points used, of both the sg-AWSCM and the sg-ALSCM is much faster than the approximation based on an isotropic sparse grid because fewer points are placed along the non-important dimensions associated with small eigenvalues. Furthermore, our new sg-AWSCM also reduces the overall complexity when compared to the sg-ALSCM approximation, and nearly matches that for the best N-term approximation. Further proof of this can be seen in Table 3 that shows a reduction in the computational complexity for computing the expected value using the sg-AWSCM, when compared to the isotropic sparse grid and sg-ALSCM, by approximately a factor of 20 and 3 respectively, to achieve a desired accuracy of 10^{-7}. In fact, for this higher-dimensional problem, the savings incurred by the sg-AWSCM compared to the sg-ALSCM and the isotropic sparse-grid approximation are much more significant than for the previous low-dimensional examples. One can expect the relative savings, compared with the isotropic sparse-grid approximation, to increase as one further increases the parameter dimension.

6 Conclusions

This work proposed a novel sparse-grid adaptive wavelet stochastic collocation method for both smooth and irregular solutions of partial differential equations with random input data. This method can be viewed as a major improvement to previous works; isotropic and anisotropic global Lagrange-type stochastic collocation based on tensor product approximations [3] or sparse grid approaches [45, 45, 58], as well as the hierarchical sparse-grid locally adaptive linear stochastic collocation method [39, 40].

The new technique consists of any standard deterministic approximation in the physical space (e.g. Galerkin finite element) and an adaptive collocation in the probability domain at sparse-grid points in the random parameter space, along with a hierarchical multi-dimensional multi-resolution linear wavelet basis. This compactly supported Riesz basis guarantees the stability of the multi-scale coefficients and leads to more efficient hierarchical sparse grid approximations. That is, we are able to guide adaptive refinement by the magnitude of the wavelet coefficient which results in a minimal number of grid points to achieve a prescribed tolerance. This alleviates the curse of dimensionality by reducing the computational complexity for problems having high stochastic dimension. Moreover, as a consequence of the interpolation property, guaranteed by the proposed lifting scheme, our approach remains completely non-intrusive and naturally allows for the solution of uncoupled deterministic problems that are trivially parallelizable, as for the Monte Carlo method.

The numerical examples included in this work provide computational verification of the advantage of our novel algorithm. The numerical results compare our new approach with several classical and heavily utilized techniques for solving stochastic problems whose solutions are both highly regular and even non-smooth with respect to the random variables. The results show that, in particular, for moderately large-dimensional problems, the sparse grid adaptive wavelet stochastic collocation approach seems to be very efficient and superior to all methods it is compared to.

Future directions of this research will include a complete convergence analysis of our new approach that will incorporate an examination of the complexity of our algorithm with respect to the number of collocation points on the sparse grid, as the dimension of the problem increases. However, as the computational results suggest, we also want to use the theoretical results to fully explain the increased stability and efficiency of these techniques when compared to previous approaches. Finally, we want to avoid solving an interpolation matrix equation for the wavelet coefficients and intend to develop fast algorithms for calculating the wavelet coefficients in sparse tensor product spaces.

Acknowledgements Max Gunzburger was supported by the US Air Force Office of Scientific Research (AFOSR) under grant number FA9550-11-1-0149. Clayton G. Webster was supported by the US AFOSR under grant number 1854-V521-12. Also supported by the Laboratory Directed Research and Development (LDRD) Program at the Oak Ridge National Laboratory (ORNL). The ORNL is operated by UT-Battelle, LLC, for the United States Department of Energy under

Contract DE-AC05-00OR22725. Guannan Zhang was supported by the US AFOSR under grant number FA9550-11-1-0149. Also supported by the Advanced Simulation Computing Research (ASCR), Department of Energy, through the Householder Fellowship at ORNL. The ORNL is operated by UT-Battelle, LLC, for the United States Department of Energy under Contract DE-AC05-00OR22725.

References

1. I.M. Babuška, R. Tempone, G.E. Zouraris, Galerkin finite element approximations of stochastic elliptic partial differential equations. SIAM J. Numer. Anal. **42**(2), 800–825 (2004)
2. I.M. Babuška, R. Tempone, G.E. Zouraris, Solving elliptic boundary value problems with uncertain coefficients by the finite element method: the stochastic formulation. Comput. Methods Appl. Mech. Eng. **194**(12–16), 1251–1294 (2005)
3. I. Babuška, F. Nobile, R. Tempone, A stochastic collocation method for elliptic partial differential equations with random input data. SIAM J. Numer. Anal. **45**(3), 1005–1034 (2007)
4. A. Barth, A. Lang, Multilevel monte carlo method with applications to stochastic partial differential equations. Int. J. Comput. Math. **89**(18), 2479–2498 (2012)
5. A. Barth, C. Schwab, N. Zollinger, Multi-level monte carlo finite element method for elliptic pdes with stochastic coefficients. Numer. Math. **119**(1), 123–161 (2011)
6. A. Barth, A. Lang, C. Schwab, Multilevel monte carlo method for parabolic stochastic partial differential equations. Bit **53**(1), 3–27 (2013)
7. J. Beck, F. Nobile, L. Tamellini, R. Tempone, Stochastic spectral Galerkin and collocation methods for PDEs with random coefficients: a numerical comparison, in *Spectral and High Order Methods for Partial Differential Equations*. Lecture Notes in Computational Science and Engineering, vol. 76 (Springer, Berlin, 2011), pp. 43–62.
8. J. Beck, F. Nobile, L. Tamellini, R. Tempone, Convergence of quasi-optimal stochastic Galerkin methods for a class of PDES with random coefficients. Comput. Math. Appl. **67**(4), 732–751 (2014)
9. H.-J. Bungartz, M. Griebel, Sparse grids. Acta Numer. **13**, 1–123 (2004)
10. J. Charrier, R. Scheichl, A.L. Teckentrup, Finite element error analysis of elliptic PDEs with random coefficients and its application to multilevel Monte Carlo methods. SIAM J. Numer. Anal. **51**(1), 322–352 (2013)
11. C. Chui, J. Wang, A general framework of compactly supported splines and wavelets. J. Approx. Theory **71**(3), 263–304 (1992)
12. A. Cohen, I. Daubechies, J. Feauveau, Biorthogonal bases of compactly supported wavelets. Commun. Pure Appl. Math. **45**(5), 485–560 (1992)
13. A. Cohen, W. Dahmen, R. DeVore, Adaptive wavelet methods for elliptic operator equations – convergence rates. Math. Comput. **70**, 27–75 (2001)
14. A. Cohen, W. Dahmen, R. DeVore, Adaptive wavelet methods for elliptic operator equations II – beyond the elliptic case. Found. Comput. Math. **2**, 203–245 (2002)
15. S. Dahlke, W. Dahmen, K. Urban, Adaptive wavelet methods for saddle point problems – optimal convergence rates. SIAM J. Numer. Anal. **40**, 1230–1262 (2002)
16. W. Dahmen, A. Kunoth, Adaptive wavelet methods for linear-quadratic elliptic control problems: convergence rates. SIAM J. Control Optim. **43**, 1640–1675 (2002)
17. I. Daubechies, Orthonormal bases of compactly supported wavelets. Commun. Pure Appl. Math. **41**(7), 909–996 (1988)
18. I. Daubechies, Wavelets - algorithms and applications. Science **262**(5139), 1589–1591 (1993)
19. D. Diaz, M. Gunzburger, A. Kunoth, An adaptive wavelet viscosity method for hyperbolic conservation laws. Numer. Math. **24**, 1388–1404 (2008)
20. T.J. Dijkema, C. Schwab, R. Stevenson, An adaptive wavelet method for solving high-dimensional elliptic PDEs. Constr. Approx. **30**(3), 423–455 (2009)

21. H. Elman, C. Miller, Stochastic collocation with kernel density estimation. Technical report, Department of Computer Science, University of Maryland, 2011
22. G. Fishman, *Monte Carlo: Concepts, Algorithms, and Applications*. Springer Series in Operations Research (Springer, New York, 1996)
23. J. Foo, X. Wan, G. Karniadakis, The multi-element probabilistic collocation method (ME-PCM): error analysis and applications. J. Comput. Phys. **227**(22), 9572–9595 (2008)
24. P. Frauenfelder, C. Schwab, R.A. Todor, Finite elements for elliptic problems with stochastic coefficients. Comput. Methods Appl. Mech. Eng. **194**(2–5), 205–228 (2005)
25. T. Gerstner, M. Griebel, Dimension-adaptive tensor-product quadrature. Computing **71**(1), 65–87 (2003)
26. R.G. Ghanem, P.D. Spanos, *Stochastic Finite Elements: A Spectral Approach* (Springer, New York, 1991)
27. M. Griebel, Adaptive sparse grid multilevel methods for elliptic PDEs based on finite differences. Computing **61**(2), 151–179 (1998)
28. M. Gunzburger, A. Kunoth, Space-time adaptive wavelet methods for optimal control problems constrained by parabolic evolution equations. SIAM J. Control Optim. **49**(3), 1150–1170 (2011)
29. J.D. Jakeman, R. Archibald, D. Xiu, Characterization of discontinuities in high-dimensional stochastic problems on adaptive sparse grids. J. Comput. Phys. **230**(10), 3977–3997 (2011)
30. A. Klimke, B. Wohlmuth, Algorithm 847: Spinterp: piecewise multilinear hierarchical sparse grid interpolation in matlab. ACM Trans. Math. Softw. **31**(4), 561–579 (2005)
31. F.Y. Kuo, C. Schwab, I.H. Sloan, Quasi-Monte Carlo methods for high-dimensional integration: the standard (weighted Hilbert space) setting and beyond. ANZIAM J. Aust. N. Z. Ind. Appl. Math. J. **53**(1), 1–37 (2011)
32. F.Y. Kuo, C. Schwab, I.H. Sloan, Quasi-Monte Carlo finite element methods for a class of elliptic partial differential equations with random coefficients. SIAM J. Numer. Anal. **50**(6), 3351–3374 (2012)
33. O.P. Le Maître, O.M. Knio, *Spectral Methods for Uncertainty Quantification* (Springer, New York, 2010)
34. O.P. Le Maître, O.M. Knio, H.N. Najm, R.G. Ghanem, Uncertainty propagation using Wiener-Haar expansions. J. Comput. Phys. **197**(1), 28–57 (2004)
35. O.P. Le Maître, H.N. Najm, R.G. Ghanem, O.M. Knio, Multi-resolution analysis of Wiener-type uncertainty propagation schemes. J. Comput. Phys. **197**(2), 502–531 (2004)
36. C.F. Li, Y.T. Feng, D.R.J. Owen, D.F. Li, I.M. Davis, A Fourier-Karhunen-Loève discretization scheme for stationary random material properties in SFEM. Int. J. Numer. Meth. Eng. (2007) www.interscience.wiley.com
37. M. Loève, *Probability Theory. I*. Graduate Texts in Mathematics, vol. 45, 4th edn. (Springer, New York, 1977)
38. M. Loève, *Probability Theory. II*. Graduate Texts in Mathematics, vol. 46, 4th edn. (Springer, New York, 1978)
39. X. Ma, N. Zabaras, An adaptive hierarchical sparse grid collocation algorithm for the solution of stochastic differential equations. J. Comput. Phys. **228**(8), 3084–3113 (2009)
40. X. Ma, N. Zabaras, An adaptive high-dimensional stochastic model representation technique for the solution of stochastic partial differential equations. J. Comput. Phys. **229**(10), 3884–3915 (2010)
41. L. Mathelin, M.Y. Hussaini, T.A. Zang, Stochastic approaches to uncertainty quantification in CFD simulations. Numer. Algorithms **38**(1–3), 209–236 (2005)
42. H.G. Matthies, A. Keese, Galerkin methods for linear and nonlinear elliptic stochastic partial differential equations. Comput. Methods Appl. Mech. Eng. **194**(12–16), 1295–1331 (2005)
43. P. Nitsche, Sparse approximation of singularity functions. Constr. Approx. **21**, 63–81 (2005)
44. F. Nobile, R. Tempone, Analysis and implementation issues for the numerical approximation of parabolic equations with random coefficients. Int. J. Numer. Methods Eng. **80**(6–7), 979–1006 (2009)

45. F. Nobile, R. Tempone, C.G. Webster, A sparse grid stochastic collocation method for partial differential equations with random input data. SIAM J. Numer. Anal. **46**(5), 2309–2345 (2008)
46. F. Nobile, R. Tempone, C.G. Webster, An anisotropic sparse grid stochastic collocation method for partial differential equations with random input data. SIAM J. Numer. Anal. **46**(5), 2411–2442 (2008)
47. B. Øksendal, *Stochastic Differential Equations: An Introduction with Applications*. Universitext, 6th edn. (Springer, Berlin, 2003)
48. P. Oswald, Hierarchical conforming finite-element methods for the biharmonic equation. SIAM J. Numer. Anal. **29**(6), 1610–1625 (1992)
49. C. Schwab, R. Stevenson, Adaptive wavelet algorithms for elliptic PDE's on product domains. Math. Comput. **77**(261), 71–92 (2008)
50. C. Schwab, R. Stevenson, Fast evaluation of nonlinear functionals of tensor product wavelet expansions. Numer. Math. **119**, 765–786 (2011)
51. C. Schwab, S. Tokareva, High order approximation of probabilistic shock profiles in hyperbolic conservation laws with uncertain initial data. Technical report, SAM Research Report No. 2011–53, 2011
52. S. Smolyak, Quadrature and interpolation formulas for tensor products of certain classes of functions. Dokl. Akad. Nauk SSSR **4**, 240–243 (1963)
53. W. Sweldens, The lifting scheme: a custom-design construction of biorthogonal wavelets. Appl. Comput. Harmon. Anal. **3**(2), 186–200 (1996)
54. W. Sweldens, The lifting scheme: a construction of second generation wavelets. SIAM J. Math. Anal. **29**(2), 511–546 (1998)
55. W. Sweldens, P. Schroder, Building your own wavelets at home. Computer **90**(1995:5), 72–107 (2000)
56. R.A. Todor, C. Schwab, Convergence rates for sparse chaos approximations of elliptic problems with stochastic coefficients. IMA J. Numer. Anal. **27**(2), 232–261 (2006)
57. N. Wiener, The homogeneous chaos. Am. J. Math. **60**, 897–936 (1938)
58. D. Xiu, J. Hesthaven, High-order collocation methods for differential equations with random inputs. SIAM J. Sci. Comput. **27**, 1118–1139 (2005)
59. D. Xiu, G.E. Karniadakis, The Wiener-Askey polynomial chaos for stochastic differential equations. SIAM J. Sci. Comput. **24**(2), 619–644 (2002)
60. G. Zhang, M. Gunzburger, Error analysis of a stochastic collocation method for parabolic partial differential equations with random input data. SIAM J. Numer. Anal. **50**(4), 1922–1940 (2012)

Robust Solutions to PDEs with Multiple Grids

Brendan Harding and Markus Hegland

Abstract In this paper we will discuss some approaches to fault-tolerance for solving partial differential equations. In particular we will discuss how one can combine the solution from multiple grids using ideas related to the sparse grid combination technique and multivariate extrapolation. By utilising the redundancy between the solutions on different grids we will demonstrate how this approach can be adapted for fault-tolerance. Much of this will be achieved by assuming error expansions and examining the extrapolation of these when various solutions from different grids are combined.

Introduction

Modern supercomputers are becoming increasingly complex having hundreds of thousands of components all working together to complete cumbersome tasks. As we head towards exascale computing the number of components in the fastest machines will significantly increase. The purpose of these components vary, from power supplies, interconnects, CPU's, hard drives, cooling and so on. Given the limited lifetime of each component, the probability that any one of these will fail within a given period of time also increases. The failure of one component will likely affect numerous neighbouring components. Of course, the origin of all these failures is a hardware component, but numerous studies of faults indicate that software errors can be just as, if not more, significant [8]. How we handle these failures whilst applications are running is a challenge that must be addressed if we are to achieve exascale computing.

B. Harding (✉) • M. Hegland
Australian National University, Canberra, ACT 0200, Australia
e-mail: brendan.harding@anu.edu.au; markus.hegland@anu.edu.au

J. Garcke and D. Pflüger (eds.), *Sparse Grids and Applications - Munich 2012*, Lecture Notes in Computational Science and Engineering 97,
DOI 10.1007/978-3-319-04537-5_7,

Current approaches to fault-tolerance tend to be checkpoint-restart in nature. These methods are infeasible at exascale because they are too memory and communication intensive leading to inefficient use of resources and high energy costs. Algorithm based fault-tolerance (ABFT) has been studied for a variety of problems and often provides a low cost solution to robustness, see for example [5, 9, 13]. Existing approaches are largely based upon the use of checksums within linear algebra calculations.

The sparse grid combination technique was introduced by Griebel et al. [11]. By solving a given problem on many regular anisotropic grids and taking a linear combination of the results one may approximate the so called sparse grid solution [6, 15]. In Sect. 1 we will introduce some notation and then discuss so called truncated combinations. We will then go on to describe what one might do if a fault affects one of the solutions making it unavailable for combination. One of the key points is that the solution on different grids can be computed independently and hence a fault need not affect the entire computation. Furthermore, there is a lot of redundancy shared between the solution of different grids which can be utilised in the event of a fault. We will demonstrate the additional errors in the solution one trades off when using our approaches with both theoretical bounds and some numerical results.

Bungartz et al. [7] describe multivariate extrapolation and combination techniques for elliptic boundary value problems. Following this, in Sect. 2 we will describe how one may compute an extrapolation when one has an assortment of grids. In particular this will lead to different ways to form extrapolations when a fault affects one of the solutions. We will demonstrate the effectiveness of these extrapolations for a variety of problems.

Finally, in Sect. 3 we will demonstrate how combination solutions and extrapolation solutions may be combined effectively making the most out of the given solutions. Furthermore, given the fault-tolerant approaches for the combination and extrapolation solutions it is straightforward to combine the two in a fault-tolerant manner. We will again demonstrate the errors with some numerical results.

1 The Combination Technique

In this section we will describe the sparse grid combination technique and introduce some notation used throughout the paper.

Let $\underline{i} = (i_1, \ldots, i_d) \in \mathbb{N}^d$ be a multi-index and $\Omega_i = \{0, h_i, 2h_i, 3h_i, \ldots, 1\}$ be a discretisation of the unit interval where $h_i := 2^{-i}$. We define a grid on the unit d-cube by

$$\Omega_{\underline{i}} := \Omega_{i_1} \times \Omega_{i_2} \times \cdots \times \Omega_{i_d}$$

Given a grid $\Omega_{\underline{i}}$ we define an associated space of d-linear functions

$$V_{\underline{i}} := \text{span}\{\phi_{\underline{i},\underline{j}} : j_t = 0, \dots 2^{i_t}, t = 1, \dots, d\}$$

where $\phi_{\underline{i},\underline{j}}$ are the usual d-linear nodal basis functions (hat functions). The hierarchical difference spaces $W_{\underline{i}}$ are given by

$$V_{\underline{i}} = W_{\underline{i}} \oplus \sum_{t=1}^{d} V_{\underline{i}-\underline{e}_t}$$

where $\underline{e}_t$ is the unit vector along the t-th axis. We can use this to write $V_n := V_{\{n,n,\dots,n\}}$ as the direct sum of difference spaces

$$V_n = \bigoplus_{i_1=0}^{n} \cdots \bigoplus_{i_d=0}^{n} W_{\underline{i}} = \bigoplus_{\|\underline{i}\|_\infty \leq n} W_{\underline{i}},$$

and of course the classical sparse grid space is given by

$$V_n^s := \bigoplus_{\|\underline{i}\|_1 \leq n} W_{\underline{i}}.$$

The sparse grid combination technique [10, 11] allows one to approximate the sparse grid solution [6] by taking linear combinations of the solution on multiple regular anisotropic grids. Suppose $f_{\underline{i}} \in V_{\underline{i}}$ denotes the solution to a given problem on the grid $\Omega_{\underline{i}}$, then the classical combination solution is given by

$$f_n^c(\underline{x}) := \sum_{q=0}^{d-1} (-1)^q \binom{d-1}{q} \sum_{|\underline{i}|_1 = n-q} f_{\underline{i}}(\underline{x}). \tag{1}$$

The combination technique can be generalised into something adaptive and flexible, see for example [12]. Given a lattice of multi-indices $\mathcal{I}$ which is a downset, we denote a combination technique onto this space with the formula

$$f_{\mathcal{I}}^c(\underline{x}) = \sum_{\underline{i} \in \mathcal{I}} c_{\underline{i}} f_{\underline{i}}(\underline{x}) \in V_{\mathcal{I}}^s := \bigoplus_{\underline{i} \in \mathcal{I}} W_{\underline{i}}, \tag{2}$$

where the $c_{\underline{i}}$ are the combination coefficients.

For $d = 2$ and $d = 3$ we will often use the notation $f_{i,j}$ and $f_{i,j,k}$ respectively to denote $f_{\underline{i}}$. Similarly for $W_{\underline{i}}$ and $V_{\underline{i}}$.

There are some classical results for the error of combination technique solutions. We assume the pointwise error splitting (see [11])

$$f - f_{i,j} = C_1(h_i)h_i^p + C_2(h_j)h_j^p + D(h_i, h_j)h_i^p h_j^p \tag{3}$$

where C_1 depends on h_i, x and y, C_2 depends on h_j, x and y, and D depends on h_i, h_j, x and y. Furthermore $|C_1|$, $|C_2|$ and $|D|$ are all bounded above by the same positive constant κ. We are primarily concerned with solutions with sufficient smoothness to permit first and second order schemes to obtain a solution satisfying the above splitting for $p = 1$ and $p = 2$ respectively.

With this assumption it is shown by Griebel et al. [11] that for $d = p = 2$ one has

$$|f - f_n^c| \le \kappa h_n^2 (1 + \frac{5}{4} n) .$$

Similarly in three-dimensions if one assumes the pointwise error splitting

$$\begin{aligned} \epsilon_{i,j,k} := f - f_{i,j,k} = & \; C_1(h_i)h_i^p + C_2(h_j)h_j^p + C_3(h_k)h_k^p \\ & + D_1(h_i,h_j)h_i^p h_j^p + D_2(h_i,h_k)h_i^p h_k^p \\ & + D_3(h_j,h_k)h_j^p h_k^p + E(h_i,h_j,h_k)h_i^p h_j^p h_k^p \end{aligned} \tag{4}$$

where $|C_1|$, $|C_2|$, $|C_3|$, $|D_1|$, $|D_2|$, $|D_3|$ and $|E|$ are bounded above by the same positive constant κ, then one may show for $p = 2$ (again see [11])

$$|f - f_n^c| \le \kappa h_n^2 (1 + \frac{65}{32} n + \frac{25}{32} n^2) .$$

These results can be easily extended to general p where h_n^2 is replaced with h_n^p and the constants are different.

1.1 Truncated Approximations

In the classical combination technique one can have solutions on extremely disproportionate (strongly anisotropic) grids contributing to the solution. It has been observed that for some problems it is better to leave some of these out of the combination, see for example [4]. This gives rise to the so called truncated combinations since the second sum of (1) is effectively truncated to exclude such solutions. In two dimensions one may express truncated combinations as

$$\tilde{f}_{i,j}^t := \sum_{\alpha=0}^{t} f_{i+\alpha,j+t-\alpha} - \sum_{\alpha=0}^{t-1} f_{i+\alpha,j+t-1-\alpha} \tag{5}$$

and similarly for three-dimensions one has

$$\begin{aligned} \tilde{f}_{i,j,k}^t := & \sum\nolimits_{\alpha+\beta=0}^{t} f_{i+\alpha,j+\beta,k+t-\alpha-\beta} \\ & -2\sum\nolimits_{\alpha+\beta=0}^{t-1} f_{i+\alpha,j+\beta,k+t-1-\alpha-\beta} \\ & +\sum\nolimits_{\alpha+\beta=0}^{t-2} f_{i+\alpha,j+\beta,k+t-2-\alpha-\beta} \end{aligned} \tag{6}$$

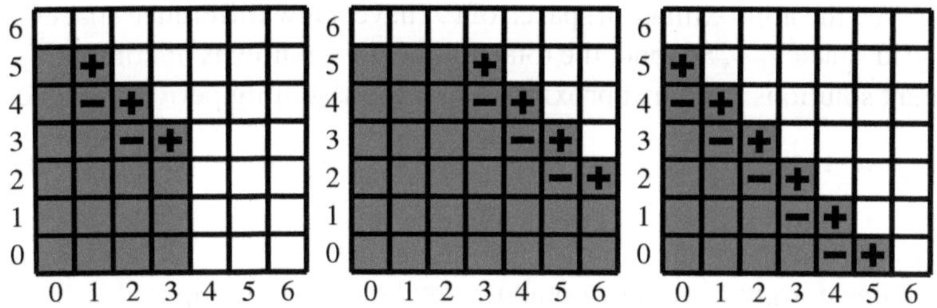

Fig. 1 Here we illustrate three examples of combinations. On the *left* we have the combination $\tilde{f}^2_{1,3}$, in the *middle* we have the combination $\tilde{f}^3_{3,2}$ and on the *right* we have the combination $\tilde{f}^5_{0,0} = f^c_5$. The *shaded squares* indicate the hierarchical spaces $W_{i,j}$ which contribute to the combined solution

where by $\sum_{\alpha+\beta=0}^{t}$ we mean $\sum_{\alpha=0}^{t}\sum_{\beta=0}^{t-\alpha}$. Some examples of these in two dimensions are depicted in Fig. 1.

In this paper we will consider truncated combinations of the form $\tilde{f}^t_{n-t,n-t}$ and $\tilde{f}^t_{n-t,n-t,n-t}$ with $2 \leq t \leq n$ for problems with two and three spatial dimensions respectively. Note that for $t = n$ the truncated combination is equivalent to the classical combination of (1). Specifically, we will discuss in detail the cases of $t = 2$ and $t = 3$. Our approach for dealing with faults in these examples will be indicative of the approach for larger t. The first examples are the truncated combinations in two spatial dimensions with $t = 2$ as given by

$$\tilde{f}^2_{n-2,n-2} = f_{n,n-2} + f_{n-1,n-1} + f_{n-2,n} - f_{n-1,n-2} - f_{n-2,n-1} \tag{7}$$

and $t = 3$ as given by

$$\tilde{f}^3_{n-3,n-3} = f_{n,n-3} + f_{n-1,n-2} + f_{n-2,n-1} + f_{n-3,n} - f_{n-1,n-3} - f_{n-2,n-2} - f_{n-3,n-1}\,. \tag{8}$$

There are a few things to note about these combinations, first of all we note that the approximation spaces for the combinations of (7) and (8) are given by

$$\bigoplus_{\|\underline{i}\|_\infty \leq n\,,\,\|\underline{i}\|_1 \leq 2n-2} W_{\underline{i}} \quad \text{and} \quad \bigoplus_{\|\underline{i}\|_\infty \leq n\,,\,\|\underline{i}\|_1 \leq 2n-3} W_{\underline{i}} \tag{9}$$

for $t = 2$ and $t = 3$ respectively. Alternatively, these may be expressed as

$$V_{n,n} = (W_{n,n} \oplus W_{n-1,n} \oplus W_{n,n-1}) \oplus \left(\bigoplus\nolimits_{\|\underline{i}\|_\infty \leq n\,,\,\|\underline{i}\|_1 \leq 2n-2} W_{\underline{i}}\right),$$
$$V_{n,n} = \left(\bigoplus\nolimits_{\|\underline{i}\|_\infty \leq n\,,\,\|\underline{i}\|_1 > 2n-3} W_{\underline{i}}\right) \oplus \left(\bigoplus\nolimits_{\|\underline{i}\|_\infty \leq n\,,\,\|\underline{i}\|_1 \leq 2n-3} W_{\underline{i}}\right)$$

where we see the approximation spaces of (9) have a few difference spaces less than the full grid space $V_{n,n}$. Second, the total number of unknowns of computing the five component solutions for the approximation of (7) assuming periodic boundaries is

$$2 \times (2^n \times 2^{n-2}) + (2^{n-1})^2 + 2 \times (2^{n-1} \times 2^{n-2}) = 3 \times 2^{2n-2} + 2 \times 2^{2n-3} = 2^{2n}.$$

The RHS is equal to the number of unknowns in the full grid space $V_{n,n}$. Therefore one does not save any computational resources compared to the classical combination technique. If computing the component solutions as in (8) one saves more but still not as much as the classical combination technique. As a result, the approaches discussed in this paper will not extend beyond three-dimensional problems as one runs into the curse of dimensionality. In three-dimensions we will consider the examples with $t = 2$ given by

$$\begin{aligned} \tilde{f}^2_{n-2,n-2,n-2} = {} & f_{n,n-2,n-2} + f_{n-1,n-1,n-2} + f_{n-2,n-1,n-1} \\ & + f_{n-1,n-2,n-1} + f_{n-2,n,n-2} + f_{n-2,n-2,n} \\ & -2(f_{n-1,n-2,n-2} + f_{n-2,n-2,n-1} + f_{n-2,n-1,n-2}) \\ & + f_{n-2,n-2,n-2}\,, \end{aligned} \tag{10}$$

and with $t = 3$ given by

$$\begin{aligned} \tilde{f}^3_{n-3,n-3,n-3} = {} & f_{n,n-3,n-3} + f_{n-1,n-2,n-3} + f_{n-2,n-1,n-3} + f_{n-3,n,n-3} + f_{n-1,n-3,n-2} \\ & + f_{n-2,n-2,n-2} + f_{n-3,n-1,n-2} + f_{n-2,n-3,n-1} + f_{n-3,n-2,n-1} \\ & + f_{n-3,n-3,n} - 2(f_{n-1,n-3,n-3} + f_{n-2,n-2,n-3} + f_{n-3,n-1,n-3} \\ & + f_{n-2,n-3,n-2} + f_{n-3,n-2,n-2} + f_{n-3,n-3,n-1}) \\ & + f_{n-3,n-3,n-2} + f_{n-3,n-2,n-3} + f_{n-2,n-3,n-3}\,. \end{aligned} \tag{11}$$

Again we note the number of unknowns in the ten component solutions of (10) is close to that of the full grid space $V_{n,n,n}$ and hence one saves very little in computational resources compared to the classical combination technique.

The advantage of these combinations is that for increasing n the solution error decreases faster than for the usual sparse grid combination technique. This is evident looking at the approximation spaces where one can see that only a few of the fine difference spaces have been excluded compared to the full grid solution space V_n. The following proposition which is similar to those in [11] gives a bound on the error.

Proposition 1. *Suppose that $f_{i,j}$ satisfies the pointwise error splitting*

$$\epsilon_{i,j} := f - f_{i,j} = C_1(h_i)h_i^p + C_2(h_j)h_j^p + D(h_i, h_j)h_i^p h_j^p$$

where $|C_1|$, $|C_2|$ and $|D|$ are bounded above by some positive constant κ. The error of $\tilde{f}^t_{n-t,n-t}$ for some fixed $t > 0$ is bounded by

$$|f - \tilde{f}^t_{n-t,n-t}| \leq 2\kappa h_n^p + \mathcal{O}((h_n^p)^2)\,.$$

Proof. We have

$$\begin{aligned}
f - \tilde{f}^t_{n-t,n-t} &= \textstyle\sum_{i=0}^t (f - f_{n-t+i,n-i}) - \sum_{i=0}^{t-1}(f - f_{n-t+i,n-1-i}) \\
&= \textstyle\sum_{i=0}^t (C_1(h_{n-t}2^{-i})h^p_{n-t}2^{-ip} + C_2(h_{n-t}2^{-t+i})h^p_{n-t}2^{(-t+i)p} \\
&\qquad + D(h_{n-t}2^{-i}, h_{n-t}2^{-t+i})h^p_{n-t}h^p_{n-t}2^{-tp}) \\
&\quad - \textstyle\sum_{i=0}^{t-1}(C_1(h_{n-t}2^{-i})h^p_{n-t}2^{-ip} + C_2(h_{n-t}2^{-t+1+i})h^p_{n-t}2^{(-t+1+i)p} \\
&\qquad + D(h_{n-t}2^{-i}, h_{n-t}2^{-t+1+i})h^p_{n-t}h^p_{n-t}2^{(-t+1)p}) \\
&= C_1(h_{n-t}2^{-t})h^p_{n-t}2^{-tp} + C_2(h_{n-t}2^{-t})h^p_{n-t}2^{-tp} \\
&\quad + \mathcal{O}(2^{-(t+1)p}(h^p_{n-t})^2) \\
&= C_1(h_n)h_n^p + C_2(h_n)h_n^p + \mathcal{O}((h_n^p)^2)
\end{aligned}$$

and therefore

$$\begin{aligned}
|f - \tilde{f}^t_{n-t,n-t}| &\leq |C_1(h_n)|h_n^p + |C_2(h_n)|h_n^p + |\mathcal{O}((h_n^p)^2)| \\
&\leq 2\kappa h_n^p + \mathcal{O}((h_n^p)^2)\,.
\end{aligned}$$

This implies that the rate of convergence is the same as that for the solution $f_{n,n}$. One can extend this result to the three-dimensional case with the error splitting of (4) to find $|f - \tilde{f}^t_{n-t,n-t,n-t}| \leq 3\kappa h_n^p + \mathcal{O}((h_n^p)^2)$.

Another advantage of these combinations is that for each n the structure of the combination is essentially the same making the analysis simpler as we begin to study the effect of faults on the combination technique.

We also remark that such combinations may work well for problems where the classical sparse grid approximation is poor. For example, since for increasing n there is an increasing minimum "fineness" of the solution along each axis one may still resolve problems with discontinuities or non-linear interactions. This minimum "fineness" also means we can try extrapolation techniques onto the largest grid common to those in the approximation, this will be discussed in Sect. 2.

Effectively we are approximating something very close to the full grid solution rather than the classical sparse grid solution, however by using multiple grids in a way similar to the combination technique we will be able to add some robustness to our computations as we will now discuss.

1.2 Combining in the Event of Faults

Suppose we have an application running on HPC architecture using MPI for communications. Generally, if a processor experiences a failure whilst the application is running an exception will be raised by MPI and the entire application will be aborted. This problem is often handled by periodically saving the state of the entire application (checkpointing), if a fault occurs the application is simply restarted from the last successful checkpoint. This is infeasible for exascale machines because checkpointing is too memory and communication intensive making it too costly in terms of computational resources, time and energy costs.

Given a problem decomposed into one of the combinations described above, we note that the computation of each solution is independent. Therefore, if a fault affects the computation of any one of these solutions, one could avoid the global abort and the computation of the remaining solutions can continue as usual. The solution which has failed can be rescheduled with the computation starting from the beginning. Since recomputation can lead to load balancing issues one may consider doing without that solution and utilising only those which remain. This approach raises the following question: how does one combine the grids if one of them is missing? This question motivates the discussion throughout the remainder of this section.

We will first consider in detail the truncated combination of (7). Given the five solutions in this combination we will go through the alternative approximations one may take when a failure affects any one of these solutions. For a failure affecting the solutions $f_{n-2,n}$ or $f_{n-2,n-1}$ one can instead compute the combination $f_{n,n-2} + f_{n-1,n-1} - f_{n-1,n-2}$. Effectively this gives an approximation in the space

$$V_{n,n-1} = W_{n,n-1} \oplus \left(\bigoplus_{i \leq n\,,\, j \leq n-1\,,\, i+j \leq 2n-2} W_{i,j} \right) .$$

Compared to the normal case as in (9) we see that some information from the difference spaces $W_{i,n}$ for $i \leq n-2$ has been lost. For large n one would expect that these have a very small contribution to the solution and therefore the loss of these should only have a small impact on the error.

Similarly for faults on $f_{n,n-2}$ or $f_{n-1,n-2}$ one can instead apply the combination $f_{n-2,n} + f_{n-1,n-1} - f_{n-2,n-1}$. Similar to before we are left with an approximation in the space

$$V_{n-1,n} = W_{n-1,n} \oplus \left(\bigoplus_{i \leq n-1\,,\, j \leq n\,,\, i+j \leq 2n-2} W_{i,j} \right) .$$

The last case to consider is a fault on $f_{n-1,n-1}$. The simplest thing to do in this case is to simply take $f_{n-2,n}$ or $f_{n,n-2}$ to be the solution. A fault affecting this solution is a worse case scenario as we will see in Proposition 2 which bounds the error of these alternate combinations. Figure 2 illustrates some of combinations described when faults affect the occur during the calculation of $\tilde{f}^2_{3,3}$.

Proposition 2. *Suppose that $f_{i,j}$ satisfies the pointwise error splitting as in Proposition 1. Let g_n be any one of the combinations above when a fault affects one of the component solutions from the combination in (7), that is,*

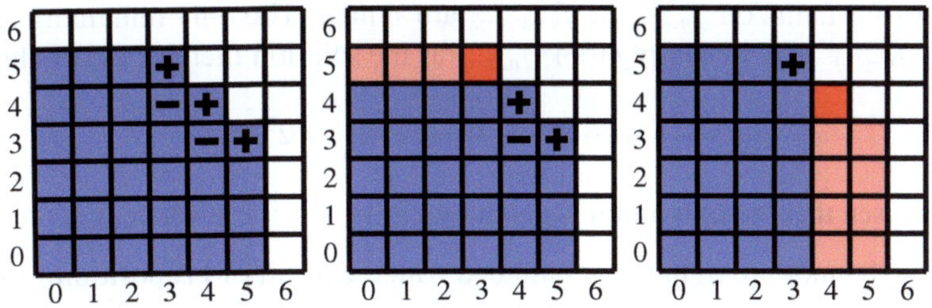

Fig. 2 Here we illustrate how a fault effects the combination $\tilde{f}^2_{3,3}$. On the *left* we have the combination when no faults occur. In the *middle* we have a possible combination if a fault occurs during the calculation of $f_{3,5}$ and/or $f_{3,4}$. On the *right* we have a possible combination if a fault occurs during the calculation of $f_{4,4}$. The *blue shaded squares* indicate the difference spaces $W_{i,j}$ which contribute to the combined solution whilst the *red shaded squares* indicate the difference spaces we lose in the modified combinations

$$g_n = \begin{cases} f_{n,n-2} + f_{n-1,n-1} - f_{n-1,n-2} & \textit{for a fault on } f_{n-2,n} \textit{ or } f_{n-2,n-1} \\ f_{n-2,n} + f_{n-1,n-1} - f_{n-2,n-1} & \textit{for a fault on } f_{n,n-2} \textit{ or } f_{n-1,n-2} \\ f_{n,n-2} \textit{ or } f_{n-2,n} & \textit{for a fault on } f_{n-1,n-1}\,. \end{cases}$$

The (pointwise) error of g_n is bounded by

$$|f - g_n| \leq \kappa h_n^p (1 + 2^{2p}) + \mathcal{O}((h_n^p)^2)\,.$$

Proof. Consider the case for a fault on $f_{n-2,n}$ or $f_{n-2,n-1}$, one has

$$\begin{aligned}
& f - (f_{n,n-2} + f_{n-1,n-1} - f_{n-1,n-2}) \\
&= (f - f_{n,n-2}) + (f - f_{n-1,n-1}) - (f - f_{n-1,n-2}) \\
&= C_1(h_n)h_n^p + C_2(h_{n-2})h_{n-2}^p + D(h_n, h_{n-2})h_n^p h_{n-2}^p \\
&\quad + C_1(h_{n-1})h_{n-1}^p + C_2(h_{n-1})h_{n-1}^p + D(h_{n-1}, h_{n-1})h_{n-1}^p h_{n-1}^p \\
&\quad - (C_1(h_{n-1})h_{n-1}^p + C_2(h_{n-2})h_{n-2}^p + D(h_{n-1}, h_{n-2})h_{n-1}^p h_{n-2}^p) \\
&= C_1(h_n)h_n^p + C_2(h_{n-1})h_{n-1}^p \\
&\quad + D(h_n, h_{n-2})h_n^p h_{n-2}^p + D(h_{n-1}, h_{n-1})h_{n-1}^p h_{n-1}^p - D(h_{n-1}, h_{n-2})h_{n-1}^p h_{n-2}^p
\end{aligned}$$

and therefore

$$\begin{aligned}
& |f - (f_{n,n-2} + f_{n-1,n-1} - f_{n-1,n-2})| \\
&\leq |C_1(h_n)|h_n^p + |C_2(h_{n-1})|2^p h_n^p \\
&\quad + |D(h_n, h_{n-2}) + D(h_{n-1}, h_{n-1}) - 2^p D(h_{n-1}, h_{n-2})|2^{2p} h_n^{2p} \\
&\leq \kappa h_n^p (1 + 2^p) + \kappa(2 + 2^p)2^{2p} h_n^{2p}) \\
&= \kappa h_n^p (1 + 2^p) + \mathcal{O}((h_n^p)^2)
\end{aligned}$$

The cases for faults on $f_{n,n-2}$ or $f_{n-1,n-2}$ are similar. The only remaining case is a fault on $f_{n-1,n-1}$. If we simply took $f_{n,n-2}$ as the solution then we clearly have

$$|f - f_{n,n-2}| \leq \kappa h_n^p(1 + 2^{2p}) + \kappa 2^{2p} h_n^{2p} .$$

Hence we see that the maximum is given by $\kappa h_n^p(1 + 2^{2p}) + \mathcal{O}((h_n^p)^2)$.

If these bounds are tight then combined with the result of Proposition 1 one may expect that in the limit of large n one has $|f - g_n| \lessapprox \frac{1+2^{2p}}{2} |f - \tilde{f}^2_{n-2,n-2}|$.

Remark 1. We stop here to note it is possible to do a little better for some of these cases. For example, given the case of a fault affecting the solution $f_{n-2,n-1}$ we note that one can still make some use of the solution $f_{n-2,n}$. We must first define the interpolation (or projection) operator $I_{\underline{i}}$. Given any $f \in V$ then

$$I_{\underline{i}} : f \in V \mapsto I_{\underline{i}} f \in V_{\underline{i}} ,$$

that is, $I_{\underline{i}} f$ interpolates (projects) f to a function defined by its values on the grid $\Omega_{\underline{i}}$. Now given an interpolation $I_{n-2,n-1} f_{n-2,n}$ of the solution $f_{n-2,n}$ onto the space $V_{n-2,n-1}$, one may then add $f_{n-2,n} - I_{n-2,n-1} f_{n-2,n}$ to the combination $f_{n,n-2} + f_{n-1,n-1} - f_{n-1,n-2}$. One sees that $f_{n-2,n} - I_{n-2,n-1} f_{n-2,n}$ extracts the hierarchical components $W_{i,n}$ with $0 \leq i \leq n-2$ from the solution $f_{n-2,n}$. As a result we are effectively recovering contributions from theses hierarchical spaces. Similarly may be done for a fault on $f_{n-1,n-2}$ via symmetry. The other case is a fault affecting $f_{n-1,n-1}$ where one may take $f_{n,n-2}$ as the solution and then add to it $f_{n-2,n} - I_{n-2,n-2} f_{n-2,n}$. Given an appropriate consistency property between the interpolation operator and approximation scheme, for example $\|f - I_{i,j} f_{k,l}\|_\infty \leq C \|f - f_{i,j}\|_\infty$ for some positive constant C and all $k \geq i$ and $l \geq j$, then one may prove error bounds similar to those of Proposition 2.

One may also enumerate the cases for the other combinations presented. For (8) one has the following:

- If $f_{n-3,n}$ and/or $f_{n-3,n-1}$ are lost due to faults one may take the combination $f_{n,n-3} + f_{n-1,n-2} + f_{n-2,n-1} - f_{n-1,n-3} - f_{n-2,n-2}$ [compare with (7)]. Here we lose contributions from the difference spaces $W_{i,n}$ for $i \leq n-3$.
- Similarly, if $f_{n,n-3}$ and/or $f_{n-1,n-3}$ are lost due to faults then one may take the combination $f_{n-3,n} + f_{n-2,n-1} + f_{n-1,n-2} - f_{n-3,n-1} - f_{n-2,n-2}$. Here we lose contributions from the difference spaces $W_{n,i}$ for $i \leq n-3$.
- If a fault affects $f_{n-1,n-2}$ or $f_{n-2,n-1}$ then one may take the combinations $f_{n-3,n} + f_{n-2,n-1} - f_{n-3,n-1}$ or $f_{n,n-3} + f_{n-1,n-2} - f_{n-1,n-3}$ respectively. For a fault on $f_{n-1,n-2}$ we see that the new combination does not have contributions from the difference spaces $W_{n-1,i}$ for $i \leq n-2$ and $W_{n,j}$ for $j \leq n-3$. Similar applies for a fault on $f_{n-2,n-1}$.
- Finally, if a fault affects $f_{n-2,n-2}$ then one may take combinations from the previous case, i.e. $f_{n,n-3} + f_{n-1,n-2} - f_{n-1,n-3}$ or $f_{n-3,n} + f_{n-2,n-1} - f_{n-3,n-1}$.

Again we have a bound on the errors for these combinations.

Proposition 3. *Suppose that $f_{i,j}$ satisfies the same pointwise error splitting as in Proposition 1. Let g_n be any one of the combinations above when a fault affects one of the component solutions from the combination in (8), that is, depending on which component the fault occurred, g_n is one of the functions*

$$g_n = \begin{cases} f_{n,n-3} + f_{n-1,n-2} + f_{n-2,n-1} - f_{n-1,n-3} - f_{n-2,n-2} \\ f_{n-3,n} + f_{n-2,n-1} + f_{n-1,n-2} - f_{n-3,n-1} - f_{n-2,n-2} \\ f_{n-3,n} + f_{n-2,n-1} - f_{n-3,n-1} \text{ or } f_{n,n-3} + f_{n-1,n-2} - f_{n-1,n-3} \\ f_{n,n-3} + f_{n-1,n-2} - f_{n-1,n-3} \text{ or } f_{n-3,n} + f_{n-2,n-1} - f_{n-3,n-1} \,. \end{cases}$$

The (pointwise) error of any of the four g_n is bounded by

$$|f - g_n| \leq \kappa h_n^p (1 + 2^{2p}) + \mathcal{O}((h_n^p)^2) \,.$$

Proof. The proof is similar to that of the previous proposition. It is straightforward to show that

$$\begin{aligned} &|f - (f_{n,n-3} + f_{n-1,n-2} + f_{n-2,n-1} - f_{n-1,n-3} - f_{n-2,n-2})| \\ &\leq |C_1(h_n)| h_n^p + |C_2(h_{n-1})| 2^p h_n^p \\ &\quad + |D(h_n, h_{n-3}) + D(h_{n-1}, h_{n-2}) + D(h_{n-2}, h_{n-1})| 2^{3p} h_n^{2p} \\ &\quad + |D(h_{n-1}, h_{n-3}) + D(h_{n-2}, h_{n-2})| 2^{4p} h_n^{2p} \\ &\leq \kappa h_n^p (1 + 2^p) + \kappa (3 + 2^{p+1}) 2^{3p} h_n^{2p}) \\ &= \kappa h_n^p (1 + 2^p) + \mathcal{O}((h_n^p)^2) \end{aligned}$$

and

$$\begin{aligned} &|f - (f_{n,n-3} + f_{n-1,n-2} - f_{n-1,n-3})| \\ &\leq |C_1(h_n)| h_n^p + |C_2(h_{n-1})| 2^{2p} h_n^p \\ &\quad + |D(h_n, h_{n-3}) + D(h_{n-1}, h_{n-2})| 2^{3p} h_n^{2p} + |D(h_{n-1}, h_{n-3})| 2^{4p} h_n^{2p} \\ &\leq \kappa h_n^p (1 + 2^{2p}) + \kappa (2 + 2^p) 2^{3p} h_n^{2p}) \\ &= \kappa h_n^p (1 + 2^{2p}) + \mathcal{O}((h_n^p)^2) \,. \end{aligned}$$

Since similar results apply to the remaining cases due to symmetry of the combinations one can see that the maximal error is $\kappa h_n^p (1 + 2^{2p}) + \mathcal{O}((h_n^p)^2)$.

Again, if the bounds are tight then in the limit of large n one may expect to observe $|f - g_n| \lesssim \frac{1+2^{2p}}{2} |f - \tilde{f}^2_{n-2,n-2}|$. One can see that if a fault occurs when using the grids of (8) then at least three of the remaining six grids can be utilised in a combination. This is somewhat nicer than the case of (7) where in one scenario only one of the remaining four grids can be utilised. Despite this the error bound is the same.

Remark 2. Similar to the previous set of combinations, one may add hierarchical components from remaining unused grids to some of the cases presented above. For example, in the case of a fault on $f_{n-1,n-3}$ one may take the combination

$f_{n-2,n-1} + f_{n-3,n} - f_{n-3,n-1}$ and add to it $f_{n,n-3} - I_{n-1,n-3} f_{n,n-3}$. In doing this one is including contributions from the difference spaces $W_{n,i}$ for $i \leq n-3$. Similar can be done for many of the remaining cases. Again one may prove error bounds similar to those above given a consistency property between the interpolation operator $I_{i,j}$ and the discrete solutions $f_{i,j}$.

In order to demonstrate how similar approaches can be applied in higher dimensions we will also enumerate the cases for the three-dimensional combination of (10).

- If $f_{n-2,n-2,n}$ is lost one may take the combination

$$\begin{aligned} g_n = {} & f_{n,n-2,n-2} + f_{n-1,n-1,n-2} + f_{n-2,n-1,n-1} + f_{n-1,n-2,n-1} + f_{n-2,n,n-2} \\ & -2(f_{n-1,n-2,n-2} + f_{n-2,n-1,n-2}) - f_{n-2,n-2,n-1} + f_{n-2,n-2,n-2}. \end{aligned} \tag{12}$$

  The information from the difference spaces $W_{i,j,n}$ for $i, j \leq n-2$ is lost. Similarly for $f_{n,n-2,n-2}$ and $f_{n-2,n,n-2}$ (one simply permutes the indices).
- If $f_{n-2,n-1,n-1}$ is lost one may take the combination

$$\begin{aligned} g_n = {} & f_{n,n-2,n-2} + f_{n-1,n-1,n-2} + f_{n-2,n-2,n} + f_{n-1,n-2,n-1} + f_{n-2,n,n-2} \\ & -2f_{n-1,n-2,n-2} - f_{n-2,n-1,n-2} - f_{n-2,n-2,n-1}. \end{aligned} \tag{13}$$

  Here we lose contributions from the difference spaces $W_{i,n-1,n-1}$ for $i \leq n-2$. Similarly for $f_{n-1,n-2,n-1}$ and $f_{n-1,n-1,n-2}$ (again, simply permute the indices).
- If $f_{n-2,n-2,n-1}$ is lost then there are many combinations that can be chosen. One such combinations which utilises the most solutions is

$$\begin{aligned} g_n = {} & f_{n-2,n-2,n} + f_{n-2,n,n-2} + f_{n,n-2,n-2} + f_{n-1,n-1,n-2} \\ & - f_{n-1,n-2,n-2} - f_{n-2,n-1,n-2} - f_{n-2,n-2,n-2}. \end{aligned} \tag{14}$$

  For this combination one loses contributions from the difference spaces $W_{i,n-1,n-1}$ for $i \leq n-2$ and $W_{n-1,i,n-1}$ for $i \leq n-2$. Similar applies for faults affecting $f_{n-1,n-2,n-2}$ or $f_{n-2,n-1,n-2}$ (again one simply permutes the indices).
- If $f_{n-2,n-2,n-2}$ is lost then again there are many options. One such is to take

$$\begin{aligned} g_n = {} & f_{n-2,n-2,n} + f_{n-2,n,n-2} + f_{n,n-2,n-2} + f_{n-2,n-1,n-1} + f_{n-1,n-1,n-2} \\ & - f_{n-1,n-2,n-2} - 2f_{n-2,n-1,n-2} - f_{n-2,n-2,n-1}. \end{aligned} \tag{15}$$

  One loses contributions from the difference spaces $W_{i,n-1,n-1}$ for $i \leq n-2$ in this case. Two similar combinations can be obtained simply by permuting the indices.

We have a bound on the error for any of these combinations.

Proposition 4. *Suppose that $f_{i,j,k}$ satisfies the pointwise error splitting of (4). Let g_n be any one of the combinations above when a fault affects one of the solutions from the combination in (10), that is, g_n is equal to one of (12), (13), (14) or (15) up to a permutation of indices. The (pointwise) error of g_n is bounded by*

$$|f - g_n| \leq \kappa h_n^p (2 + 2^p) + \mathcal{O}((h_n^p)^2) .$$

Proof. The proof is very similar to those in the two-dimensional case. The maximal error is given for the case of a fault affecting one of $f_{n,n-2,n-2}$, $f_{n-2,n,n-2}$ or $f_{n-2,n-2,n}$ and is given by

$$\begin{aligned} f - (& f_{n,n-2,n-2} + f_{n-1,n-1,n-2} + f_{n-2,n-1,n-1} + f_{n-1,n-2,n-1} + f_{n-2,n,n-2} \\ & -2(f_{n-1,n-2,n-2} + f_{n-2,n-1,n-2}) - f_{n-2,n-2,n-1} + f_{n-2,n-2,n-2}) \\ = {} & C_1(h_n)h_n^p + C_2(h_n)h_n^p + C_3(h_{n-1})h_{n-1}^p + \mathcal{O}((h_n^p)^2) \end{aligned}$$

leading to the bound $\kappa(2 + 2^p)h_n^p$.

Given this result, if the bounds are tight then in the limit of large n one may expect to observe $|f - g_n| \lessapprox \frac{2+2^p}{3}|f - \tilde{f}^2_{n-2,n-2,n-2}|$. We can see that this bound appears to be much better compared to the two-dimensional case. This is due to the extra grids which give rise to many more ways one may combine solutions. In fact, one should note that for many of the cases of faults, the bound on error using the combinations presented is $3\kappa h_n^p + \mathcal{O}((h_n^p)^2)$, i.e. it is the same as a combination without faults for the leading error terms. Given the flexibility from having extra grids we also note it is often possible to obtain good solutions even when two solutions are affected by faults.

Such results can be easily extended to the combination of (11) but we will not enumerate these cases here.

Remark 3. PDE's which have a derivative with respect to time can be more difficult to handle. Many modern solvers use automatic time stepping schemes that choose a stable time step providing results within some desired error tolerance. As a result we will not analyse the truncation errors dependant on time stepping, we will simply assume that the time stepping scheme is of higher order then the spatial discretisations. Under this assumption the error from spatial truncation errors dominates and we may apply the combinations presented above.

1.3 Results for Combinations When Faults Occur

We will demonstrate the trade off in errors for not having to recompute for a variety of problems. In particular we will consider the scalar advection equation

$$\frac{\partial f}{\partial t} + \underline{a} \cdot \nabla f = 0$$

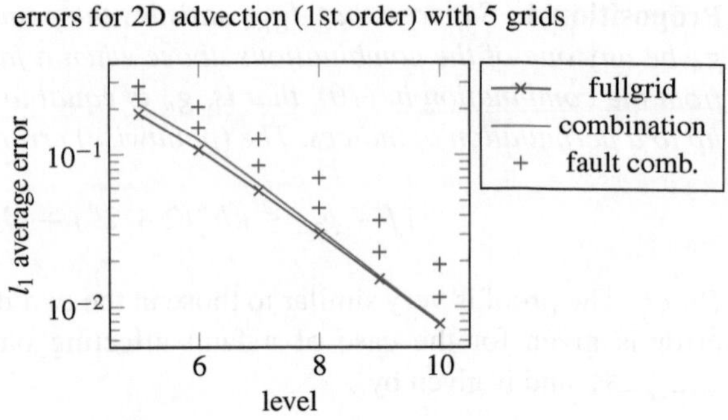

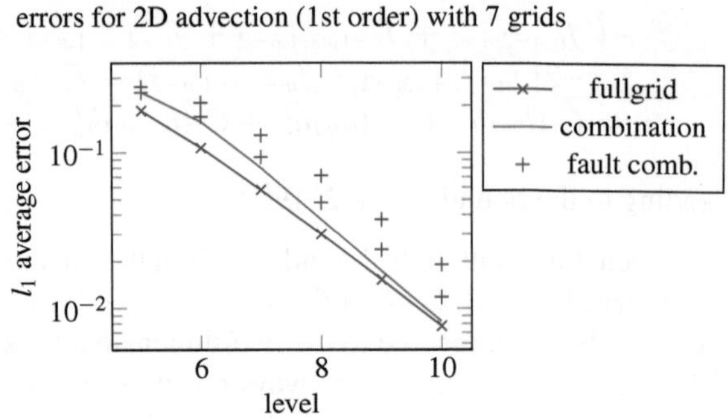

Fig. 3 We demonstrate the errors obtained after altering the combination when a fault occurs for the 2D advection problem. Given a level n, the five and seven grid combinations are $\tilde{f}^2_{n-2,n-2}$ and $\tilde{f}^3_{n-3,n-3}$ respectively. The full grid we compare with is $f_{n,n}$ and the fault combinations are those discussed in Sect. 1.2. Errors are computed on the space $V_{n,n}$ in each case

in two spatial dimensions with $\underline{a} = (1, 1)$, exact solution $f(x, y, t) = sin(2\pi(x - t))sin(2\pi(y - t))$ in the unit square $[0, 1]^2$ and periodic boundary conditions. Given $f(x, y, 0)$ we solve up to $t = 0.5$. High order Runge–Kutta methods are used for time stepping and we experiment with first order spatial discretisations. We run the solution to completion on each grid combining only once at the end.

We solve the same problem in three spatial dimensions with $a = (1, 1, 1)$, exact solution $f(x, y, z, t) = \sin(2\pi(x-t)) \sin(2\pi(y-t)) \sin(2\pi(z-t))$ in the unit cube $[0, 1]^3$ and periodic boundary conditions. Again we start with the initial condition at $t = 0$ and solve up to $t = 0.5$ using high order Runge–Kutta time stepping and second order spatial discretisations.

We will also give results for the two-dimensional Laplace problem $\Delta f = 0$ with the exact solution $f(x, y) = \sin(\pi y)\frac{\sinh(\pi(1-x))}{\sinh \pi}$ in the unit square $[0, 1]^2$ and Dirichlet boundary conditions as in [11].

Each of these problems has been implemented using PETSc [1–3] to compute each of the individual solutions and then with the combinations being done in Python.

In Figs. 3 and 4 we demonstrate the average l_1 error obtained for some different problems using many of the different combinations of solutions as discussed. In the first figure we have some results for the two-dimensional advection problem using the combination of (7). We compare the combination solution with the full grid solution ($f_{n,n}$) and also plot the error for each case of a fault affecting one of the five solutions (some of these are the same due to symmetries of the problem).

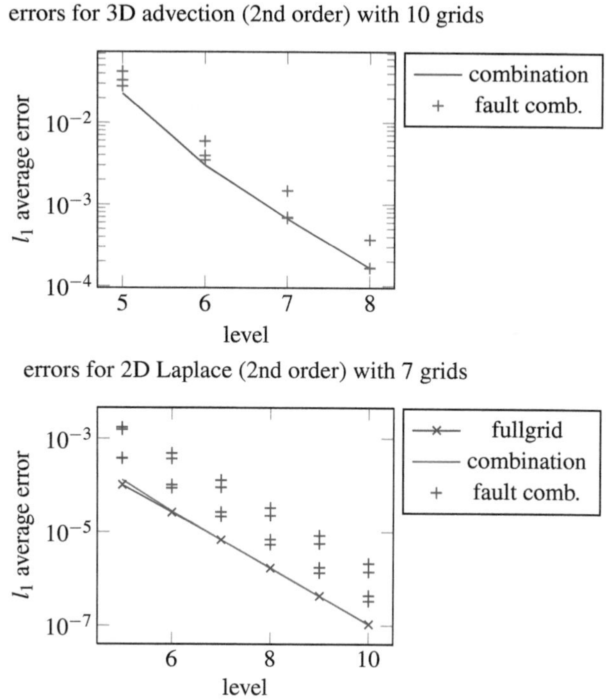

Fig. 4 We demonstrate the errors obtained after altering the combination when a fault occurs for a 3D advection problem and a 2D Laplace problem. Given a level n the combination for the 3D problem is $\tilde{f}^2_{n-2,n-2,n-2}$ and the fault combinations are those discussed in Sect. 1.2. The error in the 3D problem is computed on the space of the truncated approximation. The details for the 2D problem are the same as those of the seven grid advection problem in Fig. 3

We see that the error from the combination appears to converge towards that from the full grid solution (i.e. $f_{n,n}$). Further, the combination after a fault has occurred converges at the same rate and appears to be bounded by a constant multiple of the error from the normal combination where that constant depends on which solution the fault occurred. This is consistent with the results of Proposition 2.

The second plot is the error for the same PDE but using the solutions as in (8) as the base solutions. The observations are similar. The third plot is for the three-dimensional advection equation with second order discretisations. Here we have not included the full grid error as it is too expensive to compute but one observes similar bounds on the error when the combination is altered after a fault occurs. The last plot is the two-dimensional Laplace problem using the seven solutions as in (8) which are solved using a second order spatial discretisations. The errors in this plot after a fault has occurred appear to be quite a bit larger but are still bounded by a constant multiple of the errors when no faults occur.

2 Richardson Extrapolation and Related Ideas

Classical Richardson extrapolation involves taking two solutions $f_{n,n}$ and $f_{n+1,n+1}$ and combining them with the formula

$$\frac{2^p}{2^p-1} I_{n,n} f_{n+1,n+1} + \frac{1}{1-2^p} f_{n,n}$$

in order to obtain a higher order solution. It is important to note that this result requires a slightly different error estimate than the error splitting of (3). In particular we require the pointwise multivariate asymptotic expansion

$$\gamma_{i,j} = f - f_{i,j} = e^{p,0} h_i^p + e^{0,p} h_j^p + e^{p,p} h_i^p h_j^p + e^{2p,0} h_i^{2p} + e^{0,2p} h_j^{2p} + \cdots \quad (16)$$

where each $e^{m,n}$ term depends only upon x and y. A similar error splitting applies for three spatial dimensions. One can see the first terms are similar to those in the previous error splitting except that the constants are independent of h_i and h_j. We also assume that the interpolation operator I preserves the pth order terms of the asymptotic expansion.

In particular we are concerned with solutions smooth enough to permit first and second order schemes satisfying this expansion for $p = 1$ and $p = 2$ respectively. It is well known that with suitable assumptions on regularity then the solution of a variety of elliptic problems has such an expansion for $p = 2$. However, Richardson extrapolation is computationally expensive as one must calculate the solution on a grid four times the size in two-dimensions and eight times the size in three-dimensions.

2.1 Multivariate Richardson Extrapolation

Multivariate extrapolation has been studied by Rüde [14] and Bungartz et al. [7]. For example, given the error splitting in (16) with $p = 2$ we can extrapolate by taking the combination

$$\frac{4}{3} I_{n,n} f_{n+1,n} + \frac{4}{3} I_{n,n} f_{n,n+1} - \frac{5}{3} f_{n,n} .$$

It has been shown that multivariate Richardson extrapolation is much cheaper than classical Richardson extrapolation for increasing dimension and increasing extrapolation order [7].

In this simple two-dimensional example it is still expensive requiring two extra grids which are twice the size. We note however that one may make additional use of the information in these grids. Given the combination $g \approx f_{n,n+1} + f_{n+1,n} - f_{n,n}$ one could add the high frequency components from g to the extrapolation above. That way one has a high order extrapolation of the solution for the coarser hierarchical spaces as well as a low order approximation of contributions from some finer hierarchical spaces. Assuming the contributions from these are sufficiently small for increasing n then one should still see high order convergence with respect to n. By doing this one can make the most out of the given solutions. This is discussed further in Sect. 3.

We are concerned with having solutions from combinations like those of (7). This raises the question: what is the best way to combine these grids to obtain a similar extrapolation? Two possible choices for $p = 2$ are:

$$\frac{4}{3} I_{n-2,n-1} f_{n-1,n-1} + \frac{4}{3} I_{n-2,n-1} f_{n-2,n} - \frac{5}{3} f_{n-2,n-1}$$

or

$$\frac{4}{3} I_{n-1,n-2} f_{n,n-2} + \frac{4}{3} I_{n-1,n-2} f_{n-1,n-1} - \frac{5}{3} f_{n-1,n-2} .$$

These extrapolate onto the grids $\Omega_{n-2,n-1}$ and $\Omega_{n-1,n-2}$ respectively. Another option is to extrapolate onto the largest grid which is contained within all five grids. It is easily shown that the combination

$$-\frac{4}{9} I_{n-2,n-2} f_{n-2,n} + \frac{17}{9} I_{n-2,n-2} f_{n-1,n-1} - \frac{4}{9} I_{n-2,n-2} f_{n,n-2}$$

is a multivariate extrapolation onto the grid $\Omega_{n-2,n-2}$. There are also other possibilities where one may use all of the five grids but this is convenient computationally given it uses only 3.

In general one may obtain coefficients for such extrapolations by using the error expansion in (16) to form the equation

$$\begin{aligned} & a_1(f + (f_{n-2,n} - f)) + a_2(f + (f_{n-1,n-1} - f)) + a_3(f + (f_{n,n-2} - f)) \\ & + a_4(f + (f_{n-2,n-1} - f)) + a_5(f + (f_{n-1,n-2} - f)) \\ & = f + 0e^{p,0} + 0e^{0,p} + \mathcal{O}((h_n^p)^2) \end{aligned} \tag{17}$$

which leads to the system of equations

$$\begin{aligned} & a_1 + a_2 + a_3 + a_4 + a_5 = 1 \\ & a_1 h_{n-2}^p + a_2 h_{n-1}^p + a_3 h_n^p + a_4 h_{n-2}^p + a_5 h_{n-1}^p = 0 \\ & a_1 h_n^p + a_2 h_{n-1}^p + a_3 h_{n-2}^p + a_4 h_{n-1}^p + a_5 h_{n-2}^p = 0 . \end{aligned}$$

The problem here is that the system is under determined. One can manage this by assuming some symmetry in combination coefficients, for example $a_1 = a_3$ and $a_4 = a_5$. Alternatively one may attempt to extrapolate further terms in the error expansion, for example the $e^{2p,0}$ and $e^{0,2p}$ terms. If the problem at hand is formulated as the minimum of a functional then one could substitute the combination formula and apply the above equations as linear constraints. Another option is just to pick a few solutions one would like to work with and see if there's a suitable solution. The problem of the under determined system of equations becomes worse for the example of (8) and worse still for the three-dimensional examples of (10) and (11) since there are even more solutions and hence coefficients.

It is important to note that one must be careful since whilst one may obtain coefficients that extrapolate away the $e^{p,0}$ and $e^{0,p}$ terms, it may significantly increase the constants on the higher order terms such that the result is poor unless n is sufficiently large. It is also known that extrapolation is unstable. Despite this we will demonstrate how extrapolation may still be applied when faults occur.

2.2 *Fault-Tolerant Multivariate Extrapolation*

The idea is that when a fault occurs affecting one of the solutions then we may solve the set of linear equations similar to those derived from (17) using only the solutions which remain. For the example of the solutions computed as in (7) it is straightforward. For a fault affecting $f_{n-1,n-2}$ or $f_{n-2,n-1}$ one may still use the extrapolation $-\frac{4}{9}I_{n-2,n-2}f_{n-2,n} + \frac{17}{9}I_{n-2,n-2}f_{n-1,n-1} - \frac{4}{9}I_{n-2,n-2}f_{n,n-2}$ for $p = 2$. In general one has

$$\frac{1}{(2^p-1)^2}I_{n-2,n-2}\left(-2^p f_{n-2,n} + (2^{2p}+1)f_{n-1,n-1} - 2^p f_{n,n-2}\right) . \quad (18)$$

If a fault affects the solution $f_{n,n-2}$ then we note that one may take the extrapolation

$$\frac{2^p}{2^p-1}I_{n-2,n-1}f_{n-2,n} - \frac{2^p+1}{2^p-1}f_{n-2,n-1} + \frac{2^p}{2^p-1}I_{n-2,n-1}f_{n-1,n-1} \quad (19)$$

onto the space $V_{n-2,n-1}$. By symmetry a similar extrapolation can be obtained when a fault affects $f_{n-2,n}$. This leaves us with the last case of a fault on $f_{n-1,n-1}$. We have

$$\begin{aligned} &a_1 f_{n-2,n} + a_2 f_{n,n-2} + a_3 f_{n-2,n-1} + a_4 f_{n-1,n-2} \\ &= f + 0e^{p,0} + 0e^{0,p} + \mathcal{O}((h_n^p)^2) \end{aligned}$$

which we solve along with the assumption that $a_5 = a_4$ and $a_3 = a_1$. This results in the multivariate extrapolation formula

$$\begin{aligned} &\tfrac{2^p+2^{2p}}{2^{p+1}-2}(I_{n-2,n-2}f_{n-2,n} + I_{n-2,n-2}f_{n,n-2}) \\ &-\tfrac{2^{2p}+1}{2^{p+1}-2}(I_{n-2,n-2}f_{n-2,n-1} + I_{n-2,n-2}f_{n-1,n-2}) . \end{aligned}$$

Similar extrapolation formulas may be applied where faults occur for the solutions as in (8). In fact one can simply shift the extrapolation formula of (18) and/or (19) in order to avoid the solution that was affected by a fault.

In three-dimensions there are many different ways one may add the solutions to obtain an extrapolation. The simplest approach is to take the simple multivariate extrapolation formula

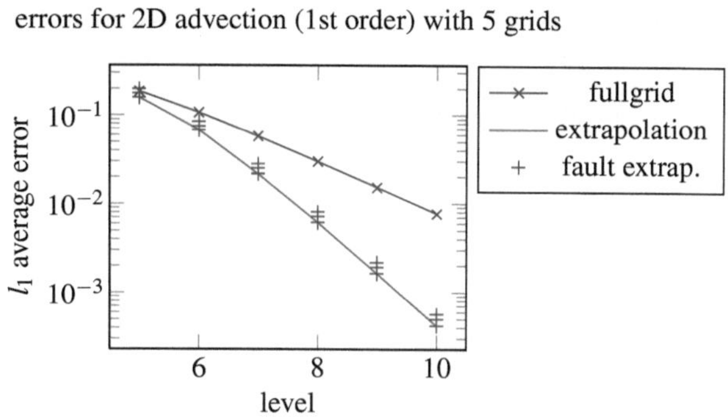

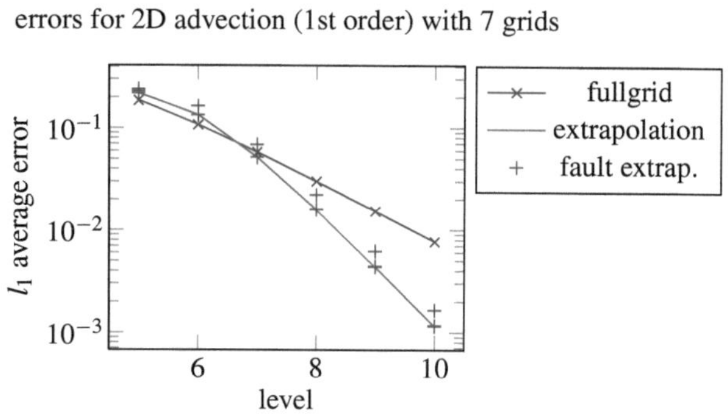

Fig. 5 We demonstrate the errors obtained after altering the extrapolation when a fault occurs for a 2D advection problem. Given a level n, for the five and seven grid plots we form an extrapolation onto the spaces $V_{n-2,n-2}$ and $V_{n-3,n-3}$ respectively. The solutions used in the extrapolation are those computed for the combinations $\tilde{f}^2_{n-2,n-2}$ and $\tilde{f}^3_{n-3,n-3}$ respectively. The full grid solution we compare with is $f_{n,n}$. The fault extrapolations are those described in Sect. 2.2. The error in the extrapolation is computed on the spaces $V_{n-2,n-2}$ and $V_{n-3,n-3}$ for the first and second plots respectively whilst for the full grid the error is computed on the space $V_{n,n}$

$$\frac{2^p}{2^p-1}(I_{i,j,k}f_{i+1,j,k} + I_{i,j,k}f_{i,j+1,k} + I_{i,j,k}f_{i,j,k+1}) - \frac{2^{p+1}+1}{2^p-1}f_{i,j,k}$$

and shift it to a combination which avoids the solution which was affected by a fault.

Remark 4. We remarked previously that it is difficult to consider truncation errors dependant on the time stepping given the common use of automatic time stepping schemes. For extrapolation we again assume that the time stepping is of an order larger than the spatial discretisations. Optimally one would like at least order $2p$ such that the extrapolation of the p order spatial errors results in an order $2p$ method.

2.3 *Results for Extrapolation*

In Figs. 5 and 6 we demonstrate the results for extrapolation using the same examples as in Sect. 1.3. For example, in the first plot of Fig. 3 we have the two-dimensional scalar advection equation where we compare the full grid solution $f_{n,n}$ with the extrapolation of the solutions computed for the combination $\tilde{f}^2_{n-2,n-2}$ onto the space $V_{n-2,n-2}$. We observe that the extrapolation is much better than the full

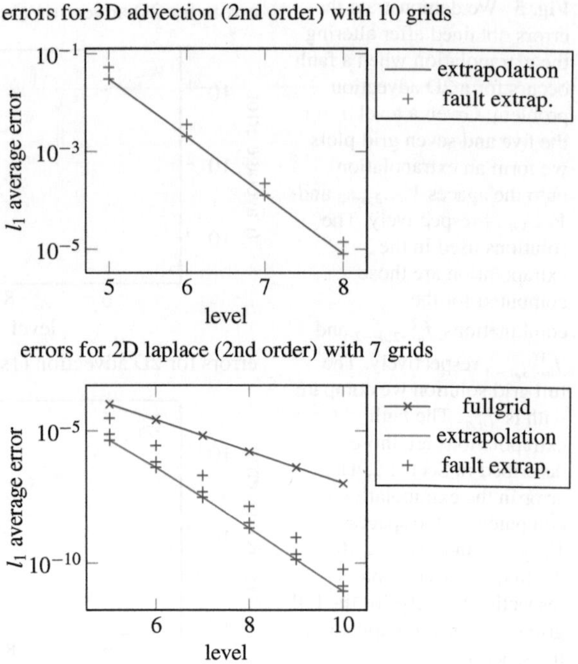

Fig. 6 We demonstrate the errors obtained after altering the extrapolation when a fault occurs for a 3D advection problem and a 2D laplace problem. The details for the 2D laplace problem are the same as those for the seven grid advection problem in Fig. 5. For the third problem, given a level n, we compute the extrapolation onto the space $V_{n-2,n-2,n-2}$ using the solutions as computed for the combination $\tilde{f}^2_{n-2,n-2,n-2}$. The error is computed on the space $V_{n-2,n-2,n-2}$

grid solution for large enough levels despite the difference in grid sizes. This is due to the extrapolation converging at a faster rate despite being on a smaller grid. This remains true when we alter the extrapolation after faults occur and we appear to have some bounds on the errors again for different cases of faults. Such bounds may be derived through a thorough analysis of the extrapolation of higher order terms in the asymptotic expansion.

3 Combining the Two Approaches

As suggested previously one may combine the two approaches. This is particularly straightforward for the case $p = 1$. We will describe this in detail for the case where one has the five solutions as in (7).

3.1 For a Standard Case (No Faults)

Consider one has the five component solutions $f_{n-2,n}$, $f_{n-1,n-1}$, $f_{n,n-2}$, $f_{n-2,n-1}$ and $f_{n-1,n-2}$ and that they satisfy the asymptotic error expansion of (16). Remember that $\tilde{f}^2_{n-2,n-2}$ denotes the combination

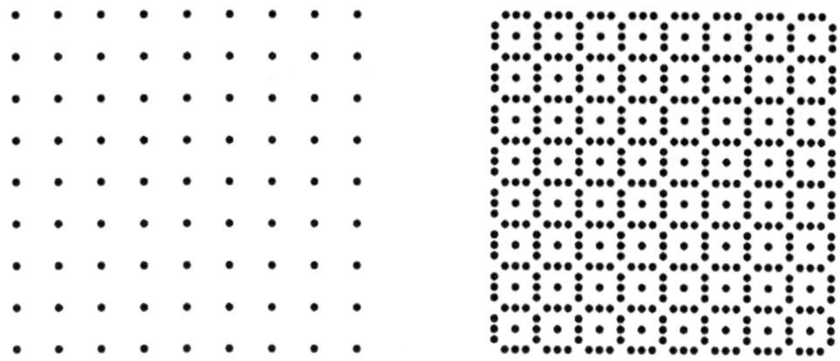

Fig. 7 This is a visual representation of the combined approach as in (20) for $n = 5$. On the *left* are the grid points which are extrapolated to a $2p$ order solution whilst those in the diagram on the *right* are obtained by adding some p order corrections

$$\tilde{f}^2_{n-2,n-2} = f_{n-2,n} + f_{n-1,n-1} + f_{n,n-2} - f_{n-2,n-1} - f_{n-1,n-2}\,.$$

Let $f^e_{n-2,n-2}$ denote an extrapolation of these solutions onto the space $V_{n-2,n-2}$, for example with $p = 1$ one can use

$$f^e_{n-2,n-2} = -2I_{n-2,n-2}f_{n-2,n} + 5I_{n-2,n-2}f_{n-1,n-1} - 2I_{n-2,n-2}f_{n,n-2}\,.$$

Now we can use both of these results by combining according to

$$I_{\mathcal{I}} f^e_{n-2,n-2} + (\tilde{f}^2_{n-2,n-2} - I_{n-2,n-2}\tilde{f}^2_{n-2,n-2}) \tag{20}$$

where $I_{\mathcal{I}} f^e_{n-2,n-2}$ is the bilinear interpolation of $f^e_{n-2,n-2}$ onto the sparse grid on which $\tilde{f}^2_{n-2,n-2}$ lives, that is $\tilde{f}^2_{n-2,n-2} \in V^s_{\mathcal{I}}$ with $\mathcal{I} = \{(i,j) \in \mathbb{N}^2 : i + j \leq 2n - 2 \text{ and } i, j \leq n\}$. The result is a solution consisting of a second order approximation from the space $V_{n-2,n-2}$ and a first order approximation from the space

$$\left(\bigoplus_{j\leq n-2} W_{n,j}\right) \oplus \left(\bigoplus_{j\leq n-1} W_{n-1,j}\right) \oplus \left(\bigoplus_{i\leq n-2} W_{i,n}\right) \oplus \left(\bigoplus_{i\leq n-2} W_{i,n-1}\right). \tag{21}$$

The points on the sparse grid that contribute to each of these function spaces for $n = 5$ are depicted in (Fig. 7). For $p > 1$ similar approaches may be taken but one needs to use a higher order interpolation (rather than the usual bilinear) to be able to effectively utilise the order $2p$ extrapolation. Future work will involve investigation into these cases.

3.2 *Fault-Tolerant Approach*

Since we have already enumerated fault-tolerant approaches for the combination and extrapolation techniques, combining the two approaches will be straightfor-

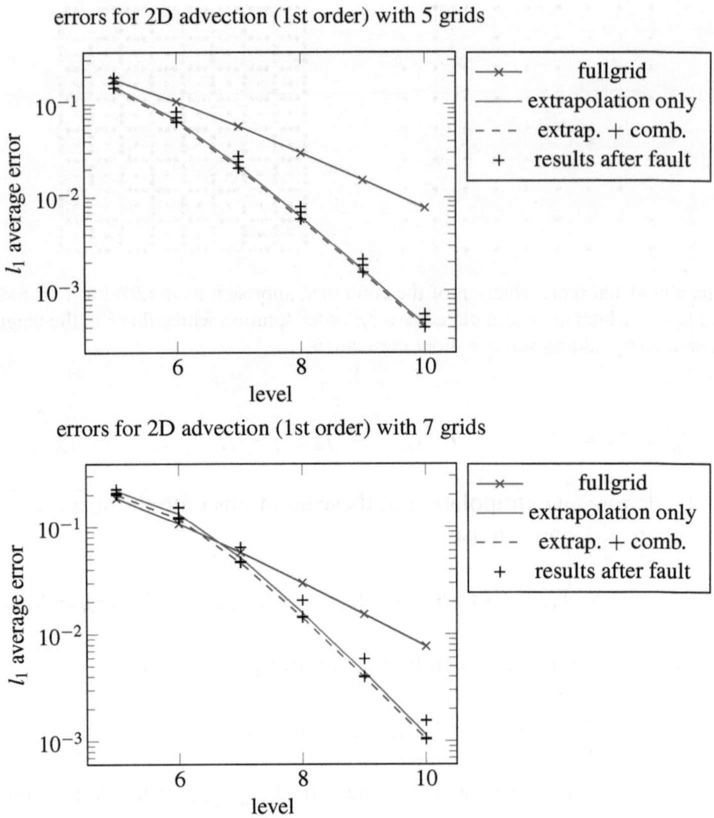

Fig. 8 We demonstrate the errors obtained when combining the two methods for the 2D scalar advection problem

ward. For example, given a fault affecting one of the five solutions of (7), let $g^e_{n-2,n-2}$ denote an extrapolation of successfully computed solutions onto the space $V_{n-2,n-2}$ and $g^c_{\mathcal{I}} \in V^s_{\mathcal{I}}$ denote some combination of the successfully computed solutions, then one simply takes

$$I_{\mathcal{I}} g^e_{n-2,n-2} + (g^c_{\mathcal{I}} - I_{n-2,n-2} g^c_{\mathcal{I}}) \, .$$

Note that in two of the cases when faults occurred with these five solutions we extrapolated onto a larger space, e.g. $V_{n-2,n-1}$ and $V_{n-1,n-2}$, however in our numerical results we extrapolate onto $V_{n-2,n-2}$ for consistency and ease of computation.

In Fig. 8 we demonstrate this combined approach with the full grid solution and the extrapolation solution for the two-dimensional scalar advection equation. We also plot the results of the combined approach when we alter it after a fault has occurred. The first plot uses the five solutions of (7) whilst the second plot uses the seven solutions from (8). The combined approach has an error slightly better than taking only the extrapolation. After a fault has occurred the results are still very good having error only a little larger then the normal combined approach.

Similar can be done for the combination of solutions given in (8), (10) and (11). The results in these cases tend to be a bit better given the extra grids available.

We conclude that the numerical results look promising for these approaches to fault-tolerance. We intend to further investigate these methods for more complex examples and higher order approximations.

Acknowledgements This research was supported under the Australian Research Council's *Linkage Projects* funding scheme (project number LP110200410). We are grateful to Fujitsu Laboratories of Europe for providing funding as the collaborative partner in this project.

References

1. S. Balay, W.D. Gropp, L.C. McInnes, B.F. Smith, Efficient management of parallelism in object oriented numerical software libraries, in *Modern Software Tools in Scientific Computing*, ed. by E. Arge, A.M. Bruaset, H.P. Langtangen (Birkhäuser, Basel, 1997), pp. 163–202
2. S. Balay, J. Brown, K. Buschelman, V. Eijkhout, W.D. Gropp, D. Kaushik, M.G. Knepley, L.C. McInnes, B.F. Smith, H. Zhang, PETSc users manual. Technical Report ANL-95/11 - Revision 3.3, Argonne National Laboratory (2012)
3. S. Balay, J. Brown, K. Buschelman, W.D. Gropp, D. Kaushik, M.G. Knepley, L.C. McInnes, B.F. Smith, H. Zhang, PETSc Web page (2012). http://www.mcs.anl.gov/petsc. Accessed 1 Nov 2012
4. J. Benk, D. Pflüger, Hybrid parallel solutions of the Black-Scholes PDE with the truncated combination technique, in *HPCS*, 2012, pp. 678–683
5. G. Bosilca, R. Delmas, J. Dongarra, J. Langou, Algorithm-based fault tolerance applied to high performance computing. J. Parallel Distrib. Comput. **69**(4), 410–416 (2009)
6. H. Bungartz, M. Griebel, Sparse grids. Acta Numer. **13**, 147–269 (2004)
7. H. Bungartz, M. Griebel, U. Rüde, Extrapolation, combination, and sparse grid techniques for elliptic boundary value problems. Comput. Methods Appl. Mech. **116**, 243–252 (1994)
8. F. Cappello, Fault tolerance in petascale/exascale systems: current knowledge, challenges and research opportunities. Int. J. High Perform. Comput. Appl. **23**(3), 212–226 (2009)
9. P. Du, A. Bouteiller, G. Bosilca, T. Herault, J. Dongarra, Algorithm-based fault tolerance for dense matrix factorizations, in *Proceedings of the 17th ACM SIGPLAN Symposium on Principles and Practice of Parallel Programming - PPoPP '12*, 2012, p. 225
10. J. Garcke, Sparse grids in a nutshell, in *Sparse Grids and Applications*, ed. by J. Garcke, M. Griebel. Lecture Notes in Computational Science and Engineering, vol. 88 (Springer, Berlin, 2013, pp. 57–80)
11. M. Griebel, M. Schneider, C. Zenger, A combination technique for the solution of sparse grid problems, in *Iterative Methods in Linear Algebra*, Brussels, 1991 (North-Holland, Amsterdam, 1992), pp. 263–281
12. M. Hegland, Adaptive sparse grids. ANZIAM J. **44**, 335–353 (2003)
13. K. Huang, J. Abraham, Algorithm-based fault tolerance for matrix operations. IEEE Trans. Comput. **c**(6), 518–528 (1984)
14. U. Rüde, Extrapolation and related techniques for solving elliptic equations. Bericht I-9135, Institut Fur Informatik, TU (1992)
15. C. Zenger, Sparse grids. Notes Numer. Fluid Mech. **31**, 241–251 (1991)

Efficient Regular Sparse Grid Hierarchization by a Dynamic Memory Layout

Riko Jacob

Abstract We consider a new hierarchization algorithm for sparse grids of high dimension and low level. The algorithm is inspired by the theory of memory efficient algorithms. It is based on a cache-friendly layout of a compact data storage, and the idea of rearranging the data for the different phases of the algorithm. The core steps of the algorithm can be phrased as multiplying the input vector with two sparse matrices. A generalized counting makes it possible to create (or apply) the matrices in constant time per row.

The algorithm is implemented as a proof of concept and first experiments show that it performs well in comparison with the previous implementation SG++, in particular for the case of high dimensions and low level.

1 Introduction

In many applications high dimensional models arise naturally, for example as a function that maps the high dimensional space $[0, 1]^d$ to the real numbers. Such a function can be represented using a regular grid. If we use a regular grid that has N points in every direction, this approach leads to N^d sample points, which can be more than what is computationally feasible.

One approach to reduce the number of sample points are sparse grids introduced by Zenger [11] in 1991. To see the potential for improvement, consider an axis parallel line through a grid point. We call the sampled points that lie on such a line a *pole*. In the regular grid, all poles consist of N points. Take two poles a and b in direction e_d that are shifted in direction e_1 by $1/N$. For a smooth function f we would expect that the restricted functions $f|_a$ and $f|_b$ are very similar, at least for

R. Jacob (✉)
ETH Zürich, Universitätstrasse 6, 8092 Zürich, Switzerland
e-mail: rjacob@inf.ethz.ch

J. Garcke and D. Pflüger (eds.), *Sparse Grids and Applications - Munich 2012*, Lecture Notes in Computational Science and Engineering 97,
DOI 10.1007/978-3-319-04537-5_8,

small $1/N$. Still, the full grid samples both functions on N points. In contrast, the sparse grid uses a different number of grid points on the two poles, say more on a than on b, and uses the samples for $f|_a$ to increase the accuracy of the representation of $f|_b$.

The sparse grid interpolation is based on hierarchical basis functions, as detailed for our concrete setting in Sect. 2. Each such basis functions has an axis parallel hyperrectangle as support and the volume of this support is $2^{-\ell}$, where the integer ℓ is the level of the basis function. Importantly, the support is not necessarily square, i.e., the side length can change drastically with the dimension. By using only basis functions of level up to $n \geq \ell$, we get some poles (actually one for each direction) that are sampled with $N = 2^{n+1} - 1$ points, whereas all other poles are sampled with fewer points. In contrast to the corresponding full grid with its N^d points, the sparse grid has only $O\left(N\binom{n+d-1}{d-1}\right) = O\left(N\left(\frac{e(n+d)}{d-1}\right)^{d-1}\right)$ sample points. The rate of convergence, i.e., how the approximation accuracy depends on increasing N, remains comparable to that of the full grid [3, 4, 11].

If we consider some examples, we see that the sparse grid with $N = 7$ points on the most densely sampled poles has for $d = 100$ only 20,401 grid points. For comparison, the corresponding full grid has $7^{100} \approx 10^{84}$ grid points, more than what is currently assumed to be the number of atoms in the universe. For $N = 7$ and $d =$ 10,000 there are 200 million sparse grid points and for $N = 15$ and $d = 100$ there are 72 million. In this situation, a different asymptotic estimate is helpful, namely $O\left(N\binom{n+d-1}{d-1}\right) = O\left(N\binom{n+d-1}{n}\right) = O\left(N\left(\frac{e(n+d)}{n}\right)^n\right)$. Concretely, we see that for $N = 7$ the number of grid points grows with $\Theta(d^2)$, for $N = 15$ with $\Theta(d^3)$, and so on. Hence high dimensions might be numerically feasible for small N. In this work, we focus as an example on one particular task, namely the hierarchization of a sparse grid (see Sect. 2). For this particular task, there is one value per grid point as input and output, and this number of values is referred to as *degrees of freedom* (DoF). Further, hierarchization is a task of relatively small computational intensity, i.e., in every round of the algorithm every variable gives rise to at most four floating point operations. Hence our algorithmic ideas are related to good memory usage. On one hand this amounts to minimizing the size of the data structures (ideally only one variable per DoF), and on the other hand we want make sure that the data access patterns are cache-friendly. This leads us to so called *memory efficient* or *I/O-efficient* algorithms. While the above points are our main focus, clearly a useful implementation must reflect other aspects of modern hardware. One noteworthy aspect is parallelism, for example in the context of several cores of the same CPU.

Many efficient algorithms known for sparse grids are based on the *unidirectional principle* [3]. It allows us to operate only on the one-dimensional sparse grids, i.e., the poles. More precisely, we iterate over the dimensions and for each dimension consider each of the poles in this dimension on its own.

The question we consider in this work is how to implement the unidirectional principle I/O-efficiently, with the example of hierarchization. Let us pictorially describe the situation by thinking of the sparse grid as a work piece that needs to be drilled from many different directions. There are two standard ways of doing this: Either you mount the working piece to the bench and move a mobile drill around it, or you mount the drill on the bench and turn the working piece. We propose to switch to the analogue of the latter method: Instead of adapting the one-dimensional hierarchization procedure to the current dimension, we move the data. For the one-dimensional hierarchization algorithm to be I/O-efficient, it would be good if each pole we work on is stored contiguously in memory. Provided that each pole fits into the cache, we immediately get an I/O-efficient algorithm: It loads the pole once and efficiently because the pole is split into few cache lines. Then it performs all operations on this pole in the cache, and writes the finished pole back to main memory to free the cache.

Because it is impossible to have a data layout of the sparse grid that stores all poles of all dimensions contiguously, we rearrange the layout of the sparse grid according to the dimension in which we currently work. More precisely, we define a *rotation* of the sparse grid that is a cyclic shift of the dimensions, i.e., maps $(x_1, \ldots, x_d) \mapsto (x_2, x_3, \ldots, x_d, x_1)$. Using this, it is sufficient to be able to perform the one dimensional hierarchization algorithm efficiently in one of the dimensions. We chose for this *working step* to operate in dimension d.

This approach has four main advantages:

- We can choose a memory layout that makes the access pattern of the working step cache-friendly
- Both phases (working and rotation) are exactly the same for all d rounds. They can be phrased as sparse matrix multiplication. Computing these matrices once is sufficient.
- There is no need to store position information (like level and index) together with a variable. This role of the variable is always implied by the position of the variable in the array representing the sparse grid. This leads to a fairly small memory-footprint, in particular in comparison to hash-based implementations.
- The algorithm can be easily and efficiently (both computation and memory access-wise) parallelized for multiple cores.

1.1 Algorithmic Background

The theory of I/O-efficient algorithms go back to the definition of the I/O-model [1]. It is based on the idea that the CPU can only work on a memory of size M (the cache). Input, output and auxiliary storage are in *external memory*, which is unbounded in size and organized in *blocks* of size B (a cache-line). The running time of an algorithm is estimated by the number of *I/O-operations* that read a block from external memory or write a block to external memory.

The differences between the I/O-model and the somewhat similar RAM or von-Neumann model can be illustrated by considering the task of permuting. Given a permutation $\pi\colon \{1, \ldots, N\} \to \{1, \ldots, N\}$, a program for *permuting* takes an input vector $(x_1, \ldots, x_N)$ and creates the output vector $(y_1, \ldots, y_N)$ according to $y_{\pi(i)} = x_i$. The naive algorithm loops over the input and writes the value to the specified position of the output. On the RAM, this takes $\Theta(N)$ operations which is trivially optimal. The naive algorithm can be adapted to the I/O-model where it causes $\Theta(N)$ I/O-operations. In contrast to the RAM this is not always optimal. The alternative algorithm is to use a B/M-way merge sort to rearrange the input array into the output array, which takes $O(\frac{N}{B} \log_{M/B} \frac{N}{B})$ I/Os. If the logarithmic term is smaller than B, this is better than the naive algorithm. When the I/O-model is used to describe a situation where the external memory is a disk and the internal memory is main-memory, the naive algorithm can easily be orders of magnitude slower. This is complemented by a lower bound stating that taking the better of the two mentioned algorithms is optimal [1]. The lower bound holds even if the algorithm can be adapted to the permutation, but this permutation is random. In contrast, there are also some permutations that can be performed easily, for example if the permutation is a cyclic shift or moves the elements not too far, e.g., $|\pi(i) - i| < M - 2B$. Many RAM algorithms have a completely unstructured memory access, similarly to the naive algorithm for permuting. Sometimes this can be avoided by rearranging the data [7]. One example of this technique is the I/O-efficient algorithm for multiplying a dense vector with a sparse matrix [2]. The first phase of the algorithm is to create all elementary products by multiplying each entry a_{ij} of the matrix with x_j. To perform this I/O-efficiently the matrix should be stored in a column major layout such that matrix entries of the same column are stored together and the columns are in the same order as the input vector. In a second phase of the algorithm the output is computed as row sums of the elementary products. For this to be I/O-efficient, they are first rearranged into a row-major layout such that all entries belonging to the same row are stored together. Here, the I/O-efficient way to rearrange might be to use the sorting algorithm.

Note that even though we phrase our result here as a multiplication with a sparse matrix and a permutation matrix, the structure of these two particular matrices usually makes the naive algorithm I/O-efficient. Still, we use the idea of working in phases and rearranging the date between the phases.

1.2 Related Work

Many different aspects of sparse grids have been investigated in the last years. The presentation here follows [3, 4, 10, 11].

Most of the proposed algorithms for sparse grids have been implemented, and some of the code is publicly available.

One easily available implementation is SG++ [10] (http://www5.in.tum.de/SGpp/). Its focus is adaptive sparse grids and it provides a lot of generality in terms of the used basis functions. Because of its availability and ease of installation, we use this for comparison in this paper.

The idea of using a layout of the complete sparse grid as a compact data structure has been proposed [8], but the layout itself differs from what we consider here. The corresponding implementation `fastsg` is described [9], where a recursive formula for the size of the sparse grid is proposed, similar to the one presented here. The code is publicly available but does not provide a hierarchization procedure for the 0-boundary case considered here.

Some of the ideas presented here (rotation, compact layout) are also the basis for a different implementation with focus on parallelism and vectorization [5]. Optimizing for this kind of parallelism favors a different layout. That code has been extended with a focus on evaluation [6].

2 Sparse Grids

Sparse grids have been designed to approximate functions in high dimensional spaces with relatively few degrees of freedom. By now, there is a body of literature discussing the mathematics of sparse grids. In contrast, this work investigates only the computer science aspects arising in the context, actually only for one particular task called hierarchization. Nonetheless, there is the need for a brief discussion of the underlying mathematical concepts, at least to explain how we phrase them in our algorithms.

In the following $(0,1) \subset \mathbb{R}$ denotes the open interval from 0 to 1, whereas $[0,1]$ denotes the closed interval. A sparse grid space as we use it here is a finite dimensional linear space. Its dimension is phrased as *degrees of freedom* (DoF) and it is given by the size of the basis we define for the space. The elements of the basis are specific continuous functions $\mathbb{R}^d \to \mathbb{R}$ with support limited to $(0,1)^d$, the so called *hierarchical basis functions* as defined in the following. An element of the *sparse grid space* is a linear combination of these basis elements, and hence also a continuous function $\mathbb{R}^d \to \mathbb{R}$ with support $(0,1)^d$.

2.1 *One-Dimensional Tree Structure*

In one dimension ($d = 1$) the hierarchical structure can be described by an annotated complete binary tree, as exemplified in Fig. 1. Each node v is annotated with an interval $I_v = (a_v, b_v) \subseteq (0,1)$ leading to the centerpoint $c_v = \frac{a_v+b_v}{2}$. The root is annotated with the open interval (0,1) and the centerpoint $1/2$. Each node v of the tree has two children l and r that are annotated with the intervals $I_l = (a_v, c_v)$ and $I_r = (c_v, b_v)$. Note that the centerpoints are unique and can be used to identify the node and the interval. Note further that two intervals of nodes u and v in the tree are either disjoint or one is contained in the other. In the latter case, if $I_u \subseteq I_v$, then v is an *ancestor* of u, i.e., the path from u to the root passes through v. For any node v, the endpoints a_v and b_v of its interval are either 0, 1, or a centerpoint of an

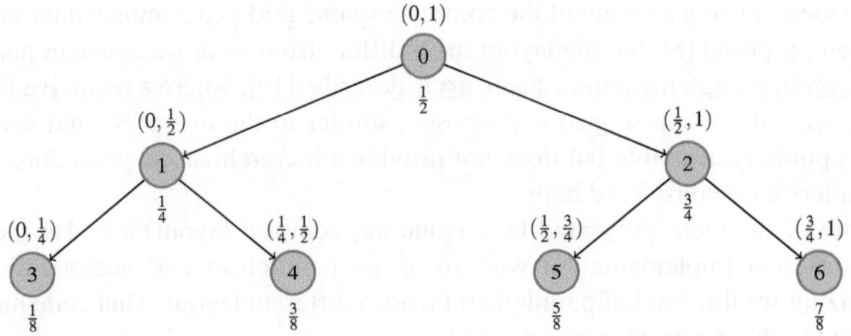

Fig. 1 The one-dimensional sparse grid of level 2: the tree T_l, the associated intervals above and the centerpoints below. The BFS-number is given inside the circle of the nodes

ancestor of v. These ancestors are called the *hierarchical predecessors* of v. There can be at most two hierarchical predecessors, namely one for a_v and another for b_v. We define the level of the root node to be 0, and the level of a node v, denoted by $l(v)$, to be its distance to the root, i.e., the number of times one has to follow a parent link to reach the root node. At level ℓ of the tree there are 2^ℓ nodes, all associated intervals have length $b_v - a_v = 2^{-\ell}$ and they partition the interval $(0, 1)$ (ignoring that centerpoints of nodes with level $< \ell$ are not element of any interval). We call the tree up to and including level ℓ the *ℓ-tree*, denoted by T_ℓ. Performing an in-order traversal of T_ℓ and collecting the centerpoints yields an equidistant grid in the interval $[0, 1]$ with spacing $2^{-(\ell+1)}$ and $2^{\ell+1} - 1$ points.

For every node v of an ℓ-tree we define a basis element of the one-dimensional sparse grid space. In this work, the one dimensional basis function f_v of a node v is piece-wise linear hat function with support (a_v, b_v) and the maximum of 1 at the centerpoint c_v. Observe that the nodes u with $f_u(c_v) > 0$ are precisely the ancestors of v.

A function f in the one-dimensional sparse grid space of level ℓ is given by coefficients λ_v, one for each node of the ℓ-tree, i.e.

$$f = \sum_{v \in T_\ell} \lambda_v f_v .$$

Such a function f is continuous and piece-wise linear on $[0, 1]$ with kinks at the centerpoints of the nodes of T_ℓ and support $(0, 1)$, and it is an element of the one-dimensional sparse grid space. Note that the value $f(c_v)$ at a centerpoint c_v is in general different from the coefficient λ_v.

Definition 1 (The Task of 1-D Hierarchization).

Input Values y_v, one for each node $v \in T_\ell$.
Output Coefficients λ_v such that the represented function $f = \sum \lambda_v f_v$ has the property $f(c_v) = y_v$ for each node v of the ℓ-tree.

The coefficients λ_v are also called *hierarchical surpluses*.

Algorithm 1: 1-D hierarchization

Input : values at the grid points, stored in $y[]$
Output : hierarchical surpluses, stored in $\lambda[]$
for i = *maxlevel* **downto** *0* **do**
    foreach node v *of* T_l with $l(v) = i$ **do**
        Let l_v be the left hierarchical predecessor of v
        Let r_v be the right hierarchical predecessor of v
        $\lambda[v] = y[v] - 0.5 * (y[l_v] + y[r_v])$

One dimensional hierarchization can be achieved by the pseudocode given in Algorithm 1. To see why, we argue that $y_v = \frac{y_l + y_r}{2} + \lambda_v$ holds. Consider any node $v \in T_\ell$. Observe that for all $u \in T_\ell, u \neq v$ the basis functions f_u falls in one of the following two cases. The first case is $f_u(a_v) = f_u(c_v) = f_u(b_v) = 0$, either because the support of f_u and f_v does not overlap, or because u is a descendant of v, i.e., in u is in the subtree below v. The second case is that f_u is linear in the complete interval $[a_v, b_v]$, which means u is an ancestor of v. Hence, the contribution of all other functions together will result in the linear interpolation between y_l at a_v and y_r at b_v, leading to the equation $\lambda_v = y_v - \frac{y_l + y_r}{2}$.

2.2 *Higher Dimensional Case*

In higher dimensions ($d > 1$), the sparse grid space is constructed using a tensor product approach, and the term sparse becomes meaningful.

We generalize the index set T_ℓ to its d-fold Cartesian product $T_\ell^d = T_\ell \times \cdots \times T_\ell$. For a vector of d tree nodes $\mathbf{v} = (v_1, \ldots, v_d) \in T_\ell^d$ the level of $\mathbf{v}$ is defined as the sum of the levels $\ell(\mathbf{v}) = \sum_{i=1}^{d} l(v_i)$. We define its d-dimensional basis function by $f_{\mathbf{v}}(x_1, \ldots, x_d) = \prod_{i=1}^{d} f_{v_i}(x_i)$. The support of $f_{\mathbf{v}}$ is $I_{v_1} \times \cdots \times I_{v_d}$ and its unique centerpoint $c_{\mathbf{v}} = (c_{v_1}, \ldots, c_{v_d}) \in (0,1)^d$ is called a grid point. Note that the d-dimensional volume of the support of $f_{\mathbf{v}}$ is $2^{-\ell(\mathbf{v})}$. The sparse grid of dimension d and level ℓ is based on the set of *sparse grid vectors* $I_\ell^d = \{\mathbf{v} \in T_\ell^d \mid \ell(\mathbf{v}) \leq \ell\}$, and the *sparse grid space* is the span of the corresponding basis functions $F_\ell^d = \{f_{\mathbf{v}} \mid \mathbf{v} \in I_\ell^d\}$. The set of *sparse grid points* is $C_\ell^d = \{c_{\mathbf{v}} \mid \mathbf{v} \in I_\ell^d\}$.

Note that in most of the established literature, the level of the one dimensional functions is counted starting from 1 for what we call the root node. This leads to the situation that the simplest d-dimensional basis function with support $(0,1)^d$ has level d, whereas in our notation it has level 0. In other words, what we call level represents directly the number of refinement steps and hence the volume of the support, independently of the dimension of the underlying space.

With these definitions, the one dimensional hierarchization task formulated in Definition 1 naturally generalizes to higher dimensions.

Algorithm 2: High dimensional hierarchization

for $i = d$ **downto** *1* **do**
  foreach pole p of the sparse grid in dimension i **do**
    perform one-dimensional hierarchization on p

Definition 2 (The Task of Hierarchization).

Input: Values $y_{\mathbf{v}}$, one for each sparse grid vector $\mathbf{v} \in I_\ell^d$
Output: Coefficients $\lambda_{\mathbf{v}}$ such that the represented function $f = \sum_{\mathbf{v} \in I_\ell^d} \lambda_{\mathbf{v}} f_{\mathbf{v}}$ has the property $f(c_{\mathbf{v}}) = y_{\mathbf{v}}$ for each $\mathbf{v} \in I_\ell^d$.

Its algorithmic solution will be the focus of this work. The well established algorithm follows the unidirectional principle [3], making use of the one dimensional algorithm. A *pole* of the sparse grid in direction i containing the sparse grid vector $\mathbf{v} = (v_1, \ldots, v_d)$ is the subset of sparse grid vectors that differ from $\mathbf{v}$ only in dimension i. All such poles have the structure of a one dimensional sparse grid (by projection to dimension i), even if some consist of only a single element. Further, the poles in direction i partition the sparse grid. The hierarchization algorithm considers the dimensions one after the other, i.e., it works in the directions $i = 1, \ldots, d$. In each iteration it runs Algorithm 1 on all poles of direction i (one-dimensional hierarchization). The pseudocode of this well established solution is given as Algorithm 2 and we do not modify it at this level of abstraction. Beyond this pseudocode, several important aspect need to be addressed:

- Which computation happens when.
- How and where the variables are stored.
- How the data is moved.

3 Memory Efficient Algorithm

This section describes the ideas that lead to a memory efficient algorithm for the hierarchization of a sparse grid of high dimension and low level.

3.1 Data Layout

At the heart of the algorithm is a specific layout of the data. From this layout the algorithm itself follows naturally.

3.1.1 Layout in One Dimension

Given the tree structure detailed in Sect. 2.1, a breadth-first-search (BFS) traversal of the nodes is well defined: It starts at the root, and then traverses the nodes of the tree with increasing level and from left to right. More precisely, we assign the BFS-number 0 to the root, BFS-number 1 to its left child, 2 to its right child, the BFS-numbers 3, 4, 5, 6 to the four nodes of level 2, and so on. These BFS-numbers are shown in Fig. 1. The BFS-number is a unique identifier of a node in the one dimensional tree T_ℓ. This BFS-layout for a complete binary tree is well understood (its perhaps most prominent use is in the heap-sort algorithm). For a node with BFS-number i, its two children have the BFS-number $2(i+1)-1$ and $2(i+1)$, and its parent has BFS-number $\lfloor (i-1)/2 \rfloor$. This simplicity of computing related BFS-numbers is one of the big advantages of the BFS-layout. Note that starting the counting at 1 would make this even simpler, but indexing the array in C-style starting from 0 is not only closer to the code, but it is also more natural when working with subarrays. Observe that the BFS-number encodes the level and the index within the level in one number. The BFS-numbers of nodes with level ℓ start with $2^{\ell+1}-1$. We define the level of a BFS-number to be the level of the corresponding node in the ℓ-tree. Further, a tree of maximum level ℓ uses precisely the BFS-numbers in $\{0, \ldots, 2^{\ell+1}-2\}$. We use the BFS-number as a position in the layout. A one dimensional sparse grid (i.e. a pole) can in this way be represented as an array (with one entry per degree of freedom).

3.1.2 Higher Dimensional Layout

Following the tensor product structure of the sparse grid, it is natural to identify a sparse grid point (defined by a vector of tree-nodes $\mathbf{v} = (v_1, \ldots, v_d)$) with a vector $\mathbf{b} \in \mathbb{N}_0^d$ of BFS-numbers. Each such vector of a sparse grid of level n has $\sum_{i=1}^{d} \ell(b_i) \leq n$. We sort the vectors of BFS-numbers lexicographically, with the significance of the positions decreasing (as is usual). This directly leads to a layout of a higher dimensional sparse grid, as exemplified in Fig. 2 in two dimensions and level 2. Because the first position is most significant, all elements with a 0 at the first position are grouped together at the beginning, then come the elements with a 1 at the first position and so on. Within these groups the sorting bundles together the elements with the same entry at the second position. And within these groups, the elements with the same entry at the third position are grouped together, and so on.

At the end of this conceptual recursion we see groups of vectors that differ only in the entry at the last position and are sorted by this entry. Hence, we see that the poles (defined in Sect. 2.2) in dimension d form subarrays in the complete lexicographical layout.

$0{:}(\frac{1}{2},\frac{1}{2})$ $1{:}(\frac{1}{2},\frac{1}{4})$ $2{:}(\frac{1}{2},\frac{3}{4})$ $3{:}(\frac{1}{2},\frac{1}{8})$ $4{:}(\frac{1}{2},\frac{3}{8})$ $5{:}(\frac{1}{2},\frac{5}{8})$ $6{:}(\frac{1}{2},\frac{7}{8})$

$7{:}(\frac{1}{4},\frac{1}{2})$ $8{:}(\frac{1}{4},\frac{1}{4})$ $9{:}(\frac{1}{4},\frac{3}{4})$

$10{:}(\frac{3}{4},\frac{1}{2})$ $11{:}(\frac{3}{4},\frac{1}{4})$ $12{:}(\frac{3}{4},\frac{3}{4})$

$13{:}(\frac{1}{8},\frac{1}{2})$

$14{:}(\frac{3}{8},\frac{1}{2})$

$15{:}(\frac{5}{8},\frac{1}{2})$

$16{:}(\frac{7}{8},\frac{1}{2})$

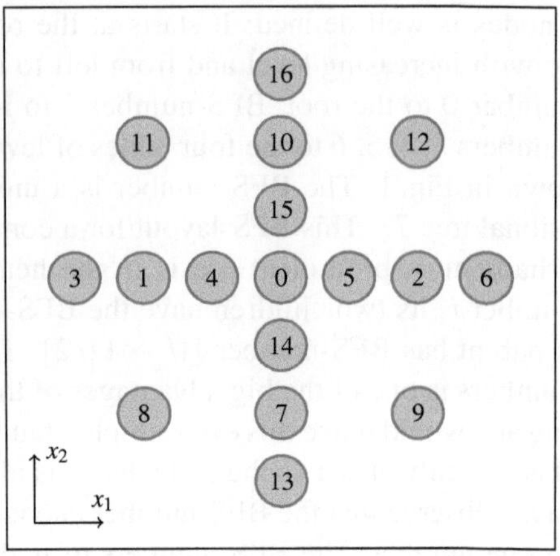

Fig. 2 The two-dimensional sparse grid of level 2: on the *right* the grid points at their positions in $[0, 1]^2$, labeled with their position in the layout. On the *left* the lexicographical layout of the grid points as triples *layout-position*:(x_1, x_2). Note the matrix structure of both figures: rows have the same x_1 coordinate, columns the same x_2 coordinate. The restriction on the level-sum yields a symmetric shape

3.2 *Numerical Work: Hierarchize All Poles*

The high dimensional hierarchization Algorithm 2 works pole-local in the inner loop. In other words, when considering dimension i, there is no interaction between variables of different poles in dimension i. Let us focus, for the moment, on the first iteration of the outer loop of Algorithm 2, where all poles of dimension d are hierarchized. This task fits particularly well to our layout of the sparse grid. Each such pole is located in a subarray, starts at a certain offset and has length $2^\ell - 1$ for a pole of level ℓ. Note that some of these poles have level 0, i.e. consist of a single element and hence require no numerical work at all.

Consider Algorithm 1 when it operates on a pole stored in BFS-layout. More precisely, consider the j-th input pole as stored in a vector $\mathbf{y}_j$ and the output in a vector $\mathbf{h}_j$, both with $N = 2^\ell - 1$ entries. We express Algorithm 1 as $\mathbf{h}_j = H_\ell \cdot \mathbf{y}_j$ for a sparse matrix H_ℓ. Consider a node v of the ℓ-tree. The variables of the input and output associated with this node are stored in the same position in the two vectors, say at position i. Hence, on the diagonal, all entries are 1. The variables associated with the hierarchical predecessors are stored before i. Hence, the matrix H_ℓ has at most two non-zero entries below the diagonal, and each of them has value $-1/2$. Note that H_k is the upper left corner of H_ℓ for $k < \ell$.

We can also express the whole first iteration of Algorithm 2 as a matrix multiplication. To this end consider all input values stored in the vector $\mathbf{y}'$ and the result in the vector $\mathbf{h}'$, both according to our layout. Let us describe the matrix W_d for which we have $\mathbf{h}' = W_d \mathbf{y}'$. The matrix W_d is a block diagonal matrix composed of many matrices H_{l_j}, where l_j is the level corresponding to the size of pole j. Hence, on the diagonal of W_d all entries are 1, above the diagonal all entries are 0, there are at most two entries of value $-1/2$ in every row and they are at distance at most 2^{ℓ} below the diagonal for a sparse grid of level ℓ.

By the above discussion it is clear that also the other hierarchization steps for dimension $i \neq d$ are linear, but the structure of the corresponding matrices would be different and harder to describe. Hence, in the following, we use a rotation to reuse W_d.

3.3 Rotation

We achieve the outermost loop of Algorithm 2 by performing rotations. In this way the one dimensional algorithm formulated as W_d in Sect. 3.2 can operate on all dimensions in turn.

Consider the shift $S(n_1, \ldots, n_d) = (n_2, n_3, \ldots, n_d, n_1)$ working on d-dimensional BFS-vectors. When following the movement of the corresponding centerpoints when applying S to each grid point, geometrically we see a rotation operating on $[0, 1]^d$. Because we are working with a non-adaptive sparse grid, this grid-point exists in our sparse grid as well. Observe that S^d is the identity, and that S^i maps the $(d - i)$-th position of the vector to the last position. In terms of our algorithm, using this rotation means that we should take a variable of our vector, understand it as a grid-point/BFS-vector, rotate it with S, and move it to the position in the vector associated with the rotated grid-point. We expresses this data movement by a permutation matrix R.

With this definition of R, we can express one execution of the outer loop of Algorithm 2 as applying the matrix RW_d, and the complete algorithm as applying $(RW_d)^d$. In other words, to transform a vector $\mathbf{y}$ of function values at the grid-points to a vector of hierarchical surpluses $\mathbf{h}$ (both in our layout), we can use the equation

$$\mathbf{h} = (RW_d)^d \mathbf{y}$$

to express Algorithm 2.

3.4 Considerations of an Implementation

With this description of the algorithm as the alternating application of two sparse matrices, it is quite natural to work with two vectors as in the pseudocode given in Algorithm 3.

Algorithm 3: Hierarchization as sparse matrix multiplication

Input : values at the grid points, stored in vector $y[]$
Output : hierarchical surpluses, stored in vector $y[]$
Let x be a vector of same size as y
for $i = 1$ **to** d **do**
  $x = W_d * y$
  $y = R * x$

This code obviously relies on the definition of the matrices W_d and R in a crucial way, and these matrices are based on the correspondence between positions in the layout, BFS-vectors, and the tree structure in the BFS-numbers. We call the translation of a BFS-vector to the position in the layout `pos`, and the reverse operation `depos`. They can be implemented quite efficiently in $O(\ell + d)$, but by the nature of having a BFS-vector with d entries as input or output, they cannot take constant time. This can be avoided by considering the BFS-vector as a data structure that changes its state according to well defined operations, in particular changing the BFS-vector such that the corresponding position in the layout is incremented. More precisely, we have an abstract data type *BFS-vector* that stores as it state a $\mathbf{b} = (b_1, \ldots, b_d)$ and supports the following operations:

`int pos():` Return $p(\mathbf{b})$, the position in the layout of $\mathbf{b}$.
`init():` set $\mathbf{b} = (0, \ldots, 0)$.
`increment():` Change $\mathbf{b}$ such that `pos` is incremented, i.e., for the current state $\mathbf{b}$ and the new state $\mathbf{b}'$ it holds $p(\mathbf{b}') = p(\mathbf{b}) + 1$.
`int shift_pos():` Return $p(S(\mathbf{b}))$, i.e. the position of the shifted current BFS-vector.
`int last_entry():` Return b_d.
`depos(x):` Change the current vector such that $p(\mathbf{b}) = x$. All of the above operations but `depos` can be implemented in $O(1)$ time using a sparse representation of the BFS-vector, as detailed in Sect. 4.

With this presentation of the hierarchization algorithm, there are several design choices to be taken. Either we can implement a carefully tuned routine that has the effect of applying W_d and R, or we can explicitly represent the two sparse matrices. Which alternative is superior depends on the circumstances and the used hardware: If several hierarchization tasks are performed, the dimension is high, and the access to the sparse matrices is fast enough, the time for precomputation of R and W_d can be amortized by sufficiently many multiplications with the matrices. If instead bandwidth to the memory is much more precious than computation on the CPU, then an on the fly computation is preferable.

Another important concern is the possibility to turn the above algorithm into a parallel one. The multiplication steps are trivial to parallelize including the load balancing. Also the creation of the matrices can easily be parallelized. Only the initialization of the individual parallel threads needs some care and uses `depos`.

4 Navigating the High Dimensional Layout

This section is concerned with computing the position of a BFS-vector $\mathbf{v} = (v_1, \ldots, v_d)$ in the layout described in Sect. 3.1.2. It relies on knowing the size of a sparse grid of dimension d' and level ℓ' for all $d' \leq d$ and $n' \leq n$.

4.1 *The Concept of Remaining Level*

In this section we work exclusively with BFS-vectors with dimension d and level $\ell \leq n$ to represent the points of the sparse grid. Remember that we use $l(b_i)$ to denote the level of the i-th digit, and that a BFS-vectors $\mathbf{b} = (b_1, \ldots, b_d)$ belongs to the sparse grid if and only if $\sum_{i=1}^{d} l(b_i) \leq n$.

Assuming that the first k entries of $\mathbf{b}$ are fixed to $b_1, \ldots, b_k$, we want to know which values are still possible for $b_{k+1}, \ldots, b_d$. This is restricted by the limit on the sum of the levels of the complete vector. Let the *remaining level* be defined as $r = n - \sum_{i=1}^{k} l(b_i)$. If $r < 0$, then the initial entries do not lead to a grid point, regardless of the remaining entries. Otherwise, the level sum of the remaining entries has to be bounded by r for the complete BFS-vector to belong to the sparse grid.

4.2 *The Number of Grid Points $v(d, \ell)$*

One important quantity is $v(d, \ell)$, the number of grid points (size) of a d-dimensional sparse grid of level ℓ, or equivalently, the number of d-dimensional BFS-vectors with level sum bounded by ℓ. In the following, we use a recursive formula for v.

For the base case, it is convenient to understand a zero-dimensional sparse grid to consist of precisely one point, regardless of the level. Hence,

$$v(0, \ell) = 1 \qquad \text{for all } \ell \in \mathbb{N}_0 . \tag{1}$$

The recursive case is given by

$$v(d, \ell) = \sum_{i=0}^{\ell} 2^i \cdot v(d-1, \ell - i) . \tag{2}$$

Here the quantity 2^i represents the number of grid points in a BFS-tree with level precisely i.

The validity of (2) follows, for example, from the layout proposed in Sect. 3.1.2: The first group has a 0 in the first position, the remaining level is ℓ, and the group is

Algorithm 4: `pos()`, the position of a BFS-vector in the layout

Input : the BFS-vector in $b[\,]$
Output : the position in p

```
p = 0
l = maxlevel
for i = 1..d do
    for x = 0..b[i]-1 do
        p = p + v(d - i, l - level(x))
    l = l - level(b[i]);
```

a sparse grid with level ℓ in $d-1$ dimensions. The next two groups have in the first digit a value of level 1, which means that the remaining level is $r = \ell - 1$ and each of them is a grid with dimension $d-1$ and level r. This argument continues with increasing level of the first digit, finally reaching the digits of level ℓ. These are 2^{ℓ} many groups, each with a remaining level 0. This means that the remaining digits must all be 0, which is consistent with our definition that a level 0 grid consists of a single grid point.

The formulas (1) and (2) form a recursion because the right hand side depends only on cases with smaller or equal level and one dimension less. For efficiency we usually compute the (relatively small) table for $v(d', n')$ using dynamic programming.

4.3 *Operations on a Full Vector*

4.3.1 `pos`: The Position in the Layout

With the help of $v(d, \ell)$ we can compute for a BFS-vector $\mathbf{b} = (b_1, \ldots, b_d)$ the position in the layout $p(\mathbf{b})$. This position can be computed by Algorithm 4, where `level`(x) denotes the level of the 1-D-tree node with BFS-number x, and `maxlevel` denotes the overall level ℓ of the sparse grid, and the BFS-vector is stored in the array $b[\,]$ (indexed with $1, \ldots, d$).

We can think of the addition to p as jumping in the layout. To account for the first entry $b_1 \in \{0, \ldots, 2^{\ell+1} - 2\}$, we jump to the group of BFS-vectors with first entry b_1, namely to the position of the BFS-vector $(b_1, 0, \ldots, 0)$. From this we can continue in the same manner for the remaining vector $(b_2, \ldots, b_d)$, interpreting this vector as an element of the sparse grid with $d-1$ dimensions and the remaining level. Note that for the last entry the increments are by $1 = v(0, \ell)$.

4.3.2 `depos`: The BFS-Vector from a Position

Now we consider the operation `depos`, the inverse to `pos`, as formulated in Algorithm 5. Given a position p, we want to find the BFS-vector $\mathbf{b} = (b_1, \ldots, b_d)$

Algorithm 5: `depos()`, the BFS-vector from the position in the layout

```
Input  : the position in p
Output : the BFS-vector in b[]
freelevel=maxlevel
b = (0, .., 0)
for i = 1..d do
    w = v(dim − i, freelevel)
    while p ≥ w do
        p = p − w
        b[i] = b[i] + 1
        w = v(dim − i, freelevel − level(b[i]))
    freelevel = freelevel − level(b[i])
```

that is mapped to p in the layout. The structure of `depos` is very similar to `pos`. Instead of jumping to the starting position of a group, we want to find the group in which p is located. In other words, the algorithm consists of d linear searches for the group of dimension i (inside the group of dimension $i-1$) in which p is located.

4.3.3 `increment`: The Next BFS-Vector in the Layout

The functions `pos` and `depos` as explained in the last two sections are not efficient enough to be used in the innermost loop of a hierarchization algorithm, a problem we circumvent by an `increment` operation. We store the BFS-vector $\mathbf{b} = (b_1, \ldots, b_d)$ in the array $b[]$, and the corresponding position p in the layout in the variable p. The `increment` operation has the purpose of incrementing p and changing $\mathbf{b}$ in a way that preserves this correspondence. This and the direct access to b_d are sufficient to create W_d (or to compute $y = W_d \cdot x$ without a representation of the matrix) serially. For a parallel version of this algorithm we assign each thread the task to create the range $p_i : p_{i+1} - 1$ of rows of W_d. Then the thread starts by setting $p = p_i$ and establishing the invariant by initializing the vector using `depos`(p_i). From then on, `depos` is no longer needed.

Assume for the moment that there is an additional array of the remaining levels $r[k] = \ell - \sum_{i=1}^{k} l(b_i)$ as defined in Sect. 4.1. If $r[d] > 0$, the current vector does not exhaust the maximum level, and the last entry $b[d]$ can be incremented without violating the constraint on the level sum. Even if $r[d] = 0$, it is possible that $b[d]$ is not the highest BFS-number of level $r[d-1]$, and hence the increment changes only $b[d] = b[d] + 1$.

If $b[d]$ is the highest BFS-number of level $r[d-1]$, we know that the increment has to change some of the entries further to the left, and the incremented vector has $b[d] = 0$. The change further left is an increment operation on the $d-1$ dimensional BFS-vector stored in $b[1], \ldots, b[d-1]$ with level-sum ℓ, and we can continue in the same way. If this process would continue to the left of $b[1]$, this means that the first entry $b[1]$ is the highest allowed BFS-number (and all other digits are 0). Then an increment is impossible because we reached the last point of the sparse grid.

After updating the digits $b[]$ as described (from right to left), we also update all $r[]$ values that might have been changed (from left to right). Note that it is actually not necessary to work with a complete vector of $r[]$-values. Instead, we can have a single variable r (`freelevel` in the code) that maintains $r[d]$, i.e., by how much the current vector does not exhaust the level ℓ.

4.3.4 `shift_pos`: The Position of the Shifted BFS-Vector

For the computation of R we can use `increment`, if we manage to update the position of the shifted BFS-vector efficiently. More precisely, if our current array $b[]$ represents the BFS-vector $\mathbf{b} = (b_1, \ldots, b_d)$, we want to compute `pos`$(S(\mathbf{b}))$. We use some additional arrays that need to be updated during `increment`.

An increment operation leaves a prefix $b_1, \ldots, b_k$ of $\mathbf{b}$ unchanged, which means that the prefix $b_2, \ldots, b_k$ of length $k-1$ of $S(\mathbf{b})$ remains unchanged. The algorithm of `pos` (Sect. 4.3.1) considers the entries of the BFS-vector from left to right. Hence, an unchanged prefix of the BFS-vector means that the initial operations of `pos` are unchanged. This can be used by recording the state of the algorithm for `pos`$(S(\mathbf{b}))$ at each entry into the body of the loops. More precisely, this gives two arrays, one for the position p and one for the remaining level l, both two-dimensional, indexed by i and x. It is sufficient for these arrays to be current in each dimension i up to the position $b[i]$.

Every increment operation increments one entry, say from b_k to $b'_k = b_k + 1$, and all entries to the left are zero, i.e., $b'_{k+1} = \cdots = b'_d = 0$. Hence, we can "jumpstart" the execution of `pos`$(S(\mathbf{b}))$ starting from what we find as state in $[k-1, b_k]$, the offset implementing the shift. From there we can complete (including recording) the execution of `pos` on the entries $(b_{k+1}, \ldots, b_d, b_1)$. Observe that the control-flow of this remaining execution is always the following: There is the last iteration of the inner loop with $i = k - 1$ and $x = b_k$ (creating a new entry in our arrays). For $i = k, \ldots, d-1$ we only update our tables at $[i, 0]$ recording the unchanged state of position p and remaining level r. The inner loop is not executed. Only for $i = d$ the inner loop iterates up to b_1, which always results in adding b_1 to p. Hence, the running time of this continuation of `pos` is proportional to the number of entries consider by `increment`, namely $d - k + 1$.

4.3.5 Performance Limitations of the Full Representation

It is well known that counting with a binary representation of the number changes amortized constant many bits per increment operation. In analogy, we might hope for a good performance of the operations `increment` and `shift_pos` in the amortized sense. This is actually not the case. In the following example the amortized cost per `increment` operation is lower bounded by $\Omega(d)$. Consider the case of level 2 and a very high dimension d. Then, almost all BFS-vectors have precisely two ones, and there are $\binom{d}{2}$ such vectors. For such a vector the cost of an increment is given by the distance of the last 1 to the right. Now all vectors that have at least $d/2$ trailing zeros require $d/2$ work, and there are roughly $\binom{d/2}{2}$ of them, i.e., roughly a fourth of all vectors. In this case enumerating all BFS-vectors with an `increment` operation takes time cubic in d, which means that the amortized time per increment is $\Omega(d)$.

4.4 Sparse Implementation

As detailed in the previous section, the increment operation on the full BFS-vector becomes problematic if there are many trailing zeros. This immediately suggests to use a sparse representation of the vector. This means storing a list of pairs (dimension, BFS-number), one for each entry that differs from zero, sorted by dimension.

Observe that the number of pairs is not only at most the dimension d, but also at most the level n of the sparse grid.

With this sparse representation, the increment takes only $O(1)$ operations, as we will argue in the following. The algorithm has to distinguish between several cases as listed below. Observe that for each case the sparse representation allows to test if it is applicable, and if so, to update the representation in constantly many operations, including the remaining level of the current BFS-vector.

- If the remaining level of the current vector is > 0, the last entry b_d can be incremented. This might extends the list by the pair $(d, 1)$.
  Hence, in all other cases we can assume that the current vector has remaining level zero.
- If the last non-zero entry can be incremented without changing the level, we do so.
- If the list consists of the single entry $(1, 2^{n+1} - 2)$ the increment operation is impossible (last BFS-vector).
- If the rightmost non-zero entry is b_i (in dimension $i > 1$ and it cannot be incremented), and $b_{i-1} = 0$, we set $b_i = 0$ and $b_{i-1} = 1$.
- Otherwise the rightmost two non-zero entries are b_{i-1} and b_i. In this case we set $b_i = 0$ and increment b_{i-1}. This is always possible because changing b_i to 0 decreases the level by at least one, whereas incrementing b_{i-1} increases the level by at most one.

It is easy to adjust the computation of `depos`, `pos` and its incremental version needed for `shift_pos` to the sparse representation. Overall this means that all operations, but `depos`, of the abstract data type formulated in Sect. 3.4 can be implemented in worst case constant time.

5 Implementation and Experiments

The described algorithms are implemented as a proof of concept and we compare running times for the hierarchization tasks with the implementation in SG++. The new approach presented in this paper is abbreviated as `rSG` (rotating Sparse Grid). SG++ is designed to deal with adaptive refinements and cannot take advantage of the simpler structure of a complete sparse grid. Further, as it is currently bundled, it is not parallelized. Another candidate for comparison would be `fastsg`, but its hierarchization procedure is only for sparse grids with boundary points, so it solves a different problem and a comparison would be meaningless. As a proof of concept, the point of the experiments is to show the potential of the new algorithmic ideas. Hence, our focus in these experiments is to show how far we can go beyond the current standard solution. Accordingly, the focus of the results are the sizes of the problems that can be tackled at all and the orders of magnitude of runtime and memory consumption. This is meant to give an indication of the strengths and weaknesses of our new approach.

The current implementation does not (yet) use the constant time algorithm for `shift_pos`. With the current setup of creating the matrices W_d and R once and then applying them several times, the possible savings in runtime are not significant enough.

5.1 Experimental Setup

The implementation is a C++ design that is parallelized with openMP, with a focus on being able to compare the performance of different algorithmic approaches. To this end, templates are used in many situations, whereas inheritance and in particular virtual classes and other constructs that are resolved at runtime are avoided. The experiments are run on a linux workstation with an Intel XEON E3-1240 V2 chip, having four cores clocked at 3.4 GHz. The total main memory of the machine is 32 Gb. The code is compiled with `gcc 4.4.7` and the compiler flags `-O3 -march=native -fopenmp -D_GLIBCXX_PARALLEL`.

The time measurements generally consider the whole execution time of the algorithm for both SG++ and `rSG`, including the initialization of data structures and setting the input vector with values. This means that in particular the phase where the two matrices W_d and R are computed (preparation phase) includes the time for using the memory for the first time (cold cache).

The time is measured inside the programm using the system call `gettimeofday()`.

The memory requirement is determined by the system utility `time` with the `%M` format, stating "Maximum resident set size of the process during its lifetime."

In this section, we describe the problem instances not only by their dimension and level but also by their number of grid points (DoF).

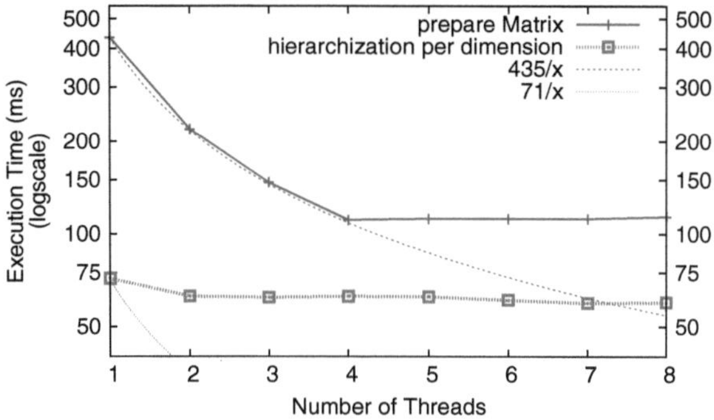

Fig. 3 Strong scaling behaviour for dimension 20, level 6

5.2 *Experiments*

5.2.1 Parallel Scaling

To evaluate the effectiveness of the ideas to parallelize the hierarchization algorithm we perform a strong scaling experiment, i.e., we solve the same problem with an increasing number of processors.

As a prototypical example we take the hierarchization of a sparse grid in 20 dimensions and level 6. This sparse grid has roughly 13 million grid points, uses 377 MB to store the grid and the execution uses in total 1.9 GB memory from the system.

Given that our machine has four cores with hyperthreading, we experiment with up to eight threads. The running times for the different numbers of threads are summarized in Fig. 3.

The preparation of the matrices scales perfectly up to four threads on the four cores, whereas more threads give no further improvement of the running time.

As a third plot in Fig. 3, we have the average hierarchization, i.e., the average time for one application of the two sparse matrices. We see that this phase scales poorly, and preparing the matrices takes roughly twice as long as applying them using four processors. Hence, with our current implementation of computing the matrices we have an algorithm that scales almost perfectly, but it is in total not fast enough to justify changing to an on the fly computation of the matrices, i.e., not creating a representation of the matrices. With different hardware or a more efficient implementation of the algorithms presented in this paper, this situation might change.

5.2.2 Running Times

In the plots given in Figs. 4 and 5 we compare the running times for the sparse grids of level 3 and level 6 for different dimensions. The range of dimensions is in both cases chosen in a way that the running times of both implementations are between

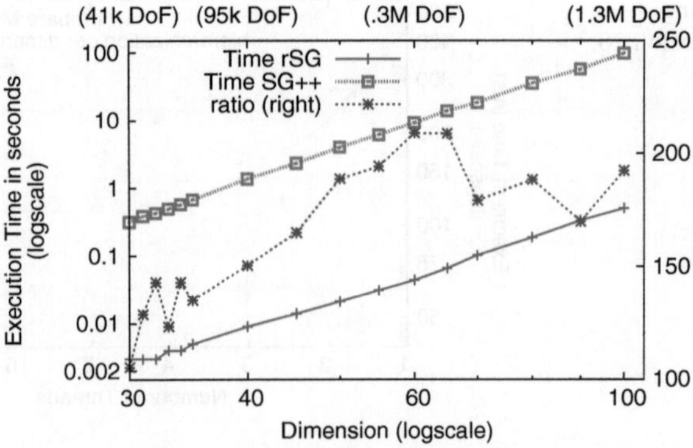

Fig. 4 Serial run time comparison for different dimensions and level 3: the x-axis gives the dimension of the sparse grid, labeled at the bottom. The dimension induces a certain DoF, as given above the figure. The y-axis on the *left* is running time in seconds for the two single core implementations with lines connecting the measurements. The y-axis on the *right* gives the ratio between the running times as plotted in the *blue dashed line*

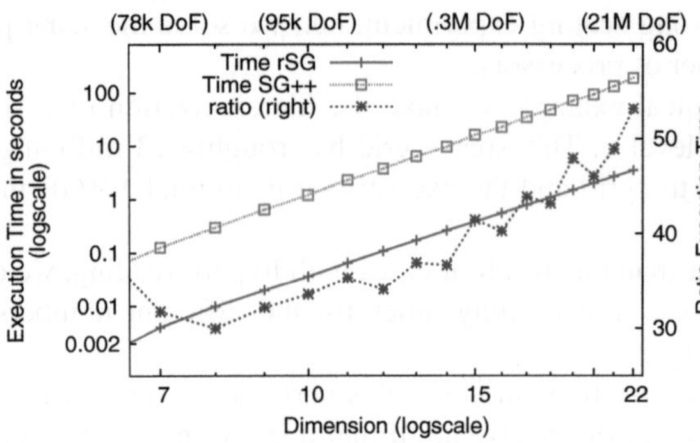

Fig. 5 Serial run time comparison for different dimensions and level 6: same setup as Fig. 4

1 ms and 3 min. In this range the implementation `rSG` (following the ideas presented in this paper) achieves speedups over SG++ as shown in the range of 100–220 for level 3 and respectively 30–53 for level 6.

5.2.3 Memory Footprint

The amount of memory used by a program is not only an important resource in itself, it also has, due to caches, a significant influence on the running time.

In Fig. 6, we plot the memory usage of the two implementations `rSG` and SG++. We see significant differences in the memory usage per DoF of the sparse grid, in particular for high dimensions and high levels. For large instances the `rSG`

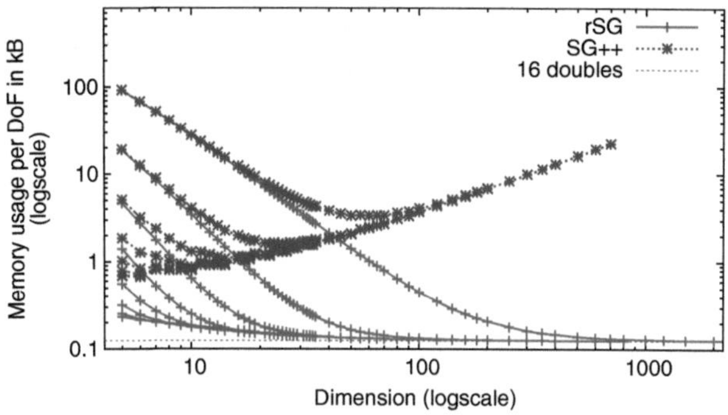

Fig. 6 Memory usage per DoF: the y-axis stands for memory usage as determined by `time -f %M` divided by the number of variables (DoF) of the sparse grid. The x-axis stands for the dimension and measurements for the same level are connected. The level-2 line is the topmost where SG++ and `rSG` have the same memory usage for small dimensions

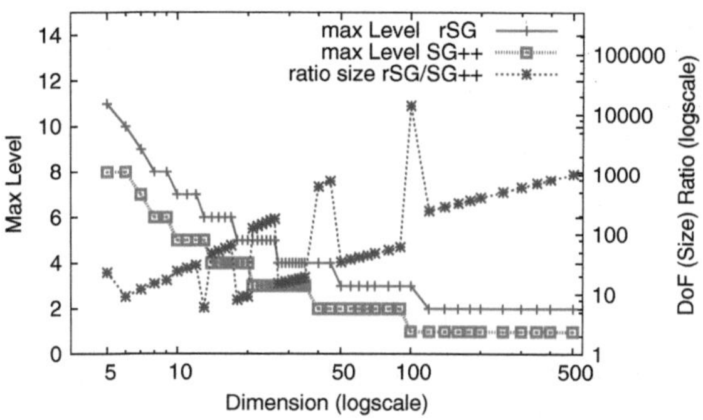

Fig. 7 Maximal sparse grid hierarchization doable serially in 1 s: for every sampled dimension, as given on the x-axis, we have the maximum level (left y-axis) that the two single core implements can hierarchize within the resource constraint. In *blue* the ratio of the DoF of the corresponding (maximal) sparse grids

implementation reaches a space usage of 16 words of 64-bit (i.e. doubles or long integers) per DoF. In contrast, the space usage of SG++ per DoF increases with the dimensionality of the problem.

5.2.4 Solvable Problem Sizes

Figures 7, 8, and 9 show the pareto front of problem sizes that can be computed with different resource limitations, namely limited computation time for the serial program and memory consumption. Remember that this comparison is somewhat unfair against SG++ because SG++ does not exploit that the grid is non-adaptive.

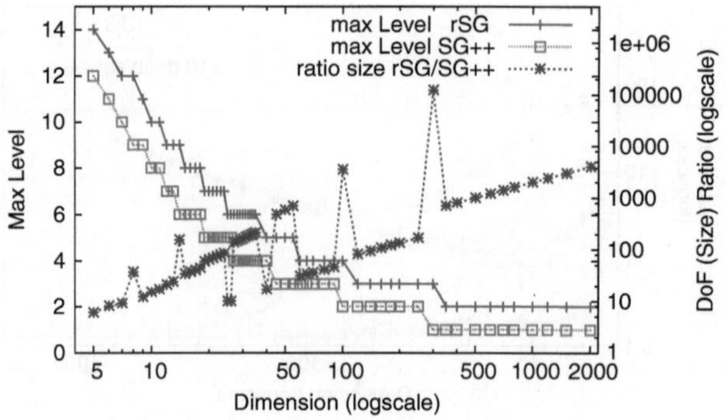

Fig. 8 Maximal sparse grid hierarchization doable serially in 1 min: same setup as Fig. 7

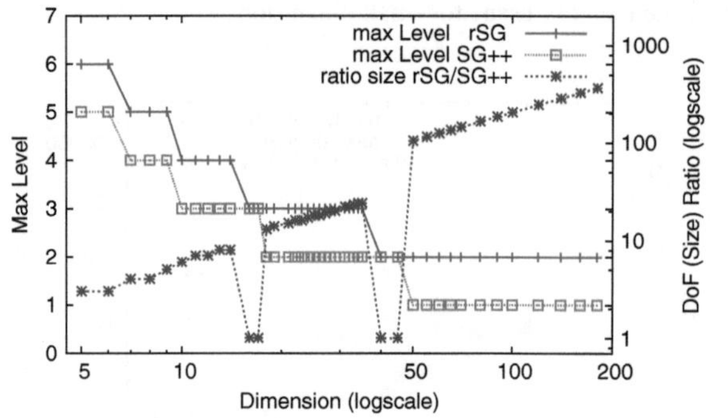

Fig. 9 Maximal sparse grid hierarchization doable with 16 GB memory: for every sampled dimension, as given on the x-axis, we have the maximum level (left y-axis) that the two implements can hierarchize within the resource constraint. In *blue* the ratio of the DoF of the corresponding (maximal) sparse grids

The first experiment sets a limit on the computation time for the hierarchization task. Figure 9 limits the used space somewhat arbitrarily to 16 GB.

In almost all considered cases, the new approach can hierarchize a sparse grid with at least one more level, sometimes three more levels, than SG++. In particular for high dimensions, the corresponding increase in degrees of freedom is a factor of up to 10000.

Observe that this comparison is not very fine grained: Increasing the level of the sparse grid by one increases the degrees of freedom by a factor of 10 to 5000. Hence, in many cases in the above comparisons, the resource constraint is by far not exhausted.

5.2.5 Relation Between Preprocessing and Hierarchization

We investigate the relation between the preprocessing time to create the matrices W_d and R, and the time to apply them once.

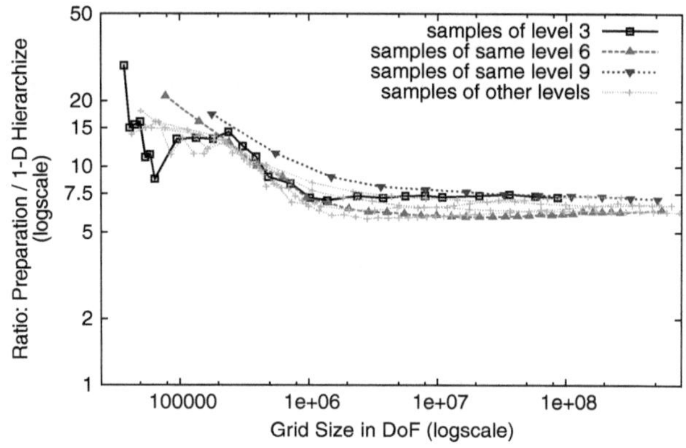

Fig. 10 Preparation vs hierarchization (serial): the y-axis shows the factor by which preparing W_d and R is slower than using them. The x-axis gives the size of the sparse grid in DoF. The measurements for the same level are connected by a line

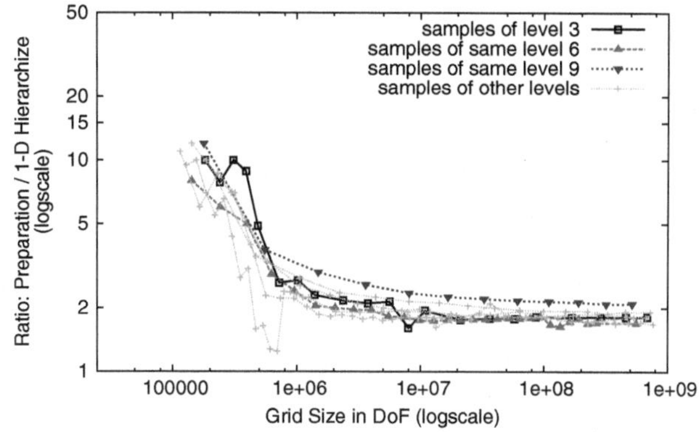

Fig. 11 Preparation vs hierarchization with four threads: with the same setup as in Fig. 10, and for comparability the same axis. For small grids some points cannot be reported because the measured time is 0

In Fig. 10 we see in the serial case that for small grids the preparation time is comparatively expensive. One explanation for this behavior is that the application step becomes memory bound for large grids: As long as the grid is small and everything fits into the cache, we see that the preparation takes a lot more computing than the application step. As soon as the grid needs to be fetched from main memory because it does not fit into the cache anymore, the application step gets slower, if there are no other effects, by a factor of roughly 3. The preparation phase is spending more time on computations on BFS-vectors, and hence the additional time for the slower memory access accounts for a smaller fraction of the total running time. Perhaps the additional memory latency is even completely hidden because the computation continues while the memory system stores the results.

We repeat this experiment with all four cores in Fig. 11. The findings of this experiment give further evidence that the application step is memory bound. It is consistent with the findings of Sect. 5.2.1 and Fig. 3. There we saw that the preprocessing step scales well with additional processors, whereas the hierarchization step improved over the serial case by at most a factor of 1.2, which suggests that the latter one is memory bound. This fits to what we can observe in Fig. 11, where we see that the ratio for big sparse grids improves to 1.7–2.

When comparing Figs. 10 and 11 directly, one notices that in the parallel case there are fewer samples for small sparse grids. This is due to the fact the computation gets faster, and that these grid are now reported with a 0 running time for hierarchization, and we cannot compute the ratio we are interested in.

Going back to the case of level 6 and dimension 20 considered in Fig. 3 (Sect. 5.2.1) we see that it is indeed prototypical. When locating this measurements in Fig. 10 by looking for the level 6 measurements with 1.28e7 DoF, we see that it actually achieves one of the smallest ratios.

One question that can be addressed by the considerations in this section is by how much the implementation of the data type for the BFS-vector needs to improve before an on the fly creation of the matrices is superior. Note that good processor scaling could provide this improvement and that the implementation does not yet use the constant time algorithm for `shift_pos`.

6 Conclusion

In this article, we have been exploring the basics of a new algorithmic idea to work with sparse grids of low level and high dimensions. The main concepts of a compact layout, rearranging the data and generalized counting present themselves as promising ideas. There are many directions in which the topic should be developed further. One direction is to consider vectorized parallel implementation, as done in [5]. Another direction is the evaluation algorithm on the data layout proposed here, as done in [6]. More directly, there is the question if a carefully tuned on-the-fly computation of the two sparse matrices can be beneficial in a parallel setting.

Acknowledgements I am grateful to Dirk Pflüger for introducing me to the topic and providing me with the SG++ source code. For many fruitful discussions I want to thank Gerrit Buse, Dirk Pflüger and Hans Bungartz. Special thanks go to Philipp Hupp for many helpful comments and discussions to improve the presentation.

References

1. A. Aggarwal, J.S. Vitter, The input/output complexity of sorting and related problems. Commun. ACM **31**(9), 1116–1127 (1988)
2. M.A. Bender, G.S. Brodal, R. Fagerberg, R. Jacob, E. Vicari, Optimal sparse matrix dense vector multiplication in the I/O-model. Theory Comput. Syst. **47**, 934–962 (2010)
3. H.-J. Bungartz, *Finite Elements of Higher Order on Sparse Grids, Habilitationsschrift* (Shaker Verlag, München, 1998)
4. H.-J. Bungartz, M. Griebel, Sparse grids. Acta Numer. **13**, 147–269 (2004)
5. G. Buse, D. Pflüger, A.F. Murarasu, R. Jacob, A non-static data layout enhancing parallelism and vectorization in sparse grid algorithms, in *ISPDC*, ed. by M. Bader, H.-J. Bungartz, D. Grigoras, M. Mehl, R.-P. Mundani, R. Potolea (IEEE Computer Society, Washington, 2012), pp. 195–202

6. G. Buse, D. Pflüger, R. Jacob, Efficient pseudorecursive evaluation schemes for non-adaptive sparse grids, in *Sparse Grids and Applications - Munich 2012* (Springer, Berlin, 2014)
7. A. Maheshwari, N. Zeh, A survey of techniques for designing I/O-efficient algorithms, in *Algorithms for Memory Hierarchies*, ed. by U. Meyer, P. Sanders, J. Sibeyn. Lecture Notes in Computer Science, vol. 2625 (Springer, Berlin, 2003), pp. 36–61
8. A. Murarasu, J. Weidendorfer, G. Buse, D. Butnaru, D. Pflüger, Compact data structure and parallel algorithms for the sparse grid technique, in *16th ACM SIGPLAN Symposium on Principles and Practice of Parallel Programming*, 2011
9. A.F. Murarasu, G. Buse, D. Pflüger, J. Weidendorfer, A. Bode, Fastsg: a fast routines library for sparse grids, in *ICCS*, ed. by H.H. Ali, Y. Shi, D. Khazanchi, M. Lees, G.D. van Albada, J. Dongarra, P.M.A. Sloot. Procedia Computer Science, vol. 9 (Elsevier, Amsterdam, 2012), pp. 354–363
10. D. Pflüger, *Spatially Adaptive Sparse Grids for High-Dimensional Problems* (Verlag Dr. Hut, München, 2010)
11. C. Zenger, Sparse grids, in *Parallel Algorithms for Partial Differential Equations*. Notes on Numerical Fluid Mechanics, vol. 31 (Vieweg, Braunschweig, 1991), pp. 241–251

Alternating Direction Method of Multipliers for Hierarchical Basis Approximators

Valeriy Khakhutskyy and Dirk Pflüger

Abstract Sparse grids have been successfully used for the mining of vast datasets with a moderate number of dimensions. Compared to established machine learning techniques like artificial neural networks or support vector machines, sparse grids provide an analytic approximant that is easier to analyze and to interpret. More important, they are based on a high-dimensional discretization of the feature space, are thus less data-dependent than conventional approaches, scale only linearly in the number of data points and are well-suited to deal with huge amounts of data. But with an increasing size of the datasets used for learning, computing times clearly can become prohibitively large for normal use, despite the linear scaling. Thus, efficient parallelization strategies have to be found to exploit the power of modern hardware.

We investigate the parallelization opportunities for solving high-dimensional machine learning problems with adaptive sparse grids using the alternating direction method of multipliers (ADMM). ADMM allows us to split the initially large problem into smaller ones. They can then be solved in parallel while their reduced problem sizes can even be small enough for an explicitly assembly of the system matrices. We show the first results of the new approach using a set of problems and discuss the challenges that arise when applying ADMM to a hierarchical basis.

V. Khakhutskyy (✉)
Institute for Advanced Study, Technische Universität München,
Lichtenbergstrasse 2a, 85748 Garching, Germany
e-mail: khakhtuv@in.tum.de

D. Pflüger
Institute for Parallel and Distributed Systems, University of Stuttgart,
Universitätsstr. 3, 70569 Stuttgart, Germany
e-mail: Dirk.Pflueger@ipvs.uni-stuttgart.de

J. Garcke and D. Pflüger (eds.), *Sparse Grids and Applications - Munich 2012*,
Lecture Notes in Computational Science and Engineering 97,
DOI 10.1007/978-3-319-04537-5_9,

1 Introduction

In recent years, sparse grids have been successfully employed for a whole range of data mining tasks such as classification, regression, time-series forecasting, and density estimation. The applications under consideration reflect the importance of data-driven problems and stem from a variety of disciplines and fields of research from astrophysics to finance, and from automotive engineering to optical digit recognition. For examples, see, e.g., [12, 13, 21].

Traditionally, two different approaches have been studied: one computes the approximation directly in a hierarchical sparse grid function space, while another one solves a problem on several full grids independently and then combines these solutions (combination technique). Each of these methods has its advantages and drawbacks.

The solution in a sparse grid function space, on the one hand, allows the usage of spatially (locally) and dimensionally adaptive refinement and can result in better performance especially on clustered data. But it results in highly nested and multi-recursive algorithms (UpDown scheme) which are difficult to parallelize, and the main effort so far focused on shared memory systems and vectorizations.

The combination technique, on the other hand, is restricted to dimensionally adaptive refinement and cannot refine single grid points in a locally adaptive manner. It also can exhibit divergent behavior in its direct version. From the algorithmic point of view, however, the combination technique offers multiple layers of parallelism: it is embarrassingly parallel with respect to the single solutions to be computed, while the (typically sparse) systems of linear equations for the individual grids can be efficiently parallelized using canonical techniques.

The alternating direction method of multipliers (ADMM) allows a new perspective on the approximation problem in the direct and spatially adaptive formulation, while additionally offering distributed parallelization of the optimization process. ADMM has been known under different names for years in the optimization community, and good convergence results have been reported for different applications [3, 5, 22, 23]. But to the best of our knowledge, the method has not been studied for hierarchical bases and sparse grids so far.

This paper intends to fill this gap. It gives an introduction to the method and insights into the interplay of alternating directions for sparse grid approximation problems used for data mining. Our experiments show interesting insights and the results can be used to build further upon.

ADMM was introduced in the 1970s by Glowinski and Marocco [14] and Gabay and Mercier [10] and was applied for approximations using conventional finite elements. It was rigorously studied in the optimization community thereafter ([4, 6, 8], and [9] to name a few). For a detailed review of the method we refer to the survey by Boyd et al. [3].

With respect to the parallelization of sparse grid methods, the parallelization of the combination technique for data mining applications was described by Garcke and Griebel [11] as well as by Garcke et al. [13]. Studies of parallelization for spatially adaptive sparse grid algorithms using shared memory parallelization, vectorization, and hardware-accelerators were conducted and published by Heinecke and Pflüger [18, 19].

This paper is organized as follows: Sect. 2 formally defines the class of data mining problems we consider in this paper. It introduces the alternating direction method of multipliers and describes the transformation of the original optimization problem into the constrained optimization problem that is required for ADMM. The study of ADMM for data mining with hierarchical sparse grids is presented in Sect. 3. The results suggest that ADMM for sparse grids is competitive with the state-of-the-art implementation on shared memory systems, although its application on massively parallel systems requires further improvement of the algorithms. Section 4 summarizes and concludes this paper.

2 Theoretical Background

The alternating direction method of multipliers (ADMM) is a further development of the dual ascent algorithm by Arrow et al. [1]. It solves constrained optimization problems of the form

$$\begin{aligned} \min_{\mathbf{x},\mathbf{z}} \quad & h(\mathbf{x}) + g(\mathbf{z}) \\ \text{subject to} \quad & A\mathbf{x} + B\mathbf{z} = \mathbf{c}, \end{aligned} \tag{1}$$

with variables $\mathbf{x} \in \mathbb{R}^n$, $\mathbf{z} \in \mathbb{R}^m$, linear constraints given by vector $\mathbf{c} \in \mathbb{R}^p$ and matrices $A \in \mathbb{R}^{p \times n}$, $B \in \mathbb{R}^{p \times m}$, and convex cost functions $h : \mathbb{R}^n \to \mathbb{R}$, $g : \mathbb{R}^m \to \mathbb{R}$.

The problem is then solved iteratively using the augmented Lagrangian function

$$L_\rho(\mathbf{x}, \mathbf{z}, \mathbf{u}) = h(\mathbf{x}) + g(\mathbf{z}) + \mathbf{u}^T(A\mathbf{x} + B\mathbf{z} - \mathbf{c}) + \frac{\rho}{2}\|A\mathbf{x} + B\mathbf{z} - \mathbf{c}\|_2^2, \tag{2}$$

with the Lagrangian multiplier $\mathbf{u} \in \mathbb{R}^p$ and a positive penalization coefficient $\rho \in \mathbb{R}^+$.

In each iteration of the ADMM algorithm, the $\mathbf{x}$ and $\mathbf{z}$ variables are calculated as minimizers of the augmented Lagrangian followed by the update of the Lagrangian multiplier variable $\mathbf{u}$:

$$\mathbf{x}^{k+1} := \arg\min_{\mathbf{x}} L(\mathbf{x}, \mathbf{z}^k, \mathbf{u}^k) , \tag{3}$$

$$\mathbf{z}^{k+1} := \arg\min_{\mathbf{z}} L(\mathbf{x}^{k+1}, \mathbf{z}, \mathbf{u}^k) \tag{4}$$

$$\mathbf{u}^{k+1} := \mathbf{u}^k + \rho(A\mathbf{x}^{k+1} + B\mathbf{z}^{k+1} - \mathbf{c}) . \tag{5}$$

Obviously, the method can be considered as the splitting of one large vector of variables into two vectors $\mathbf{x}$ and $\mathbf{z}$ updated in a Gauss-Seidel pass. This splitting allows one to exploit the properties of the functions h and g (e.g., differentiability) independent from each other.

The usage of the penalization term $\rho/2\|\cdot\|_2^2$ in the augmented Lagrangian function improves the convergence of the dual ascent algorithm [6]. It can be easily seen that for feasible solutions $(\mathbf{x}, \mathbf{z})$ the penalization term becomes 0. Assuming that the saddle point of the Lagrangian function exists, one can furthermore prove that the saddle point of the Lagrangian function without the penalization term is a saddle point of L_ρ and vice versa [10].

The proof of convergence of the ADMM for closed, proper convex functions $h : \mathbb{R}^n \to \mathbb{R} \cup \{+\infty\}$ and $g : \mathbb{R}^m \to \mathbb{R} \cup \{+\infty\}$ and $\rho \geq 0$ was shown in early days [10]. Recently, He and Yuan were able to give a convergence rate using a variational inequality (VI) reformulation of problem (1) [17]. The authors proved that after T iterations of the ADMM algorithm the average $\tilde{\mathbf{w}}_t = 1/T \sum_{k=1}^{T} \tilde{\mathbf{w}}^k$, with $\tilde{\mathbf{w}}^k = (\mathbf{x}^k, \mathbf{z}^k, \mathbf{u}^k)^T$, would approximate the solution of VI with accuracy $\mathcal{O}(1/T)$.

The data mining problem we consider in this work—regression—is formulated as follows: given a set of (training) data S drawn from a distribution $\mathcal{S}$,

$$S = \{(\mathbf{v}_i, y_i) \in \mathbb{R}^d \times \mathbb{R}\}_{i=1}^n, \tag{6}$$

with target values y_i that presumably were generated by some unknown function $f \in V$ from the inputs $\mathbf{v}_i$ (which, again, can be affected by some noise). The goal is to recover this function f from the data: find a function $\hat{f} \in V$ so that (besides approximating already known values $\hat{f}(\mathbf{v}_i) \approx y_i$, $i = 1, \ldots, m$) it is able to generalize and predict new data $\hat{f}(\mathbf{v}) \approx y$ for all new pairs $(\mathbf{v}, y)$ drawn from the distribution $\mathcal{S}$. Compared to classical function approximation problems in computational science, lower prediction accuracies are sufficient for many data mining applications due to the high degree of noise contained in the available knowledge S.

While V can theoretically be infinite-dimensional, we use sparse grid discretization to find the best approximation of f in a finite dimensional subspace $V_n \subset V$. The function f is then approximated by a linear combination of basis functions $\{\phi_j\}_{j=1}^n$ that span V_n:

$$\hat{f}_n(\mathbf{v}) = \sum_{j=1}^{N} x_j \phi_j(\mathbf{v}).$$

Figure 1 illustrates the so-called modified linear basis functions [21] in one dimension and up to level 3. The first level contains a single constant basis function ϕ_1, the second level two functions ϕ_2 (linear on the interval $[0, 0.5]$ and 0 otherwise) and ϕ_3 (linear on the interval $[0.5, 1]$ and 0 otherwise). Finally, the third level has four functions: linear functions ϕ_4 and ϕ_7 and the classical hat functions ϕ_5

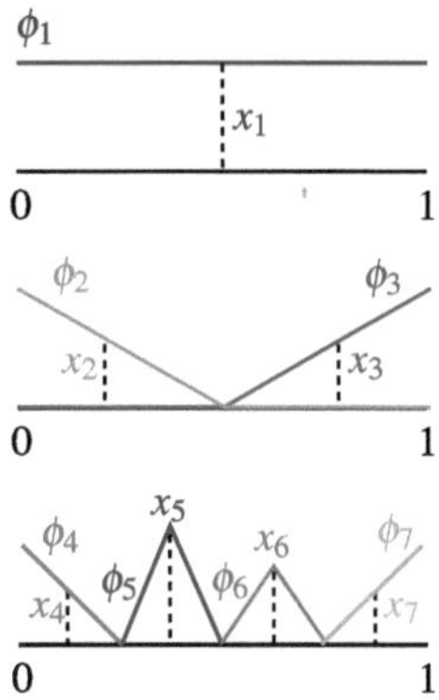

Fig. 1 The weighted modified one-dimensional basis functions ϕ_i which are extrapolating towards the boundary; coefficients x_5 and x_6 are larger than others for illustration

$$B = \begin{pmatrix} \phi_1(\mathbf{v}_1) & \phi_2(\mathbf{v}_1) & \phi_3(\mathbf{v}_1) & \phi_4(\mathbf{v}_1) \\ \phi_1(\mathbf{v}_2) & \phi_2(\mathbf{v}_2) & \phi_3(\mathbf{v}_2) & \phi_4(\mathbf{v}_2) \\ \phi_1(\mathbf{v}_3) & \phi_2(\mathbf{v}_3) & \phi_3(\mathbf{v}_3) & \phi_4(\mathbf{v}_3) \end{pmatrix} \begin{pmatrix} x_1 \\ x_2 \\ x_3 \\ x_4 \end{pmatrix} = \underbrace{\begin{pmatrix} \phi_1(\mathbf{v}_1) & \phi_2(\mathbf{v}_1) \\ \phi_1(\mathbf{v}_2) & \phi_2(\mathbf{v}_2) \\ \phi_1(\mathbf{v}_3) & \phi_2(\mathbf{v}_3) \end{pmatrix} \begin{pmatrix} x_1 \\ x_2 \end{pmatrix}}_{B_1\mathbf{x}_1} + \underbrace{\begin{pmatrix} \phi_3(\mathbf{v}_1) & \phi_4(\mathbf{v}_1) \\ \phi_3(\mathbf{v}_2) & \phi_4(\mathbf{v}_2) \\ \phi_3(\mathbf{v}_3) & \phi_4(\mathbf{v}_3) \end{pmatrix} \begin{pmatrix} x_3 \\ x_4 \end{pmatrix}}_{B_2\mathbf{x}_2}$$

Fig. 2 Illustration of matrix and vector decomposition for ADMM

and ϕ_6. This idea carries on for the next levels: the two basis functions closest to the boundary are linear, extrapolating towards the boundary, while all the inner ones are the classical hat functions. In Fig. 1, the scaling coefficients of the basis functions are depicted as $x_1, \dots, x_7$.

Approximating function $\hat{f}_n$ is hence reduced to solving the regularized least-squares problem for coefficients $\{x_j\}_{j=1}^n$ [21],

$$\min_{\mathbf{x}\in\mathbb{R}^n} \frac{1}{2}\|B\mathbf{x} - \mathbf{y}\|_2^2 + \lambda\|\mathbf{x}\|_2^2, \tag{7}$$

with $B = \begin{pmatrix} \phi_1(\mathbf{v}_1) & \phi_2(\mathbf{v}_1) & \dots & \phi_N(\mathbf{v}_1) \\ \vdots & \vdots & \ddots & \vdots \\ \phi_1(\mathbf{v}_n) & \phi_2(\mathbf{v}_n) & \dots & \phi_N(\mathbf{v}_n) \end{pmatrix}$, coefficient vector $\mathbf{x} = \begin{pmatrix} x_1 \\ \vdots \\ x_N \end{pmatrix}$, and regularization parameter λ. Note that this choice of regularization is only reasonable for hierarchical basis functions.

Following [3], we split the coefficient vector $\mathbf{x}$ into P subvectors $\mathbf{x} = (\mathbf{x}_1^T, \dots, \mathbf{x}_P^T)^T$ and break down the matrix-vector product $B\mathbf{x}$ into the sum of smaller matrix-vector products $B_i\mathbf{x}_i$ as Fig. 2 illustrates. Thus we can rewrite the problem (7) in the form

$$\min_{(\mathbf{x}_1^T,\dots,\mathbf{x}_P^T)^T\in\mathbb{R}^n} \frac{1}{2}\|\sum_{i=1}^P B_i\mathbf{x}_i - \mathbf{y}\|_2^2 + \sum_{i=1}^P \lambda\|\mathbf{x}_i\|_2^2 \tag{8}$$

so that by splitting the problem into a sum of two functions we obtain an optimization problem in the form required for ADMM:

$$\begin{aligned} &\min \ g(\sum_{i=1}^{P} \mathbf{z}_i) + \sum_{i=1}^{P} h_i(\mathbf{x}_i) \\ &\text{subject to} \ \ B_i \mathbf{x}_i = \mathbf{z}_i \quad \forall i = 1, \ldots, P \ , \end{aligned} \tag{9}$$

with convex functions $g(\mathbf{z}) = \frac{1}{2}\|\mathbf{z} - \mathbf{y}\|^2$ and $h_i(\mathbf{x}) = \lambda\|\mathbf{x}\|^2$. We now solve it using the ADMM algorithm in three steps:

$$\mathbf{x}_i^{k+1} := \arg\min_{\mathbf{x}_i} \left\{ h_i(\mathbf{x}_i) + \frac{\rho}{2}\|B_i\mathbf{x}_i - \mathbf{z}_i^k + \frac{1}{\rho}\mathbf{u}_i^k\|_2^2 \right\} , \tag{10}$$

$$\mathbf{z}^{k+1} := \underset{\mathbf{z}=(\mathbf{z}_1^T,\ldots,\mathbf{z}_P^T)^T}{\arg\min} \left\{ g(\sum_{i=1}^{P}\mathbf{z}_i) + \frac{\rho}{2}\sum_{i=1}^{P}\|B_i\mathbf{x}_i^{k+1} - \mathbf{z}_i + \frac{1}{\rho}\mathbf{u}_i^k\|_2^2 \right\} , \tag{11}$$

$$\mathbf{u}_i^{k+1} := \mathbf{u}_i^k + \rho\left(B_i\mathbf{x}_i^{k+1} - \mathbf{z}_i^{k+1}\right). \tag{12}$$

It can be proven (see, e.g., [3]) that solving the minimization step (11) is equivalent to solving

$$\bar{\mathbf{z}}^{k+1} = \arg\min_{\mathbf{z}} \ g(P\bar{\mathbf{z}}) + \frac{\rho}{2}P\|\mathbf{z} - \frac{1}{P}\sum_{i=1}^{P}B_i\mathbf{x}_i^{k+1} - \frac{1}{P\rho}\sum_{i=1}^{P}\mathbf{u}_i^{k+1}\|_2^2 \tag{13}$$

and

$$\mathbf{z}_i^{k+1} = B_i\mathbf{x}_i^{k+1} + \frac{1}{\rho}\mathbf{u}_i^{k+1} + \bar{\mathbf{z}}^{k+1} - \frac{1}{P}\sum_{i=1}^{P}B_i\mathbf{x}_i^{k+1} - \frac{1}{P\rho}\sum_{i=1}^{P}\mathbf{u}_i^{k+1} . \tag{14}$$

Finally, by substituting (13) and (14) into (10)–(12), one can see that the different Laplacian multiplier vectors $\mathbf{u}_i$ become equal and we finally obtain the equations

$$\mathbf{x}_i^{k+1} := \arg\min_{\mathbf{x}_i} \left\{ \lambda\|\mathbf{x}_i\|_2^2 + \frac{\rho}{2}\|B_i\mathbf{x}_i - B_i\mathbf{x}_i^k - \bar{\mathbf{z}}^k + \frac{1}{P}\sum_{i=1}^{P}B_i\mathbf{x}_i^k + \frac{1}{\rho}\mathbf{u}^k\|_2^2 \right\} , \tag{15}$$

$$\bar{\mathbf{z}}^{k+1} := \arg\min_{\mathbf{z}} \left\{ \frac{1}{2}\|P\cdot\mathbf{z} - \mathbf{y}\|_2^2 + \frac{P\cdot\rho}{2}\|\mathbf{z} - \frac{1}{P}\sum_{i=1}^{P}B_i\mathbf{x}_i^{k+1} - \frac{1}{\rho}\mathbf{u}^k\|_2^2 \right\} , \tag{16}$$

$$\mathbf{u}^{k+1} := \mathbf{u}^k + \rho\left(\frac{1}{P}\sum_{i=1}^{P}B_i\mathbf{x}_i^{k+1} - \bar{\mathbf{z}}^{k+1}\right) . \tag{17}$$

Algorithm 1 ADMM algorithm for sparse grids

{Initialize variables:}
$\mathbf{x}_i \leftarrow \mathbf{0}$;
$\mathbf{z} \leftarrow \frac{1}{P}\mathbf{y}$;
$\mathbf{u} \leftarrow \mathbf{0}$;
precompute B_i;
repeat
  update $\mathbf{x}_i$ by solving (18) simultaneously in all processes;
  `Allreduce`($B_i\mathbf{x}_i$, $B\mathbf{x}$, SUM); {sum up the vectors $B_i\mathbf{x}_i$ from all processes}
  update $\mathbf{z} \leftarrow \frac{1}{P+\rho}\left(\mathbf{y} + \frac{\rho}{P}B\mathbf{x} + \mathbf{u}\right)$;
  update $\mathbf{u} \leftarrow \mathbf{u} + \frac{\rho}{P}B\mathbf{x} - \rho\mathbf{z}$;
until $\|B\mathbf{x}-\mathbf{y}\|_2^2 \leq$ convergence threshold; {training error is one possible convergence criterion}
`Gatherv`($\mathbf{x}_i$, $\mathbf{x}$); {collect partial results}
return $\mathbf{x}$;

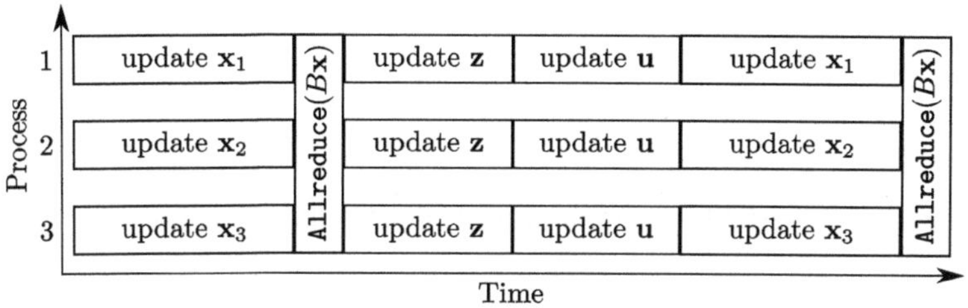

Fig. 3 Illustration of the ADMM algorithm

The update steps (16) and (17) are performed with simple linear algebra operations, while the update step (15) requires the solution of a system of linear equations:

$$\left(\frac{\lambda}{2}I + \rho B_i^T B_i\right)\mathbf{x}_i = \rho B_i^T\left(B_i\mathbf{x}_i^k + \bar{\mathbf{z}}^k - \frac{1}{P}\sum_{i=1}^{P} B_i\mathbf{x}_i^{k+1} - \frac{1}{\rho}\mathbf{u}^k\right). \tag{18}$$

At this point we would like to give a couple of remarks about the implementation of the ADMM regression. Since the last step requires the repeated solution of a linear system with the same system matrix and different right-hand side vectors, it is efficient to precompute the Cholesky factorization of the matrix $(\lambda/2I + \rho B_i^T B_i)$ at the initialization step of the algorithm. If ADMM runs on a distributed system and subproblems are solved on different machines, the processes need to communicate the sum $\sum_{i=1}^{P} B_i\mathbf{x}_i^{k+1}$ amongst each other after the update step (15). This can be done by an MPI operation `Allreduce`. No other communication between processes is required, as steps (16) and (17) are not computationally costly and can be performed independently by each process. After ADMM converges, the sub-vectors $\mathbf{x}_i$ need to be gathered only once. The overall method is summarized in Algorithm 1.

The sequence diagram in Fig. 3 further illustrates the structure of the ADMM algorithm. After the update of their vector x_i, processes synchronize their internal values of the vector $B\mathbf{x}$ using collective communication. No other synchronization is required.

3 Results

In the following, we show numerical results for the application of ADMM for regression problems with the direct sparse grid method. We are employing regular sparse grids with the modified linear basis functions shown before. This choice of basis functions allows us to approximate even those points that lie close to the boundaries without having to spend extra degrees of freedom on the boundary. Unless mentioned otherwise, we measure performance for regular sparse grids with level 6 for our experiments. We focus primarily on the optimization algorithm and do not discuss the optimal regularization parameter or convergence criteria to prevent overfitting on the dataset.

As testing data we picked the Friedman #2 data set [7], a classical benchmark problem. It is an artificial four-dimensional data set that simulates the impedance in an alternating current circuit. In most experiments, we used 20,016 data points for training,[1] generated by the function

$$y = x_1^2 + \sqrt{x_2 x_3 - (x_2 x_4)^{-2}} + \varepsilon, \tag{19}$$

where the random variables x_i are uniformly distributed in the ranges

$$0 < x_1 < 100, \quad 20 < \frac{x_2}{2\pi} < 280, \quad 0 < x_3 < 1, 1 < x_4 < 11,$$

and ε is additional uniformly distributed noise with mean 0 and variance 125.

Solving the Friedman problem with ADMM, we observe the peculiarity of the algorithm: The memory footprint of the individual system matrices drops quadratically with the growing number of processors (e.g., if the number of CPUs doubles, the size of the system matrix is reduced by the factor of 4). And, hence, the *time of one ADMM iteration* also decreases nearly quadratically.[2] At the same time, the *number of ADMM iterations* until convergence will raise, since more subproblems need to agree on the common solution. This rise, however, is difficult to estimate theoretically.

It is apparent that for large problems the total time is approximated by the duration of one ADMM iteration times the number of ADMM iterations. Hence, as long as individual iterations are dominated by back-substitution and the number of iterations grows less than quadratically, adding more processors leads to speedup.

Moreover, we would like to stress that decreasing memory footprint of subproblems would allows the solution of some high-dimensional problems in the

---

[1] The implementation of sparse grids algorithms used in this paper requires the dataset size to be a multiple of 48 in order to improve data padding. Therefore we are using 20,016 data points instead of 20,000.

[2] The complexity of individual ADMM operations is dominated by back-substitution in the x-update step (15), which is linear in the size of the system matrix.

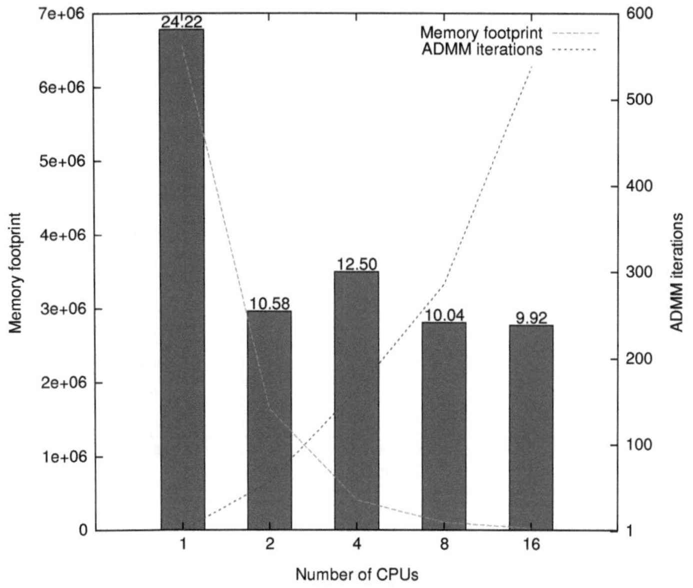

Fig. 4 Wall-clock time, number of iterations and memory footprint of the individual system matrices for ADMM with different number of processors trained with 20,016 data entries and a sparse grid of level 6. The more processes, the less time is spent per ADMM iteration, but the more iterations until convergence are required

first place as those problems would not fit into memory of any single computational node.

Figure 4 illustrates this behavior: While the serial version of the ADMM algorithm naturally requires only one ADMM-iteration, it also has the highest memory requirement to store the system matrix. Using ADMM with two processors, and thus subtasks, allows the algorithm to converge in 57 steps, whereas the required memory per processor is reduced by the factor of 4. These trends proceed as the number of processors grows: The realization with four processors requires 163 iterations to converge to the same error on the training data, and even though each single individual ADMM iteration takes less time for four subtasks, the overall runtime is higher for the same test-error convergence criterion.

While the algorithm requires further improvement, the comparison with the implementation of sparse grid regression in the software package SG^{++} [20] already shows the advantage of the ADMM implementation. Figure 5 provides the wall-clock time for the same learning task of both implementations. Note that we ensured that SG^{++} and ADMM iterate until the same value of the regularized loss functional is reached. SG^{++} uses a matrix-free conjugate gradient method. On the left, we show the results on an AMD Opteron™ 6128HE system equipped with eight processors with 2 GHz and on the right the results on an Intel i7-2600 running four processors with 3.4 GHz. We turned on the intrinsics supported by SG^{++}: SSE3 on

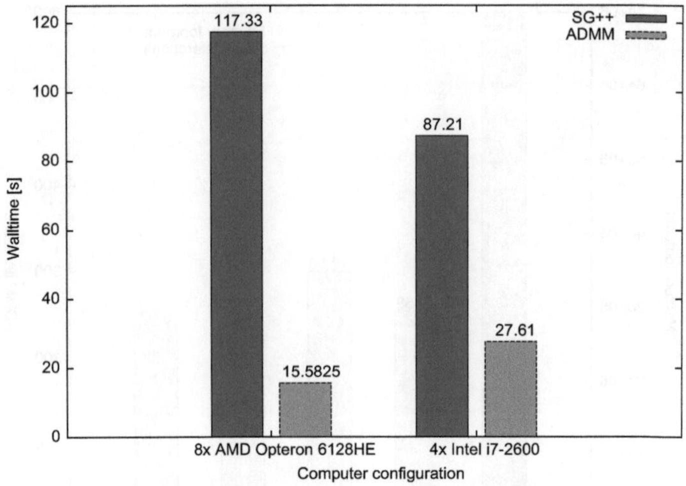

Fig. 5 Comparison of the ADMM method with a state-of-the-art implementation trained with 20,016 data points and sparse grid level 7. On both systems, ADMM shows its computational advantage

the AMD system and AVX on the Intel system.[3] On both systems, we are faster using ADMM than with the matrix-free conjugate gradient method.

Even though the number of processors on the Intel system is half that of the AMD system, the AVX implementation of the algorithms used in SG^{++} allows to process twice as many arithmetic operations in one clock cycle as the SSE3 implementation used on the older AMD systems. This and the difference between the CPU frequencies on these two machines are responsible for the fact that SG^{++} is faster on the Intel system.

The implementation of ADMM on the Intel system uses the GotoBLAS2 library [16], which does not support AVX. Therefore, we observe an increase of the computational time for 4-core Intel system compared to the 8-core Opteron. This increase, however, is less than by a factor of 2 due to the difference in the CPU frequency and the different number of ADMM iterations that is required for convergence on different numbers of processors as described above.

The ADMM implementation needs significantly less time to solve the same minimization problem as the current SG^{++} implementation since the small submatrices B_i can be efficiently stored in memory and thus directly used for computation. These results are promising, though they do not illustrate the scaling properties of different implementations. If the number of ADMM iterations that is required is very high, this computational advantage will disappear. In the remainder of this section we therefore focus on the methods to improve the convergence.

[3]For further details on the parallelization of sparse grid regression using intrinsics we refer to [18].

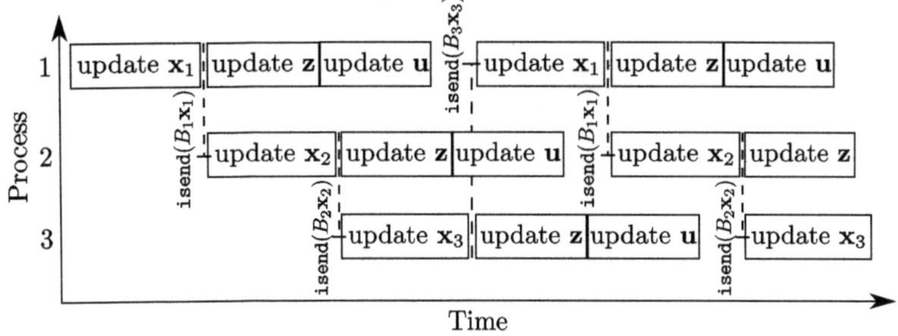

Fig. 6 Gauss-Seidel implementation of the **x**-update step with asynchronous communication

One drawback of the ADMM algorithm used with hierarchical basis functions in contrast to conventional bases is the problem that initially every process will tend to approximate the solution for scaled target values $\mathbf{z}^0 := 1/P \cdot \mathbf{y}$. As all values of $\mathbf{x}$ and $\mathbf{u}$ are set to 0 at the beginning, step (15) is reduced to finding a regularized least squares approximation of $\mathbf{z}^0$. This leads to wrong initial values of $\mathbf{x}_i$ and to slower convergence towards the correct ones. This explains the increase in the number of iterations until convergence for a higher number of subtasks.

ADMM thus better suits to a set of independent approximants, whose contributions to the overall result are of the similar size, than to the decomposition of hierarchical basis functions into subsets. For the hierarchical basis, we would rather require that the functions with larger support, those on the first levels, approximate the target values coarsely, and that then those on the higher levels take care of the residual.

The desired operation can be achieved if the $\mathbf{x}$-update step (15) is carried out in Gauss-Seidel manner ($\mathbf{x}_i^{k+1}$ is updated using new values $\mathbf{x}_0^{k+1}, \ldots, \mathbf{x}_{i-1}^{k+1}$ and old values $\mathbf{x}_{i+1}^k, \ldots, \mathbf{x}_P^k$) instead of in Jacobi manner ($\mathbf{x}_i^{k+1}$ is updated using new values $\mathbf{x}_0^k, \ldots, \mathbf{x}_{i-1}^k, \mathbf{x}_{i+1}^k, \ldots, \mathbf{x}_P^k$).

If we order the basis functions such that level sums are ascending and then partition them into subproblems, processor 1 will obtain the basis functions with the largest support and impact onto the whole approximation. At the beginning of its iteration processor 2 receives the new value of $B_1\mathbf{x}_1$ and estimates $B_2\mathbf{x}_2$ correcting for residual and so forth.

This Gauss-Seidel pass can be implemented using asynchronous communication of the vector $\sum_{i=1}^{P} B_i\mathbf{x}_i$ between processors in a ring. In the iteration (k+1) processor j receives the sum $\sum_{i=1}^{j-1} B_i\mathbf{x}_i^{k+1} + \sum_{i=j}^{P} B_i\mathbf{x}_i^k$, estimates the new coefficient vector $\mathbf{x}_j^{k+1}$ and sends the updated sum $\sum_{i=1}^{j} B_i\mathbf{x}_i^{k+1} + \sum_{i=j+1}^{P} B_i\mathbf{x}_i^k$ to the next processor ($j + 1 \mod P$). Figure 6 illustrates the time line of the asynchronous algorithm for three processes (compare to Fig. 3 with synchronous communication).

Figure 7 shows how the usage of Gauss-Seidel pass improves convergence. We further observed that if we rescale the variables $\mathbf{z}$ and $\mathbf{u}$ (e.g., using the factor

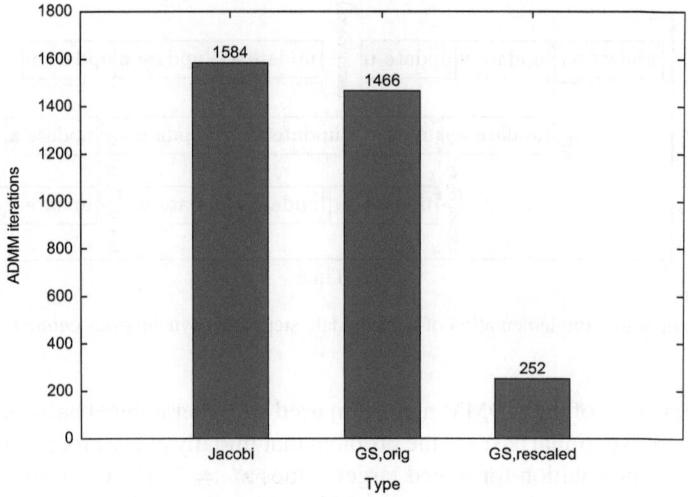

Fig. 7 The number of ADMM iterations for different values of the scaling parameter, trained on 20,016 data points for sparse grid level 6 using eight processors. The algorithmic realization with the Gauss-Seidel pass shows generally better convergence than the classical Jacobi pass. For the Gauss-Seidel realization, the scaling factor of 1 leads to better convergence

1 instead of P in Eqs. (15)–(17)) the convergence of the Gauss-Seidel algorithm further improves.[4] With original scaling factors the first subproblem approximates only at the order of $1/P\mathbf{y}$ which contradicts the intuition of residual correction described above and leads to larger number of ADMM iterations.

Furthermore, we observed that the convergence of the Gauss-Seidel implementation is invariant to the choice of the penalization parameter ρ. Choosing the right value for the classical ADMM algorithm and the parallel Jacobi implementation is rather tricky and was often discussed in the literature [3, 15, 17]. However, the theoretical treatment of the impact of the penalization parameter was conducted mainly for the original version of ADMM algorithm without parallelization. The behavior of the parallel version differs [2] and in our experience the heuristics for dynamic penalty adjustment do not lead to desired results.

While the parallel Jacobi implementation suffers from the poor choice of ρ, the parallel Gauss-Seidel implementation shows a very stable behavior. Figure 8 contrasts the number of iterations for the different values of ρ for both algorithms. While the number of iterations for Jacobi implementation grows rapidly once the value of the penalization parameter leaves the interval $[0.1, 0.2]$, the number of iterations for Gauss-Seidel implementation remains the same.

Unfortunately, even with non-blocking communication techniques the Gauss-Seidel algorithmic implementation can lead to dramatic computation/communication imbalance. Since the $\mathbf{x}$-update step dominates iterations and $\mathbf{z}$- and

[4]The similar rescaling of equations for a Jacobi pass leads to divergence of the algorithm.

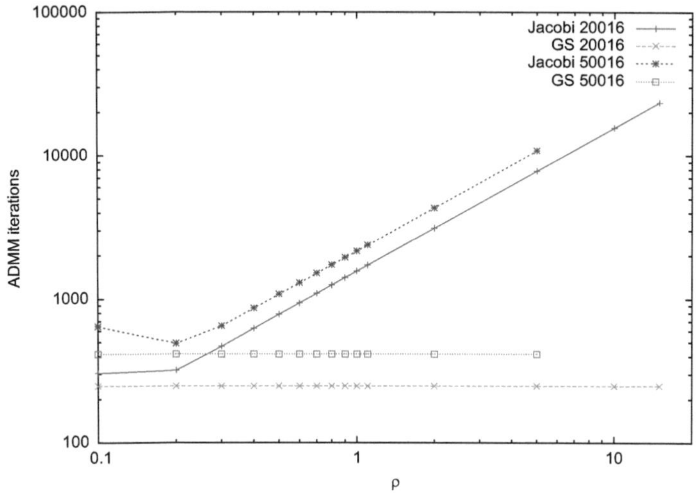

Fig. 8 Comparison of the Jacobi and Gauss-Seidel implementation for approximation of Friedman 2 dataset with 20,016 points and sparse grid level 6 on 8 processors as well as with 50,016 points and sparse grid level 6 on 16 processors. Gauss-Seidel implementation is stable across different values of the penalization parameter ρ

u-updates can be negligibly small, in the worst case the time for communication and interruption could take up to $P - 1$ times that for the computation. This imbalance would cancel any convergence improvements.

We therefore combine the both methods to accelerate the convergence with a few expensive Gauss-Seidel steps at the beginning followed by faster Jacobi steps. Figures 9 and 10 illustrate the advantages of this approach. They compare the number of ADMM iterations and the total time required to solve same problems using only Jacobi iterations with the cases where they are preceded by 1, 5, or 10 Gauss-Seidel iteration steps. Although, the number of iterations continues to decrease the more Gauss-Seidel iterations we allow, the run with only one Gauss-Seidel step takes the least time.

Finally, another way to improve convergence for hierarchical basis functions is to enforce that subproblems share the functions with larger support on the first levels. This reflects the property of the hierarchical basis that the basis functions on the first levels are more important than those on higher levels. We illustrate this principle using the example of an one-dimensional grid with level three (compare, again, Fig. 1).

Suppose that the calculations are to be split between two processors, then process 1 would calculate the coefficients for the functions on the top two levels, ϕ_1, ϕ_2, and ϕ_3, as well as for two on level 3, e.g., ϕ_4 and ϕ_5. Process 2, would then calculate the scaling coefficients for the same functions on the top two levels as well as for the other two on level 3, namely ϕ_6 and ϕ_7. Hence, the coefficient vector of the process

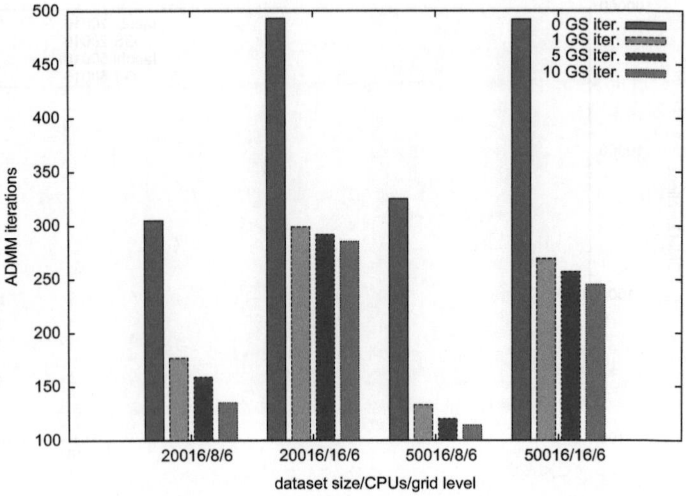

Fig. 9 The number of ADMM iterations for training of sparse grids with levels 6 and 8 on 20,016 and 50,016 data points. The algorithms runs for a given number of Gauss-Seidel iterations steps and then switches to Jacobi iterations steps. The convergence improves with the number of Gauss-Seidel steps, but only insignificantly

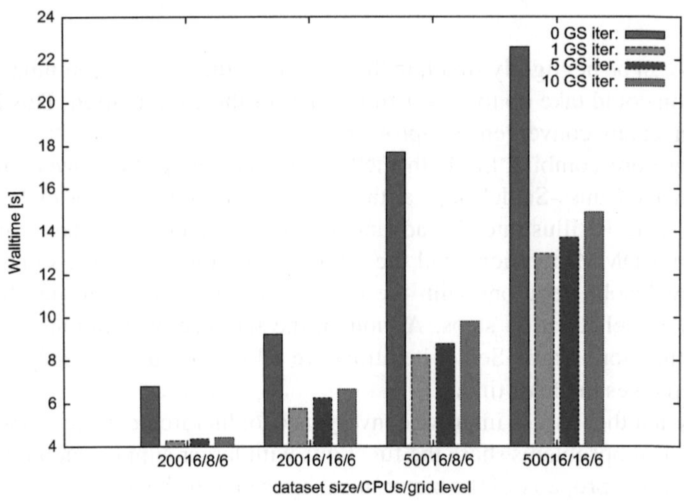

Fig. 10 The number of ADMM iterations for training of sparse grids with levels 6 and 8 on 20,016 and 50,016 data points. The algorithms runs for a given number of Gauss-Seidel iterations steps and then switches to Jacobi iterations steps. The run with only one Gauss-Seidel step takes the least time

1 would be $\mathbf{x}_1 = (x_1, x_2, x_3, x_4, x_5)^T$, while the coefficient vector of the process 2 would be $\mathbf{x}_2 = (x_1, x_2, x_3, x_6, x_7)^T$. Correspondingly, the matrix B_1 would contain columns 1, 2, 3, 4, and 5 of the original matrix B, while the matrix B_2 would contain columns 1, 2, 3, 6, and 7:

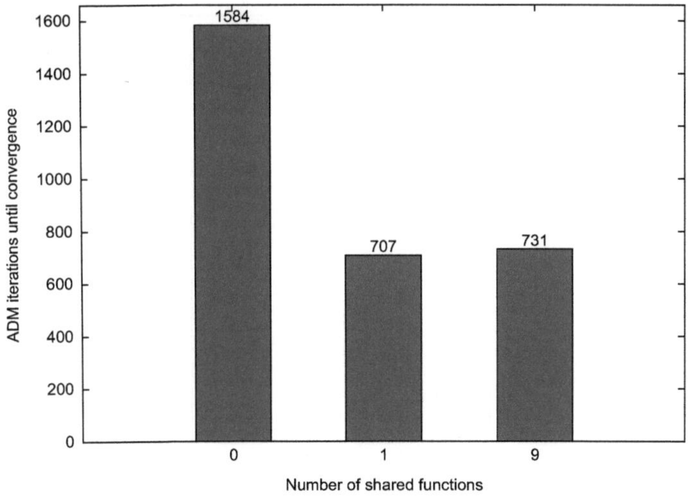

Fig. 11 ADMM convergence (number of iterations) when sharing the first basis functions trained with 20,016 data entries and sparse grid level 6. One shared function corresponds to level 1, 9 shared functions correspond to levels 1 and 2. Sharing the function of the first levels significantly improves convergence behavior

$$B_1 = \begin{pmatrix} \phi_1(\mathbf{v}_1) & \phi_2(\mathbf{v}_1) & \phi_3(\mathbf{v}_1) & \phi_4(\mathbf{v}_1) & \phi_5(\mathbf{v}_1) \\ \phi_1(\mathbf{v}_2) & \phi_2(\mathbf{v}_2) & \phi_3(\mathbf{v}_2) & \phi_4(\mathbf{v}_2) & \phi_5(\mathbf{v}_2) \\ \phi_1(\mathbf{v}_3) & \phi_2(\mathbf{v}_3) & \phi_3(\mathbf{v}_3) & \phi_4(\mathbf{v}_3) & \phi_5(\mathbf{v}_3) \end{pmatrix},$$

$$B_2 = \begin{pmatrix} \phi_1(\mathbf{v}_1) & \phi_2(\mathbf{v}_1) & \phi_3(\mathbf{v}_1) & \phi_6(\mathbf{v}_1) & \phi_7(\mathbf{v}_1) \\ \phi_1(\mathbf{v}_2) & \phi_2(\mathbf{v}_2) & \phi_3(\mathbf{v}_2) & \phi_6(\mathbf{v}_2) & \phi_7(\mathbf{v}_2) \\ \phi_1(\mathbf{v}_3) & \phi_2(\mathbf{v}_3) & \phi_3(\mathbf{v}_3) & \phi_6(\mathbf{v}_3) & \phi_7(\mathbf{v}_3) \end{pmatrix}.$$

If we denote the shared columns of a matrix B_i by B_i^{shared} (these would be the columns 1, 2, and 3 in the example above) and the shared components of a vector $\mathbf{x}_i$ by $\mathbf{x}_i^{\text{shared}}$ (the coefficients x_1, x_2, and x_3 in the example above), then the evaluation of the sparse grid function calculated as $\sum_{i=1}^{P} B_i \mathbf{x}_i$ and used in steps (15)–(17) can be approximated by

$$B\mathbf{x} \approx \sum_{i=1}^{P} B_i \mathbf{x}_i - \frac{d-1}{d} \sum_{i=1}^{P} B_i^{\text{shared}} \mathbf{x}_i^{\text{shared}}. \tag{20}$$

Figure 11 shows the number of ADMM iterations until convergence is reached for problem splittings with different number of shared functions. The results show that especially sharing the function on the first level allows a significant convergence speedup, while the additional computational cost remains negligible.

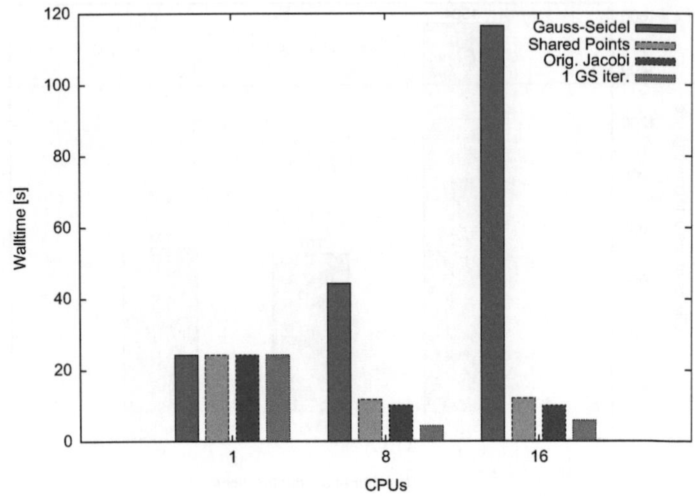

Fig. 12 Total time to train a sparse grid with level 6 on 20,016 data points in parallel using 8 and 16 processors. The algorithm extension with an Gauss-Seidel initialization step performs best

At last, we compare the original implementation of the ADMM method for sparse grids with the extensions described in this paper. Figure 12 shows the time required to train a sparse grid with level 6 on 20,016 data points in parallel using 8 or 16 processors.

The pure Gauss-Seidel implementation takes the longest due to the long waiting times. Because of an additional ALLREDUCE call, the shared point implementation also requires more time than the original Jacobi, despite the reduced number of iterations. The combination of one Gauss-Seidel initialization step followed by simultaneous Jacobi steps shows the best results.

4 Conclusions

We introduced ADMM as a new approach to sparse grid regression with promising potential for parallelization. However, and in contrast to nodal bases as they are considered for ADMM elsewhere, working with a hierarchical basis raised a few new issues. The local updates have stronger and more direct global influence, and these numerical dependencies result in a strong increase in ADMM iterations with increasing processor count.

To remedy this, we have introduced a Gauss-Seidel style pass for the **x**-update only in the first iteration. Introducing sequential dependencies significantly improved the convergence rate. Apart from the first iteration, we do not introduce communication overhead and employ Jacobi-style passes. In the future we endeavor to relax the inter-process dependency in the first iteration using pre-initialization of individual solutions and introducing one-sided and asynchronous communication patterns.

In addition, we have shown that a similar effect can be achieved if all processes share the basis functions on the first levels and then use an approximation of $B\mathbf{x}$ during synchronization. They are the basis functions with large supports and have thus rather global influence. Despite the algorithmic changes and improvements and new parallel dependencies, the computational time of a single iteration is almost unaffected.

In our implementation we assembled the system matrices $(\lambda/2 \cdot I + \rho B_i^T B_i)$ in (18) explicitly. This is not a drawback as these matrices are significantly smaller than the original system matrix from problem (7). Hence, they can fit into memory even for huge problems (data points and/or grid size) and can be efficiently solved for many relevant problems. This is another advantage of our approach.

Furthermore, preliminary results show that ADMM with sparse grids is competitive with other state-of-the-art sparse grids methods in our data mining setting on shared memory systems with moderate number of processors. These results, however, are to be investigated more in detail and in a broader scope as part of further research. In the future, we also want to address the major problem of ADMM—the increasing number of iterations with growing processor numbers—in more detail. This will be a necessary step towards massively parallel data mining with ADMM in the hierarchical basis.

References

1. K. Arrow, L. Hurwicz, H. Uzawa, *Studies in Linear and Non-linear Programming* (Stanford University Press, Stanford, 1958)
2. D.P. Bertsekas, J.N. Tsitsiklis, *Parallel and Distributed Computation: Numerical Methods* (Prentice-Hall, Upper Saddle River, 1989)
3. S. Boyd, N. Parikh, E. Chu, B. Peleato, J. Eckstein, Distributed optimization and statistical learning via the alternating direction method of multipliers. Found. Trends Mach. Learn. **3**(1), 1–122 (2010)
4. J. Eckstein, M. Fukushima, Some reformulations and applications of the alternating direction method of multipliers, in *Large Scale Optimization: State of the Art* (Springer, US, 1994), pp. 115–134
5. M.A.T. Figueiredo, J.M. Bioucas-Dias, Restoration of Poissonian images using alternating direction optimization. IEEE Trans. Image Process. Publ. IEEE Signal Proc. Soc. **19**(12), 3133–3145 (2010)
6. M. Fortin, R. Glowinski, Augmented Lagrangian methods in quadratic programming, in *Augmented Lagrangian Methods: Applications to the Numerical Solution of Boundary-Value Problems*. Studies in Mathematics and its Applications, vol. 15 (Springer, Berlin, 1983), pp. 1–46
7. J. Friedman, Multivariate adaptive regression splines. Ann. Stat. **19**(1), 1–67 (1991)
8. M. Fukushima, Application of the alternating direction method of multipliers to separable convex programming problems. Comput. Optim. Appl. **1**(1), 93–111 (1992)
9. D. Gabay, Applications of the method of multipliers to variational inequalities, in *Augmented Lagrangian Methods: Applications to the Numerical Solution of Boundary-Value Problems*,

ed. by M. Fortin, R. Glowinski. Studies in Mathematics and Its Applications, vol. 15 (Elsevier, New York, 1983), pp. 299–331

10. D. Gabay, B. Mercier, A dual algorithm for the solution of nonlinear variational problems via finite element approximation. Comput. Math. Appl. **2**(1), 17–40 (1976)
11. J. Garcke, M. Griebel, *On the Parallelization of the Sparse Grid Approach for Data Mining* (Springer, Berlin, 2001), pp. 22–32
12. J. Garcke, M. Griebel, M. Thess, Data mining with sparse grids. Computing **67**(3), 225–253 (2001)
13. J. Garcke, M. Hegland, O. Nielsen, Parallelisation of sparse grids for large scale data analysis, in *Computational Science — ICCS 2003*, ed. by P.M.A. Sloot, D. Abramson, A.V. Bogdanov, Y.E. Gorbachev, J.J. Dongarra, A.Y. Zomaya. Lecture Notes in Computer Science, vol. 2659 (Springer, Berlin, 2003), pp. 683–692
14. R. Glowinski, A. Marroco, Sur l'approximation, par éléments finis d'ordre un, et la résolution, par pénalisation-dualité d'une classe de problèmes de dirichlet non linéaires. Revue française d'automatique, informatique, recherche opérationnelle. Analyse numérique **9**(2), 41–76 (1975)
15. R. Glowinski, P.L. Tallec, *Augmented Lagrangian Methods for the Solution of Variational Problems*, Chap. 3 (Society for Industrial and Applied Mathematics, Philadelphia 1989), pp. 45–121
16. K. Goto, R. Van De Geijn, High-performance implementation of the level-3 BLAS. ACM Trans. Math. Softw. **35**(1), 1–4 (2008) [Article 4]
17. B. He, H. Yang, S. Wang, Alternating direction method with self-adaptive penalty parameters for monotone variational inequalities. J. Optim. Theory Appl. **106**(2), 337–356 (2000)
18. A. Heinecke, D. Pflüger, Multi- and many-core data mining with adaptive sparse grids, in *Proceedings of the 8th ACM International Conference on Computing Frontiers* (ACM, New York, 2011), pp. 29:1–29:10
19. A. Heinecke, D. Pflüger, Emerging architectures enable to boost massively parallel data mining using adaptive sparse grids. Int. J. Parallel Prog. **41**(3), 357–399 (2013)
20. D. Pflüger, SG++ (2013). http://www5.in.tum.de/SGpp
21. D. Pflüger, *Spatially Adaptive Sparse Grids for High-Dimensional Problems* (Verlag Dr. Hut, München, 2010)
22. G. Steidl, T. Teuber, Removing multiplicative noise by Douglas-Rachford splitting methods. J. Math. Imaging Vis. **36**, 168–184 (2010)
23. X. Zhang, M. Burger, S. Osher, A unified primal-dual algorithm framework based on Bregman iteration. J. Sci. Comput. **46**(1), 20–46 (2010)

An Opticom Method for Computing Eigenpairs

Christoph Kowitz and Markus Hegland

Abstract Solving the linearized and five-dimensional gyrokinetic equations can already be used to retrieve the microinstabilities driving the microturbulence of a hot magnetized plasma. The microinstabilities can be computed by calculating the eigenvalues of the linear gyrokinetic operator, which have a positive real part. The growth rate and frequency of the instability is given by the computed eigenvalue and its structure by the respective eigenvector. Due to the moderately high dimensionality, the sparse grid combination technique is used to tackle the curse of dimensionality for solving the gyrokinetic eigenvalue problem with the already highly optimized plasmaturbulence code GENE. Whereas the classical combination technique can retrieve approximation of the searched eigenvalues, the combination of the eigenvectors requires the application and reformulation of the optimized combination technique (OptiCom). The reformulation and an algorithm for solving the reformulated system to compute the eigenpairs is proposed in this paper. A first analytical test problem is solved and the applicability of the method is shown.

1 Introduction

In the field of fusion energy research, the hot magnetized plasmas can be simulated by using a variety of approaches including single particle descriptions, magnetohydrodynamics (MHD) or kinetic models. Each of the models has a certain application

C. Kowitz (✉)
Institute for Advanced Study, Technische Universität München,
Lichtenbergstrasse 2a, D-85748 Garching, Germany
e-mail: kowitz@in.tum.de

M. Hegland
Australian National University, Canberra, ACT, Australia
e-mail: markus.hegland@anu.edu.au

J. Garcke and D. Pflüger (eds.), *Sparse Grids and Applications - Munich 2012*, Lecture Notes in Computational Science and Engineering 97,
DOI 10.1007/978-3-319-04537-5_10,

area. The five-dimensional gyrokinetic equations can be utilized to simulate the microscopic turbulence effects, leading to anomalous transport in current magnetic confinement devices such as tokamaks and stellarators. Ab initio simulations of microturbulence can be done using gyrokinetics and the transport of heat and particles can be directly observed in a simulation. One gyrokinetic code being widely used in the fusion community, is the massively parallel plasma turbulence code GENE [9], which is developed in the group of Prof. Frank Jenko at the Max-Planck-Institute for Plasmaphysics (IPP) in Garching, Germany. Despite its high performance and its good scaling properties [10], GENE is influenced by the curse of dimensionality due to the underlying five-dimensional model. Even setups with moderate resolutions lead to a high number of degrees of freedom, since it is a five-dimensional problem.

As one remedy to this problem, sparse grids [3] have been identified. Due to the fact, that GENE cannot be easily refactored to use a hierarchical basis required by sparse grids, the sparse grid combination technique [11] is applied to create sparse grid solutions of gyrokinetic problems. Whereas a previous study [17] evaluated the application of the classical combination technique with its integer valued combination coefficients for the linear gyrokinetic initial value problem, this paper focusses on the application of the optimized combination technique (OptiCom) to the gyrokinetic eigenvalue problem. The classical combination could be used to combine the eigenvalues, but it cannot be applied to the corresponding eigenvectors. Thus a new approach which uses the OptiCom for eigenvalue problems will be proposed, which shall be used in the context of linear gyrokinetics. An analytic test problem and the performance of the algorithm will be presented.

1.1 The Linear Gyrokinetic Eigenvalue Problem

The gyrokinetic equation is a five-dimensional description of a magnetized plasma, which is derived from the six-dimensional Vlasov-Equation

$$\frac{\partial f_s}{\partial t} + \mathbf{v}\frac{\partial f_s}{\partial \mathbf{x}} + \left(\frac{q_s}{m_s}\left(\mathbf{E}(f_s) + \mathbf{v}\times\mathbf{B}(f_s)\right)\right)\frac{\partial f_s}{\partial \mathbf{v}} = \mathrm{C}(f_s)\,, \tag{1}$$

describing the behaviour of the six-dimensional distribution function $f_s(\mathbf{x}, \mathbf{v}; t)$ of different charged particle species s in an electric field $\mathbf{E}$ and a magnetic field $\mathbf{B}$. The normalized distribution function is giving the probability to find a particle of species s (ion or electron) with mass m_s and charge q_s at a certain position in phase space. With $\mathbf{E}$ and $\mathbf{B}$ being influenced and generated by the distribution function in the plasma due to Maxwells equations

$$\nabla\mathbf{E} = 4\pi\rho \qquad \nabla\times\mathbf{E} = -\frac{1}{c}\frac{\partial\mathbf{B}}{\partial t} \qquad \nabla\mathbf{B} = 0 \qquad \nabla\times\mathbf{B} = \frac{1}{c}\frac{\partial\mathbf{E}}{\partial t} + \frac{4\pi}{c}\mathbf{j} \tag{2}$$

and the moments of the distribution function

$$\rho(\mathbf{x}) = q_s \int f_s(\mathbf{x}, \mathbf{v}) d\mathbf{v} \qquad \mathbf{j}(\mathbf{x}) = q_s \int \mathbf{v} f_s(\mathbf{x}, \mathbf{v}) d\mathbf{v} \tag{3}$$

the equation is nonlinear. Note that the term $\frac{1}{c}\frac{\partial \mathbf{E}}{\partial t}$, which is describing electromagnetic waves in Maxwells equations, is not regarded in GENE. The Vlasov-equation (1) is not only governing the interaction of the particles due to long range field interactions on the lefthandside, but also describing their collision by the collision operator C on the righthandside of the equation.

The Vlasov equation (1) fully resolves the gyration of the charged particles around the magnetic field lines, which determines its characteristic time-scale. In the gyrokinetic equation, the fast gyration of the charged particles around the magnetic field lines is not resolved anymore and only the magnetic moment μ, i.e. the gyration velocity, is part of the model. That not only reduces the dimensionality to five, but it also removes the fastest motion from the model. The characteristic time-scale of gyrokinetics is now the movement of the center of gyration (guiding center) parallel and perpendicular to the magnetic field. This time-scale is significantly longer than the characteristic time-scale in the original Vlasov equation (1) and thus decreases the computational effort tremendously. Further approximations [2], which are not explained in this paper, yield the gyrokinetic equation. For the transformed distribution function g it reads in a general form as

$$\frac{\partial g}{\partial t} = \mathcal{L}(g) + \mathcal{N}(g) \tag{4}$$

with $\mathcal{L}$ being a linear integro-differential operator and $\mathcal{N}$ being a nonlinear operator on g governing the $E \times B$ drift of the charged particles, driving the turbulent behaviour.

In linear gyrokinetics only the linear operator $\mathcal{L}$ is regarded, since it already describes the instabilities leading to microturbulence. The linear gyrokinetic equation

$$\frac{\partial g}{\partial t} = \mathcal{L}(g) \tag{5}$$

is a set of ODE's, since the operator $\mathcal{L}$ can also be expressed as a matrix L, when g is discretized on a grid. Since its solution is determined by the eigenvectors and eigenvalues of $\mathcal{L}$ and the initial condition is a superposition of all of them, the time development of g is completely characterized by the spectrum of $\mathcal{L}$. Only eigenmodes having eigenvalues with a positive real part are thus growing modes. In general, the spectrum of $\mathcal{L}$ contains only few eigenvalues with positive real part (as seen in Fig. 1), which means only a few modes actually destabilize a plasma. An analysis of these unstable modes allows then an estimation of the turbulent transport by quasilinear transport models [1]. Furthermore it allows feasible parameter scans for suitable nonlinear simulation regimes due to their

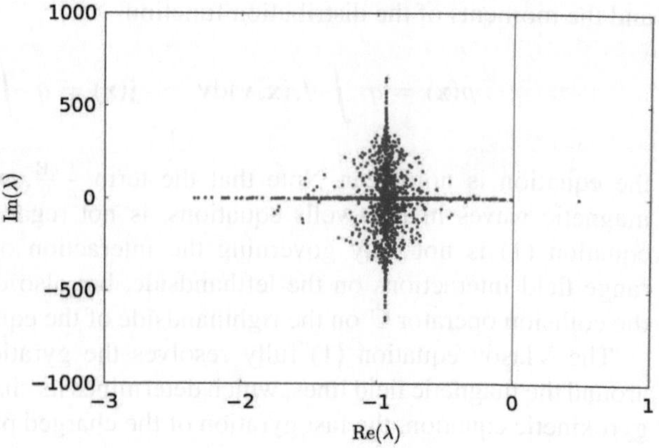

Fig. 1 The spectrum of the linear gyrokinetic operator of a small test problem. Nearly all eigenvalues have negative real parts, which results in a damping of the corresponding eigenmodes. Only eigenmodes having an eigenvalue with a positive real part are growing and thus unstable modes (*rightmost dot*)

much smaller demands in computational resources than nonlinear simulations [10]. In GENE two possibilities exist to do linear computations: a time-integration to compute the single most unstable mode and a computationally more demanding eigenvalue computation of all unstable modes for problems of about a few hundred thousands unknowns [18, 20]. The latter one will be the focus of this paper, since the currently used preconditioned Jacobi-Davidson method only scales up to 64 processors [15]. A further improvement of the performance could allow to do a detailed linear analysis on even larger linear gyrokinetic problems.

1.2 The Combination Technique

The sparse grid combination technique can be used to compute a sparse grid approximation f_c of a problem at a position $\mathbf{x} \in \mathbb{R}^d$ by a proper summation of approximations of it on coarser, possibly anisotropic grids using piecewise linear ansatz functions. All these regular cartesian grids span over the same domain, but do have different resolutions, which is denoted by level vector $\mathbf{l} \in \mathbb{N}^d$. It is giving the level of resolution in each dimension i of the grid by $2^{l_i} + 1$ being the number of grid-points in the ith of d dimensions and with $l_i > 0$. The general form of the combination technique is thus given by

$$f_c(\mathbf{x}) = \sum_{\mathbf{l} \in \mathcal{S}_\mathbf{k}} c_\mathbf{l} f_\mathbf{l} \tag{6}$$

with $\mathcal{S}_\mathbf{k} \subset \{\mathbf{l} \in \mathbb{N}^d\}$ being the set of coarse grid resolutions used for the combination and the k_i being the maximum level of resolution in the ith dimension. The $\mathbf{c}$ is the vector containing the combination coefficients $c_\mathbf{l}$ for each of the different grid resolutions $\mathbf{l}$, governing if a solution $f_\mathbf{l}$ is added or subtracted.

1.2.1 Classical and Dimension Adaptive Combination Scheme

The classical combination technique relies on combinatorics to retrieve a suitable index set S with the appropriate combination coefficients [11]

$$f_c(\mathbf{x}) = \sum_{q=0}^{d-1}(-1)^q \binom{d-1}{q} \sum_{\sum_i l_i = n+(d-1)-q} f_{\mathbf{l}}(\mathbf{x}) \,. \tag{7}$$

Whereas this approach can be used to find sparse grid approximations for problems on domains requiring an isotropic grid layout, other approaches for defining a suitable index set exist [6, 13] which can adapt the sparse grid approximation by refining or coarsening the resolution dimensionally. These approaches are optimizing the grid structure and index set towards a better approximation, but they do not adapt the combination coefficients $\mathbf{c}$ to better suit the underlying problem to be solved.

1.2.2 OptiCom

In the previous section, the classical combination scheme with its integer valued combination coefficients has been introduced. But problems exist, where these coefficients do not lead to sufficiently accurate results or no results at all. They then need to be adapted to the problem to improve or enable an approximation by the combination technique. This method of optimizing the combination coefficients (OptiCom) [14] is based on the minimization of the functional

$$J(\mathbf{c}) = \left\| P_{\mathbf{k}} f_{\mathrm{f}} - \sum_{i=1}^{m} c_i P_{\mathbf{l}_i} f_{\mathrm{f}} \right\|^2 \tag{8}$$

with $P_{\mathbf{n}}, \mathbf{n} \in \mathbb{N}^d$ being projection operators projecting the solution f into a solution space, which is characterized by the underlying cartesian grid with $2^{n_i}+1$ points in the ith of d dimensions. It is basically the minimization of the difference between the solution on a full grid $P_{\mathbf{k}}$ and the combined solution $\sum_{i=1}^{m} c_i P_{\mathbf{l}_i} f_{\mathrm{f}}$ by changing the combination coefficients. If the projection operators are orthogonal, the optimization can be done by solving

$$\begin{bmatrix} \| P_{\mathbf{l}_1} f_{\mathrm{f}} \|^2 & \dots & (P_{\mathbf{l}_1} f_{\mathrm{f}}, P_{\mathbf{l}_m} f_{\mathrm{f}}) \\ \vdots & \ddots & \vdots \\ (P_{\mathbf{l}_m} f_{\mathrm{f}}, P_{\mathbf{l}_1} f_{\mathrm{f}}) & \dots & \| P_{\mathbf{l}_m} f_{\mathrm{f}} \|^2 \end{bmatrix} \begin{bmatrix} c_1 \\ \vdots \\ c_n \end{bmatrix} = \begin{bmatrix} \| P_{\mathbf{l}_1} f_{\mathrm{f}} \|^2 \\ \vdots \\ \| P_{\mathbf{l}_m} f_{\mathrm{f}} \|^2 \end{bmatrix} \tag{9}$$

to retrieve the optimal combination coefficients by computing the norms $\|\cdot\|$ and scalar products $(\cdot,\cdot)$ in the embedding spaces [6]. These norms and scalar product have to be chosen according to the projection operators P.

1.2.3 Applicability of OptiCom for the Gyrokinetic Eigenvalue Problem

Applying the sparse grid combination technique for eigenvalue problems cannot be done straightforwardly. If the underlying eigenvalue problem is symmetric, the combination of eigenvectors according to (7) can be used to compute the respective eigenvalues by using the Rayleigh quotient and the typical error cancellation and extrapolation behavior occurs [4]. Since the linear gyrokinetic operator is neither symmetric nor hermitian, this approach is not applicable for GENE. Nevertheless applying the classical combination technique directly for the eigenvalues can lead to an approximation [16]. The approximation of the eigenvectors on the other hand requires an additional scaling before. Some problems allow a computation of an appropriate scaling before combination [8], but in general it is unknown. Thus the OptiCom can be used in order to calculate the scaling implicitly since the scaling is done in the optimization process. There is an approach to compute eigenpairs, which relies on the access of the matrix-vector product, i.e. the application of the operator on a test-vector [7]. Besides that, it is also based on the discretization of the operator by a Galerkin approach. Unfortunately, the method is not applicable to the gyrokinetic eigenvalue problem using GENE. Whereas there is access to its matrix-vector product, the linear gyrokinetic operator in GENE is not a result of a Galerkin formulation. Thus we need to change the approach to suit our non-symmetric and non-hermitian eigenvalue problem.

2 Method

In this section the method to compute eigenvalues and eigenvectors from previously computed approximations on coarser grids is presented. It reformulates the basic idea of the OptiCom and connects it with a nonlinear optimization of the combination coefficients.

2.1 Reformulation of the OptiCom

Since the method to compute eigenvalues described in Sect. 1.2.3 cannot be applied to the gyrokinetic eigenvalue problem, an alternative is proposed here. Instead of formulating the underlying minimization principle in the form of (8), it can be tailored towards an eigenvalue problem with the general description

$$\mathsf{L}\mathbf{g}_0 = \lambda_0 \mathbf{g}_0 \tag{10}$$

which can be reformulated as a minimization of the function

$$J(\mathbf{g}, \lambda) = \|\mathsf{L}\mathbf{g} - \lambda\mathbf{g}\|^2 = \|(\mathsf{L}\mathsf{I} - \lambda\mathsf{I})\mathbf{g}\|^2 \tag{11}$$

with zero being its minimum, which is reached for $\lambda = \lambda_0$ and $\mathbf{g} = \mathbf{g}_0$. The vector $\mathbf{g}_0$ is a righthand eigenvector on the full grid with its respective eigenvalue λ_0, both being the eigenpair of the linear operator L of a regular full grid, i.e. the operator $\mathcal{L}$ discretized on the full grid of resolution $\mathbf{n}$. The eigenpair is unknown and thus the functional in (11) can give an estimate how close the current estimates of $\mathbf{g}$ and λ are to the correct solutions $\mathbf{g}_0$ and λ_0 respectively. Following the concept of the OptiCom, the eigenvector $\mathbf{g}$ shall be approximated by a weighted sum of eigenvector approximations on the set of combination grids $\mathcal{S}$

$$\mathbf{g} \approx \sum_{\mathbf{l} \in \mathcal{S}} c_{\mathbf{l}} \mathbf{g}_{\mathbf{l}} = \mathbf{g}_c \quad \text{with} \quad \mathsf{L}_{\mathbf{l}} \hat{\mathbf{g}}_{\mathbf{l}} = \lambda_{\mathbf{l}} \hat{\mathbf{g}}_{\mathbf{l}} \quad \text{and} \quad \mathbf{g}_{\mathbf{l}} = P_{\mathbf{n}} \hat{\mathbf{g}}_{\mathbf{l}} \tag{12}$$

with $\mathbf{g}_{\mathbf{l}}$ being the eigenvector of a smaller resolution operator matrix $\mathsf{L}_{\mathbf{l}}$, linearly prolongated onto the full grid of resolution $\mathbf{n}$ by some prolongation operator $P_{\mathbf{n}}$. They can thus be easily added and a combination approximation $\mathbf{g}_c$ on the full grid of resolution $\mathbf{n}$ can be created. The prolongation and storage of $\mathbf{g}_{\mathbf{l}}$ is time consuming but allows the direct evaluation of the eigenpair on the full grid problem later on and is thus justified. This combination $\mathbf{g}_c$ might be able to uniquely represent the eigenvector $\mathbf{g}_{\mathbf{n}}$, but usually it will just be an approximation of it and thus the $\mathbf{c}$ will not be unique.

Putting all interpolated partial solutions of set $\mathcal{S}$ into a rectangular matrix $\mathsf{G} = [\mathbf{g}_{\mathbf{l}_1} \dots \mathbf{g}_{\mathbf{l}_n}]$, $\mathbf{g}_c$ can be represented as the matrix-vector product

$$\mathbf{g}_c = \mathsf{G}\mathbf{c} \tag{13}$$

so that the functional (11) then reads as

$$J_{\text{EV}}(\mathbf{c}, \lambda) = \|\mathsf{L}\mathbf{g}_c - \lambda \mathbf{g}_c\|^2 = \left\| (\mathsf{L} - \lambda \mathsf{I}) \sum_{\mathbf{l} \in \mathcal{S}} c_{\mathbf{l}} \mathbf{g}_{\mathbf{l}} \right\|^2 = \|(\mathsf{L} - \lambda \mathsf{I})\mathsf{G}\mathbf{c}\|^2 \tag{14}$$

which has to be minimized as well.

In contrast to (8), the function J_{EV} is does not only depend on the combination coefficients $\mathbf{c}$ but additionally also on the respective approximation of the eigenvalue λ. The advantage of formulating the OptiCom minimization principle like this is that it does not involve any projection operators and does not rely on any method or basis in which the partial solutions $\mathbf{g}_{\mathbf{l}}$ are formulated in. Also this formulation shows that at the end, the combination coefficients for combining the eigenpair of the full resolution eigenvalue problem (10) is found. Previous approaches using OptiCom for eigenvalue computations did not take the full resolution operator into account, because they did not need to.

Now having the function $J_{\text{EV}}(\mathbf{c}, \lambda)$ for minimization, a straightforward method would be the application of the least squares method by setting up the normal equations

$$[(\mathsf{LI} - \lambda \mathsf{I})\mathsf{G}]^* [(\mathsf{LI} - \lambda \mathsf{I})\mathsf{G}]\, \mathbf{c} = \mathsf{K}(\lambda)\mathbf{c} = 0 \tag{15}$$

with $\mathsf{K}(\lambda)$ now being a function of λ. The matrix $\mathsf{K}(\lambda)$ will be singular as soon as with $\lambda = \lambda_0$ the correct eigenvalue has been found. Any solution $\mathbf{c}_0$ to the singular system $\mathsf{K}(\lambda_0)\mathbf{c} = 0$ will then give the combination coefficients to compute the eigenvector $\mathbf{g}_0$ on the fullgrid by

$$\mathbf{g}_0 = \mathsf{G}\mathbf{c}_0. \tag{16}$$

Since it can not be guaranteed, that $\mathbf{g}_0$ can be exactly represented in the basis of combination grid solutions G by $\mathsf{G}\mathbf{c}$, the minimum of the functional J_{EV} might not evaluate to 0, even if with $\lambda = \lambda_0$ the correct eigenvalue has been found.

2.2 *The Method of Osborne*

Solving the linear eigenvalue problem in (10) has now changed into solving a nonlinear eigenvalue problem in (15), since there the λ_0 has to be found, which leads to a singular system. For retrieving this value, a method [12] is used which was originally proposed for general eigenvalue problems [19]. In there, an eigenvalue problem

$$\mathsf{A}(\lambda)\mathbf{g}(\lambda) = 0, \text{ with } \mathsf{A}(\lambda) = \mathsf{B}_1 - \lambda \mathsf{B}_2 \tag{17}$$

is formulated, where A is not restricted to be linear in λ. In the description of the method here, we restrict B_2 to be the identity. Again the matrix $\mathsf{A}(\lambda)$ is singular if with $\lambda = \lambda_0$ the eigenvalue has been found. The solution $\mathbf{g}$ is then the corresponding eigenvector, which can have an arbitrary scaling. In order to find the eigenpair (λ_0 and $\mathbf{g}_0$), the system (17) is embedded into a slightly larger set of equations of the form

$$\begin{pmatrix} \mathsf{A}(\lambda) & \mathbf{x} \\ \mathbf{s}^* & 0 \end{pmatrix} \begin{pmatrix} \mathbf{g} \\ \beta \end{pmatrix} = \begin{pmatrix} 0 \\ 1 \end{pmatrix} \tag{18}$$

with $\mathbf{x}$ and $\mathbf{s}$ being arbitrary nonzero vectors and $\mathbf{s}$ giving a scaling of a retrieved eigenvector. Since the system will not be singular for any λ, a $\mathbf{g}$ and β can always be computed. In the case of $\lambda = \lambda_0$ the matrix A is singular and thus there is only a solution of the system with $\beta = 0$ since there cannot be a unique solution for a $\beta \neq 0$. Due to the scaling condition $\mathbf{s}$, there is only one solution for $\mathbf{g}$ in this case, which is then the eigenvector $\mathbf{g}_0$ belonging to λ_0. The value of β is thus decreasing as λ is approaching the eigenvalue. Thus the computation of an eigenpair has now been reformulated into finding roots of the nonlinear function $\beta(\lambda)$.

In order to find the roots, Newtons method is employed and an iteration formula λ_i is formulated by [12]

$$\lambda_{i+1} = \lambda_i - \frac{\mathbf{s}^*\mathbf{g}}{\mathbf{s}^*\mathsf{A}^{-1}\frac{\mathrm{d}\mathsf{A}}{\mathrm{d}\lambda}\mathbf{g}}, \tag{19}$$

requiring an initial guess in the proximity of the eigenvalue. There are also update formulas for the respective vectors $\mathbf{s}$ and $\mathbf{x}$ since their appropriate choice speeds up the convergence of Newton's method. The numerical intensive part is then the inversion of A for the update of the current approximation of the eigenvalue.

2.3 Combining Both Approaches

In order to use the method of Sect. 2.2 for the function $\mathsf{K}(\lambda)$ in (15) in Sect. 2.1, small adjustments have to be made. First of all the function $\mathsf{K}(\lambda)$ is now a complex function. Since only the derivative $\frac{\mathrm{d}\mathsf{K}}{\mathrm{d}\lambda}$ is required, the update of (19) using the Newton step described in [12] can be computed by

$$\lambda_{i+1} = \lambda_i - \frac{\beta}{\frac{\mathrm{d}\beta}{\mathrm{d}\lambda}} = \lambda_i - \frac{\mathbf{s}^*\mathbf{c}}{\mathbf{s}^*\mathsf{K}(\lambda_i)^{-1}\frac{\mathrm{d}\mathsf{K}}{\mathrm{d}\lambda}\mathbf{c}} \tag{20}$$

to iterate towards the root of $\beta(\lambda)$. An example for that approach can be seen in Sect. 3. Note that we now find $\mathbf{c}$, which are the combination coefficients to compute an approximation the eigenvector of L by $\mathbf{g}_0 = \mathsf{G}\mathbf{c}$.

The applicability of this Newton iteration heavily depends on the partial solutions G since they influence the structure of K. As long as it is possible to represent the eigenvector $\mathbf{g}_0$ accurately with the partial solutions G, the function $\beta(\lambda_0)$ is zero. That will usually not be the case, since the partial approximations $\mathbf{g}_\mathrm{l}$ may not contain parts of the frequencies of the full grid eigenvector. As soon as the $\mathbf{g}_\mathrm{c}$ can just represent an approximation of the full grid eigenvector, the function $\beta(\lambda)$ does not have roots at the eigenvalues of L anymore and a Newton iteration will not converge. Since the deviations are assumed to be small, the minimum of $|\beta|$ will still indicate an approximation eigenvalue of L, since its real and complex part will be closest to zero in this case. The corresponding set of combination coefficients will thus give also the best approximation of the respective eigenvector.

To allow the computation of the minimum another approach than the straightforward Newton iteration can be taken. One could also use the Newton algorithm to find the minima of a function, but for that a second derivative of β would be required, which is difficult for the special form of K of our OptiCom-eigenvalue problem. The easiest approach to find the minimum is a gradient descent method which we will present here. For the gradient descent, the required gradient of β can be computed by rearranging the second term in (20) to get

$$\frac{d\beta}{d\lambda} = \beta \frac{\mathbf{s}^* \mathsf{K}(\lambda_i)^{-1} \frac{d\mathsf{K}}{d\lambda} \mathbf{c}}{\mathbf{s}^* \mathbf{c}}. \tag{21}$$

This has the same computational effort as the Newton step. Having that gradient, one can use the gradient descent method to find the minimum. Unfortunately the method has a really slow convergence, so that other optimization methods requiring the first derivative might be used by applications using this algorithm.

2.4 Computational Effort

The computational effort of the proposed scheme lies in several subproblems, which are: the computation of the approximation of the eigenvector and the eigenvalue on the combination grids, the projection of these up onto the full grid to get the matrix G, computing the matrix-matrix products LG and $\mathsf{G}^*\mathsf{G}$, solving the inversion of K^{-1} and the minimization of $\beta(\lambda)$ using Newtons or the gradient descent method.

Solving the eigenvalue problems (12) is not done by the proposed algorithm itself but completely by the application, which will in our case be GENE. It is assumed, that it will be a lot faster than solving the eigenvalue problem on the full grid, due to a tremendously reduced grid size. The prolongation of the results $\hat{\mathbf{g}}_{\mathbf{l}}$ up onto the grid of the operator L to get $\mathbf{g}_{\mathbf{l}}$ is time consuming and might require a large amount of time compared to the previous step, since it is filling a full resolution grid. The computation of the matrix-matrix products in the next step is then definitely requiring some computational effort. The size of the operator $\mathsf{L} \in \mathbb{C}^{n \times n}$ with n being the number of grid points of the full grid is indeed rather large. But for evaluating the product LG, it just has to be applied m times, with m being the number of grids used for the combination. Solving the eigenvalue problem on the full grid will require a lot more applications of the operator L onto some vector, since the matrix is assumed to be too large for being explicitly accessible in memory and thus an iterative eigenvalue solver is used. In GENE for example, the underlying solver for solving the eigenvalue problem on the large full grid, requires a lot of iterations [18] so that a single evaluation the linear gyrokinetic operator $\mathcal{L}$ on each of the m vectors will be much less demanding. The evaluation of $\mathsf{G}^*\mathsf{G}$ will not require much effort, since these matrices are explicitly stored in memory. The results of the matrix-matrix products are themselves of the same size as the matrix $\mathsf{K} \in \mathbb{C}^{m \times m}$ and do thus not consume a lot of memory.

With K being a rather small matrix, executing its inversion does not require a lot of effort. Due to that, the execution time of a single iteration, including the computation of K, the inversion of the system and following that the computation of the gradient of β in (21) is done very quickly.

Since a single iteration is not time consuming, the choice of the optimization method is not critical. The Newton method has a much higher rate of convergence than the gradient descent method. But due to the small size of the system and the fast computation of a single iteration, the latter still convergences fast compared to the time required to compute the initial combination grid approximations of the eigenpair and the connected projection onto the full resolution grids.

One issue which might also come up in future applications of the method, is that the computed approximations of the eigenvector on the different combination grids do not belong to the same eigenvalue [8]. Thus the retrieved eigenvector approximations might have to be checked for their similarity, which can be a demanding task. Especially the envisioned larger sets of combination grids will then lead to a large overhead. In the gyrokinetic eigenvalue problem, the eigenvalues of interest can sometimes be mutually close, so that step will be required.

3 First Results: Analytic Example

We applied the proposed algorithm to an analytic problem on finding an eigenfunction of a simple ODE. The chosen analytic problem is finding the periodic eigenfunction u_{f} for

$$L_u u = \lambda u \quad \text{with} \quad L_u = \frac{\mathrm{d}}{\mathrm{d}t} \tag{22}$$

in the unit interval $\Omega = [0, 1]$, which is

$$u_{\mathrm{f}}(t) = \mathrm{e}^{\lambda_0 t} \quad \text{with} \quad \lambda_0 = 2\pi k \mathrm{i}, \tag{23}$$

with $k \in \mathbb{N}^+$. To perform an analytic study of the problem using the proposed algorithm, we emulated the combination technique by applying a semidiscretization. For that, the domain is split in two parts. Each half of the domain can be represented by either the correct eigenfunction u_{f} (23) or by a linear approximation

$$u_{\mathrm{c}}(t) = \begin{cases} (1-4t) & t \leq \frac{1}{2} \\ (4t-3) & t > \frac{1}{2}\,. \end{cases} \tag{24}$$

Similar to the combination technique, solutions of different approximation qualities have to be created. For that we set up three different approximations:

$$u_1(t) = \begin{cases} u_{\mathrm{f}}(t) & t \leq \frac{1}{2} \\ u_{\mathrm{c}}(t) & t > \frac{1}{2} \end{cases} \qquad u_2(t) = \begin{cases} u_{\mathrm{c}}(t) & t \leq \frac{1}{2} \\ u_{\mathrm{f}}(t) & t > \frac{1}{2} \end{cases} \qquad u_3(t) = \begin{cases} u_{\mathrm{c}}(t) & t \leq \frac{1}{2} \\ u_{\mathrm{c}}(t) & t > \frac{1}{2} \end{cases}$$

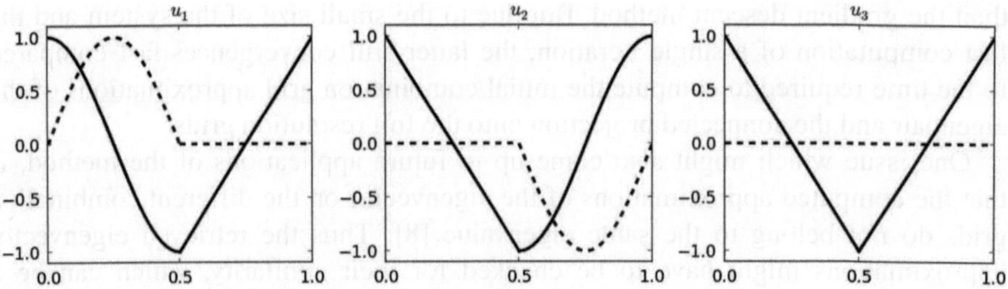

Fig. 2 The approximations of the eigenfunction u_i with the real parts (*solid*) and the imaginary parts (*dashed*)

which can be seen in Fig. 2. None of them gives the eigenfunction u but the sum

$$u_{\mathrm{f}}(t) = \sum_{i=1}^{3} c_i u_i(t) \quad \text{with} \quad \begin{aligned} c_1 &= 1 \\ c_2 &= 1 \\ c_3 &= -1 \end{aligned} \tag{25}$$

actually gives the correct solution. This is similar to the combination technique where the approximation of different, even anisotropic, resolutions $\mathbf{g}_\mathbf{l}$ are combined. In this one-dimensional problem the anisotropy is emulated by using different resolutions, i.e. approximation qualities, in the two halves of the domain instead of different dimensions. The function u_f represents the solution on an infinitely fine grid, thus the analytical solution, whereas u_c is only an approximation on a very coarse grid of two grid points (a linear approximation). This approach is thus basically a one-dimensional combination scheme with only two scales [5].

Having done this approach, the method proposed in Sect. 2.3 can be applied. The matrix $\mathsf{K}_u(\lambda)$ is then constructed as described in (15)

$$\mathsf{K}_u(\lambda) = [(L_u - \lambda)\mathsf{G}_u]^*[(L_u - \lambda)\mathsf{G}_u] \tag{26}$$

with G_u being the row vector $[u_1\ u_2\ u_3]$ and L_u being the operator applying the first derivative onto a function u_i. That leads to a matrix $\mathsf{K}_u \in \mathbb{C}^{3\times 3}$ with the entries

$$K_{u i,j} = \overline{(L_u u_i - \lambda u_i)}(L_u u_j - \lambda u_j). \tag{27}$$

Since we are using analytic functions here, we use an integral expression as scalar product so that

$$K_{u i,j} = \int_0^1 \overline{(L_u u_i - \lambda u_i)}(L_u u_j - \lambda u_j)\mathrm{d}t. \tag{28}$$

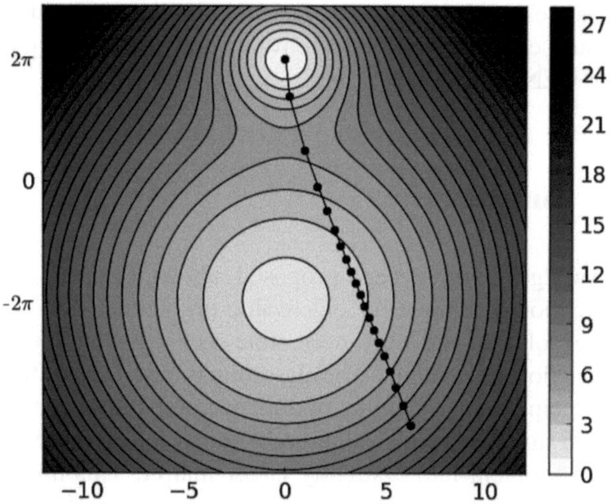

Fig. 3 The absolute value of $\beta(\lambda)$ in the region around the eigenvalue $\lambda_0 = 2\pi\mathrm{i}$ on the complex plane. The Newton iteration steps do converge directly to the root at $2\pi\mathrm{i}$ (*dotted line*). A second local minimum can be observed at $-2\pi\mathrm{i}$, which might be problematic for the gradient descent method

Note that the derivative $\frac{\mathrm{d}\mathsf{K}_u}{\mathrm{d}\lambda}$ is computed by

$$\frac{\mathrm{d}\mathsf{K}_u}{\mathrm{d}\lambda} = \frac{\mathrm{d}}{\mathrm{d}\lambda}\left([(L_u - \lambda)\mathsf{G}_u]^*[(L_u - \lambda)\mathsf{G}_u]\right) \tag{29}$$

$$= -[(L_u - \lambda)\mathsf{G}_u]^*\,\mathsf{G}_u \tag{30}$$

so that we get

$$\left(\frac{\mathrm{d}\mathsf{K}_u}{\mathrm{d}\lambda}\right)_{i,j} = -\int_0^1 \overline{(L_u u_i - \lambda u_i)} u_j \,\mathrm{d}t. \tag{31}$$

Now the Newton iteration according to (19) can be computed.

The iteration is converging in a few steps towards the correct combination coefficients mentioned in (25). Since the sum $u_1 + u_2 - u_3$ can exactly represent the eigenfunction u_{f} in the domain, the Newton iteration converges because the root of $\beta(\lambda)$ is existing. But besides the Newton iteration the slower converging gradient descent method is also leading to the solution. The root of β is also a minimum of $|\beta|$ and thus the gradient descent will also converge. Its step-size is determined by the absolute value of β. Nevertheless, the gradient-descent bears the risk of only finding a local minimum of β which is actually not the eigenvalue. It can be seen in Fig. 3. Thus a sufficiently good initial guess for the eigenvalue is required to retrieve the combination coefficients leading to an approximation of an eigenvector using the gradient descent method.

The classical combination technique would not be able to solve the eigenvalue problem, since it is a purely extrapolating technique. Whereas it would give the same combination coefficients, it would not be able to handle any perturbations. The proposed method on the other hand uses the partial solutions as basis functions and scales them to solve the underlying eigenvalue problem. The method could thus be applied to other eigenvalue problems as the gyrokinetic eigenvalue problem in GENE.

4 Summary

An algorithm has been proposed, allowing to use an optimized sparse grid combination technique for eigenvalue problems, where the coarse approximations of the solution are only accessible as is and no information regarding the basis functions is available. The algorithm is based on a reformulation of the minimization principle underlying the optimized combination technique. This formulation leads to a nonlinear eigenvalue problem, which is solved employing an embedding of the problem into a slightly larger matrix. This allows to use a Newton iteration or a gradient descent method to iterate towards the desired eigenvalue. Having found the eigenvalue, the eigenvector is computed. Depending on the quality of the initial approximations, the computed eigenvalue might not exactly be the solution to the full resolution eigenvalue problem. But the embedding of the optimization problem also gives an estimate of the accuracy of the retrieved eigenpair, so that an optimization of the chosen index set for combination can be triggered. The effort to compute an eigenpair using the proposed algorithm has been studied and it appears to be able to reduce the computational effort to find eigenpairs in high dimensional settings. It has been tested on a one-dimensional analytical problem and exhibited the expected behaviour in finding the correct solution. Pitfalls of using the gradient descent method have been identified, which can be circumvented by using a rather close initial guess for the iteration towards the eigenvalue.

The method thus seems to be a viable tool for identifying microinstabilities using the gyrokinetic code GENE, since eigenpair approximations of lower resolution, which can be computed rather cheaply, can be combined to give a high resolution approximation of the eigenpair. This would not be possible using the classical combination technique or the method based on the Rayleigh quotient [7] and thus it is one step towards using the combination technique for computing eigenpairs in GENE. Compared to the currently applied Jacobi-Davidson method, an improved runtime is expected.

In future work the algorithm will be applied to GENE to give exact measurements of the performance, since the time for the computation of the partial solutions, the interpolation onto the full resolution grid and the evaluation of the matrix-matrix products cannot be calculated exactly here. Using this method might lead to improved scaling properties, since all of this steps can be done in parallel. Together with a proper initial choice of the coarse combination grids, the it could be a computationally more efficient alternative to the currently used iterative eigenvalue solver.

References

1. C. Angioni, A.G. Peeters, F. Jenko, T. Dannert, Collisionality dependence of density peaking in quasilinear gyrokinetic calculations. Phys. Plasmas **12**(11), 112310 (2005)
2. A. Brizard, T. Hahm, Foundations of nonlinear gyrokinetic theory. Rev. Mod. Phys. **79**(2), 421–468 (2007)
3. H.-J. Bungartz, M. Griebel, Sparse grids. Acta Numer. **13**, 147 (2004)
4. H.-J. Bungartz, M. Griebel, U. Rüde, Extrapolation, combination, and sparse grid techniques for elliptic boundary value problems. Comput. Methods Appl. Mech. Eng. **116**(1–4), 243–252 (1994)
5. Y. Fang, *The Combination Approximation Method*. Ph.D. thesis, Australian National University, 2012
6. J. Garcke, A dimension adaptive sparse grid combination technique for machine learning. ANZIAM J. **48**, C725–C740 (2007)
7. J. Garcke, An optimised sparse grid combination technique for eigenproblems. PAMM **7**(1), 1022301–1022302 (2007)
8. J. Garcke, M. Griebel, On the computation of the eigenproblems of hydrogen and helium in strong magnetic and electric fields with the sparse grid combination technique. J. Comput. Phys. **165**(2), 694–716 (2000)
9. GENE Development Team (2014). http://www.ipp.mpg.de/~fsj/gene/
10. T. Görler, X. Lapillonne, S. Brunner, T. Dannert, F. Jenko, F. Merz, D. Told, The global version of the gyrokinetic turbulence code GENE. J. Comput. Phys. **230**(18), 7053–7071 (2011)
11. M. Griebel, M. Schneider, C. Zenger, A combination technique for the solution of sparse grid problems, in *Iterative Methods in Linear Algebra*, ed. by P. de Groen, R. Beauwens (North Holland, Amsterdam, 1992), pp. 263–281
12. D. Harrar II, M. Osborne, Computing eigenvalues of ordinary differential equations. ANZIAM J. **44**, C313–C334 (2003)
13. M. Hegland, Adaptive sparse grids. ANZIAM J. **44**, C335–C353 (2003)
14. M. Hegland, J. Garcke, V. Challis, The combination technique and some generalisations. Linear Algebra Appl. **420**(2–3), 249–275 (2007)
15. C. Kowitz, Preconditioning for fast linear computations with the plasma turbulence code GENE. Master's thesis, Technische Universität München, 2010
16. C. Kowitz, M. Hegland, The sparse grid combination technique for computing eigenvalues in linear gyrokinetics, in *ICCS*, 2013, pp. 449–458
17. C. Kowitz, D. Pflüger, F. Jenko, M. Hegland, The combination technique for the initial value problem in linear gyrokinetics, in *Sparse Grids and Applications*, ed. by M. Griebel, J. Garcke (Springer, Berlin, 2013), pp. 205–222
18. F. Merz, C. Kowitz, E. Romero, J. Roman, F. Jenko, Multi-dimensional gyrokinetic parameter studies based on eigenvalue computations. Comput. Phys. Commun. **1**, 1–9 (2011)
19. M.R. Osborne, A new method for the solution of eigenvalue problems. Comput. J. **7**(3), 228–232 (1964)
20. J.E. Roman, M. Kammerer, F. Merz, F. Jenko, Fast eigenvalue calculations in a massively parallel plasma turbulence code. Parallel Comput. **36**(5–6), 339–358 (2010)

Classification with Probability Density Estimation on Sparse Grids

Benjamin Peherstorfer, Fabian Franzelin, Dirk Pflüger, and Hans-Joachim Bungartz

Abstract We present a novel method to tackle the multi-class classification problem with sparse grids and show how the computational procedure can be split into an Offline phase (pre-processing) and a very rapid Online phase. For each class of the training data the underlying probability density function is estimated on a sparse grid. The class of a new data point is determined by the values of the density functions at this point. Our classification method can deal with more than two classes in a natural way and it provides a stochastically motivated confidence value which indicates how to rate the respond to a new point. Furthermore, the underlying density estimation method allows us to pre-compute the system matrix and store it in an appropriate format. This so-called Offline/Online splitting of the computational procedure allows an Online phase where only a few matrix-vector products are necessary to learn a new, previously unseen training data set. In particular, we do not have to solve a system of linear equations anymore. We show that speed ups by a factor of several hundred are possible. A typical application for such an Offline/Online splitting is cross validation. We present the algorithm and the computational procedure for our classification method, report on the employed density estimation method on sparse grids and show by means of artificial and real-world data sets that we obtain competitive results compared to the classical sparse grid classification method based on regression.

B. Peherstorfer (✉) • H.-J. Bungartz
Department of Informatics, Technische Universität München, Boltzmannstr. 3, 85748 Garching, Germany
e-mail: pehersto@in.tum.de

F. Franzelin • D. Pflüger
Institute for Parallel and Distributed Systems, University of Stuttgart, Universitätsstr. 38, 70569 Stuttgart, Germany

J. Garcke and D. Pflüger (eds.), *Sparse Grids and Applications - Munich 2012*, Lecture Notes in Computational Science and Engineering 97,
DOI 10.1007/978-3-319-04537-5_11,

1 Introduction

We consider the classification problem arising in the context of data mining. We want to reconstruct an unknown function $f : \mathbb{R}^d \to \{1, \ldots, n\}$ which assigns class labels $1, \ldots, n$ to points in the d-dimensional space $\mathbb{R}^d$. Given is only a set of training data

$$S = \left\{(\mathbf{x}_i, y_i) \in \mathbb{R}^d \times \{1, \ldots, n\}\right\}_{i=1}^{M},$$

where $\mathbf{x}_i$ is a point in the d-dimensional space $\mathbb{R}^d$ and y_i the class label corresponding to $\mathbf{x}_i$. Usually the set S also contains noise, thus, in general, interpolating the training points in S does not generalize well enough to new data points. Classification methods try to find an approximation of the unknown function f by using only the training data set S. There are many classification methods available, see, e.g., [3, 12].

Here we are interested in sparse-grid-based classification techniques. What we call the classical sparse grid classification method employs regression where the regression function is discretized on a sparse grid. This approach has been shown to be competitive with the best existing methods for moderately high-dimensional problems, see, e.g., [9–11, 14, 21, 23]. The key advantage of such a grid-based method is that it scales only linearly with the number of training data points [11] because the ansatz functions are not associated to the data points but stem from a grid. Since common (full) grids suffer from the curse of dimensionality, i.e., the number of grid points grows exponentially with the number of dimensions, sparse grids are employed.

In contrast to the classical approach with sparse grid regression, our proposed classification method is based on probability density functions which are discretized on sparse grids. We assume that each class of the training data has been generated separately, i.e., the data points in each class have a different underlying density function. We approximate these density functions by splitting the training data set into its classes and by estimating the individual density functions on sparse grids. The density estimation is based on the non-parametric approach introduced in [13] which has already been successfully used in the context of clustering [19]. To assign a class label to a new data point, we evaluate the estimated density functions at the new point and respond with the class label associated to the density function which yields the highest value. Note that this approach is highly related to Bayesian classification methods [16, 17]. Just like the classical sparse grid classification, our method scales only linearly with the number of training data points.

With the proposed method we can cope with more than two classes in a natural way and we can also derive a natural confidence value of the respond by comparing the values of all density functions: If we assign a label to a data point for which one density function yields a much higher value than the others, it is very likely that the data point is correctly classified. In contrast, if several density functions yield about the same value, the data point most probably lies in a region where different classes overlap and thus the class label cannot be assigned with high certainty. However, the information that a response should be considered more like an educated guess than a founded statement is already valuable for certain applications. For example, for the detection of non-coding RNAs, classifiers are trained in [24] to have a pre-screening first before a biochemical test is performed to verify the result. Due to the large number of non-ncRNAs and due to the difficulty, time, and cost of a biochemical test, a classifier should minimize the false positives (i.e., non-ncRNA classified as ncRNA) and rather miss a few true positives (i.e., correctly classified points). We refer to [24] for an extensive discussion of this biochemical classification problem.

Both the classical and our proposed classification method requires us to solve a system of linear equations to construct a classifier for a given training data set. However, the system matrix of our method is independent from the training data points. Thus, we can employ a so-called Offline/Online splitting of the computational procedure: In the Offline phase, we pre-compute the system matrix and store it in an appropriate format. Usually, this is a computationally very expensive and time consuming task. In the Online phase, when we are given a new training data set, we load the system matrix and solve the corresponding system of linear equations. Since we already have the system matrix this is cheap. Furthermore, if the system matrix has been decomposed (e.g. LU decomposition or diagonalization) during the Offline phase we do not even have to solve the system in the Online phase again but only need a few matrix-vector products to construct a classifier. Thus, the computational procedure in the Online phase of our method is in $\mathcal{O}(N^2)$ rather than in $\mathcal{O}(N^3)$ as for the classical sparse grid classification based on regression.

Such an Offline/Online scheme pays off if we compensate the costly Offline phase by repeating the Online phase many times or if our (real-time) application requires a classifier immediately after new training data has been provided. Such scenarios can be found in e.g. online learning or data stream mining [2,7,8]. Another very common example is parameter selection. If we employ cross validation to find a good parameter configuration for the current data set at hand, we have to train many classifiers with different parameters on different training data sets. Hence, we have to repeat the Online phase very often and can so afford an expensive Offline computation.

In the following Sects. 2 and 3, we state some properties of sparse grids and the classification approach based on sparse grid regression. In Sect. 4 we discuss how to estimate probability density functions using sparse grid discretization. In Sect. 5 we continue with an in-depth discussion of our proposed classification method based on density estimation. The Offline/Online splitting is introduced in Sect. 6 and the results for various artificial as well as real-world data sets are presented in Sect. 7. In terms of accuracy, we do not only obtain competitive results with our proposed method, but can also improve on the results of the classical approach. As far as the runtime of our method is concerned, with the Offline/Online splitting we gain a factor of up to several hundred in the Online phase compared to the prediction step of the method without the Offline/Online splitting.

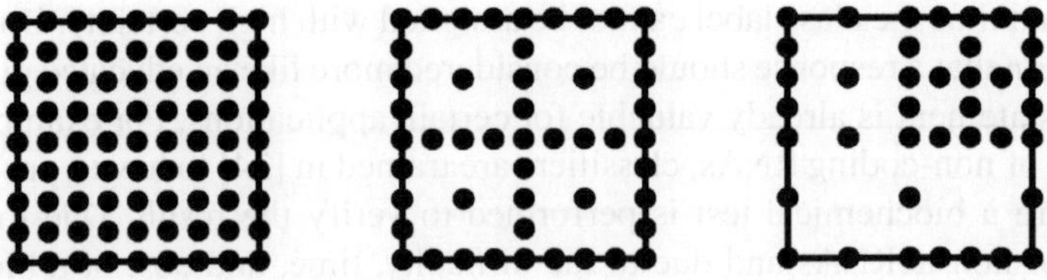

Fig. 1 On the *left* a full grid, in the *middle* a regular sparse grid, and on the *right* an adaptively refined sparse grid

2 Sparse Grids

In contrast to many other methods in data mining which employ ansatz functions associated to the training data points, we pursue a grid-based approach. This means, we have ansatz functions associated to grid points rather than to data points. Sparse grids are necessary because a straightforward discretization with 2^ℓ grid points in each direction would suffer the curse of dimensionality: The number of grid points is in $\mathcal{O}(2^{\ell d})$ and depends exponentially on the dimension d. Here, the dimension d is the dimension of the usually high-dimensional space where our data points $\{\mathbf{x}_1, \ldots, \mathbf{x}_M\}$ of the training data set S come from. Thus, a full grid typically becomes computationally infeasible already for four or five dimensions. For sufficiently smooth functions, sparse grids enable us to reduce the number of grid points by orders of magnitude to only $\mathcal{O}(2^\ell \ell^{d-1})$ while keeping a similar accuracy as in the full grid case. Figure 1 shows a full and a sparse grid. Following [4] we denote with $V_\ell^{(1)}$ the sparse grid space of level ℓ and dimension d. It is also possible to adaptively refine sparse grids, see Fig. 1 (right). Many error indicators exist which help to select appropriate grid points for refinement. More details about adaptivity in the context of classification can be found in [21, 22]. For more details on sparse grids in general, their applications and further reading, we refer to [4].

3 Classification with Sparse Grid Regression

In this section, we briefly describe the classical sparse grid classification method which employs sparse grid regression. For more details, see [11, 21, 22] and the references therein.

Let $S = \{(\mathbf{x}_i, y_i) \in \mathbb{R}^d \times \{1, \ldots, n\}\}_{i=1}^M$ be the training data set. We are then looking for a function $f \in V$ such that

$$f = \arg\min_{u \in V} \left(\frac{1}{M} \sum_{i=1}^{M} (y_i - u(\mathbf{x}_i))^2 + \lambda \|\mathcal{L}u\|_{L^2}^2 \right). \tag{1}$$

The first term of (1) ensures closeness of f to the training data and the second one imposes a certain smoothness on f in order to generalize to new, previously unseen data. The regularization parameter λ controls the trade-off between fidelity and smoothness. The regularization operator $\mathcal{L}$ is typically ∇, but simpler and computationally more efficient choices are possible [21, 22].

In the following, we always set the space V to a sparse grid space $V_\ell^{(1)}$ of level ℓ and dimension d. Thus we can write the function f as linear combination

$$f(\mathbf{x}) = \sum_{i=1}^{N} \alpha_i \phi_i(\mathbf{x}), \tag{2}$$

with hierarchical coefficients $\alpha_1, \ldots, \alpha_N$ and hierarchical basis functions $\phi_1, \ldots, \phi_N$. The optimization problem (1) can be solved by plugging (2) into (1) and setting the derivatives with respect to the hierarchical coefficients to zero [11]. This leads to a system of linear equations

$$\left(\frac{1}{M} BB^T + \lambda C\right) \boldsymbol{\alpha} = \frac{1}{M} B\mathbf{y}, \tag{3}$$

where $B_{ij} = \phi_i(\mathbf{x}_j)$, $C_{ij} = (\mathcal{L}\phi_i, \mathcal{L}\phi_j)_{L^2}$ and $\mathbf{y} = (y_1, \ldots, y_M)$. After f has been computed, the class label of a new data point $\mathbf{x} \in \mathbb{R}^d$ is determined by evaluating the function f at $\mathbf{x}$. In the case of a binary classification problem, i.e., if we have only two labels ± 1, we can define the label y of a point $\mathbf{x}$ as

$$y = \begin{cases} -1, & \text{if } f(\mathbf{x}) < 0, \\ 1, & \text{if } f(\mathbf{x}) \geq 0. \end{cases}$$

The method can also be extended to more than two classes by constructing multiple classifiers, see, e.g., [21]. Evaluating f at point $\mathbf{x}$ means evaluating the sum (2). Clearly, the sum (2) is independent from the number of training data points M and scales only linearly with the number of grid points N.

From a computational point of view the system of linear equations (3) can be solved with the conjugate gradient method such that only a procedure for the matrix-vector product with the matrices BB^T and C is required. Whereas the matrix-vector product with BB^T is straightforward, the computational procedure for the product with C might become very complicated depending on the regularization operator $\mathcal{L}$. In [21] it is argued that the regularization term $\|\mathcal{L}f\|_{L^2}^2$ can be replaced with $\sum_i \alpha_i^2$ in the minimization problem (1) which corresponds to the identity matrix I instead of the general matrix C. We also stick to this choice in the following.

It is important to note that the system matrix $\frac{1}{M}BB^T + \lambda C$ of (3) has dimension $N \times N$ and thus is independent from the number of data points M. However, since usually the matrix BB^T is not explicitly formed but only a procedure for the matrix-vector product with B and B^T is provided, the product with BB^T still depends on the number of data points M because B has dimension $N \times M$ and thus a matrix-vector product with a vector of size M has to be performed in each iteration of the CG method. For the regularization term $\sum_i \alpha_i^2$, the product with the weighted regularization matrix $\lambda C = \lambda I$ is in $\mathcal{O}(N)$, see [21, 22]. Overall, a matrix-vector product with the system matrix $\frac{1}{M}BB^T + \lambda C$ is in $\mathcal{O}(NM + N)$ and thus clearly depends on the number of data points M. This is a drawback of the classification method based on sparse grid regression when it comes to large data sets. In contrast, the minimization problem corresponding to our classification method based on sparse grid density estimation relies on a system matrix which is independent from the data points. Therefore, the matrix-vector product with the system matrix is completely decoupled from the number of data points M.

4 Density Estimation with Sparse Grids

Before we introduce our classification method based on sparse grid density estimation we first discuss how to actually perform density estimation on sparse grids. We follow the density estimation method introduced in [13] but instead of full grids we employ sparse grids and we replace the regularization term with a computationally more efficient one.

We want to estimate the density function corresponding to the data points in the training data set $S = \{\mathbf{x}_1, \dots, \mathbf{x}_M\} \subset \mathbb{R}^d$. Note that S does not contain any labels anymore but only data points. Let f_ϵ be an initial guess of the density function of the data in S. We then want to find an estimated density function f such that

$$f = \arg\min_{u \in V} \int_\Omega (u(\mathbf{x}) - f_\epsilon(\mathbf{x}))^2 \, \mathrm{d}\mathbf{x} + \lambda \|\mathcal{L}u\|_{L^2}^2. \tag{4}$$

Again we have a regularization or penalty term $\|\mathcal{L}u\|_{L^2}^2$ to impose a smoothness constraint and a regularization parameter λ to control the trade-off between error and smoothness. We set the initial guess f_ϵ to

$$f_\epsilon = \frac{1}{M} \sum_{i=1}^{M} \delta_{\mathbf{x}_i} \tag{5}$$

where $\delta_{\mathbf{x}_i}$ is the Dirac delta function centered on $\mathbf{x}_i$. We refer to [13] for a motivation and justification of (5). For us this is a simple and sufficient choice. However, other choices for f_ϵ might become necessary in certain cases, see [13] and the examples therein. In [13] we then find the necessary transformations to obtain the variational equation

$$\int_\Omega u(\mathbf{x}) s(\mathbf{x}) \mathrm{d}\mathbf{x} + \lambda \int_\Omega \mathcal{L}u(\mathbf{x}) \cdot \mathcal{L}s(\mathbf{x}) \mathrm{d}\mathbf{x} = \frac{1}{M} \sum_{i=1}^{M} s(\mathbf{x}_i), \qquad \forall s \in V, \tag{6}$$

with the test functions $s \in V$. Note that for the following classification method the statistical properties (unit integrand, moments, etc.) of the estimated density functions are not important.

Instead of employing the finite element method with full grids, as proposed in [13], we employ sparse grids to discretize f. Let $V_\ell^{(1)}$ be the sparse grid space of level ℓ and $\Phi = \{\phi_1, \ldots, \phi_N\}$ the set of the corresponding (hierarchical) basis functions. We set the test space to the sparse grid space $V_\ell^{(1)}$ and follow the usual Galerkin approach and obtain the system of linear equations

$$(P + \lambda C)\boldsymbol{\alpha} = \mathbf{b}, \tag{7}$$

where $P_{ij} = (\phi_i, \phi_j)_{L^2}$, $C_{ij} = (\mathcal{L}\phi_i, \mathcal{L}\phi_j)_{L^2}$ and $b_i = \frac{1}{M}\sum_{j=1}^{M} \phi_i(\mathbf{x}_j)$. In order to preserve moments a special regularization operator $\mathcal{L}$ is developed in [13]. Since for our purposes we do not require our estimated density functions to have such statistical properties, we can employ the simple but very effective regularization term $\sum_i \alpha_i^2$ as introduced in [21] and described in Sect. 3. Then, again, the matrix C becomes the identity matrix I.

Just as in the case of the classical classification method based on sparse grid regression, the system of linear equations (7) can be solved with the conjugate gradient method. Thus, we only have to provide the matrix-vector product with the matrix P because the matrix C has become the identity matrix I. Efficient and parallel algorithms exist to compute the product in $\mathcal{O}(2^d N)$ for the hierarchical basis Φ of a sparse grid space [21]. Just as for the regression problem (3), the system matrix $P + \lambda I$ of the density estimation problem is of dimension $N \times N$ where N is the number of sparse grid points and thus is independent from the number of data points M. However, whereas for the regression problem the product with the matrix BB^T still depends on the number of data points M, the product with the system matrix of the density estimation problem is truly independent from M. Moreover, the system matrix corresponding to the density estimation is not only independent from the number of data points but even from the data points themselves. Whereas an entry B_{ij} for the regression problem is the hierarchical basis function ϕ_i evaluated at the data point $\mathbf{x}_j$, an entry P_{ij} for the density estimation problem is the L^2 dot product of the two hierarchical basis functions ϕ_i and ϕ_j which does not depend on the data points in S. As far as the density estimation problem is concerned, the data points influence only the right hand side but not the system matrix of the system of linear equations (7). We show in Sect. 6 that this gives us the opportunity to split the computational procedure in an (expensive) Offline and a (cheap) Online phase.

5 Classification with Sparse Grid Density Estimation

In this section, we introduce the multi-class classification problem with density estimation. We employ the density estimation method discussed in the previous section.

Let $S = \left\{(\mathbf{x}_i, y_i) \in \mathbb{R}^d \times \{1, \ldots, n\}\right\}_{i=1}^{M}$ be the training data set. Note that we now have labels again. We have M pairs of a data point $\mathbf{x}_i \in \mathbb{R}^d$ and a class label

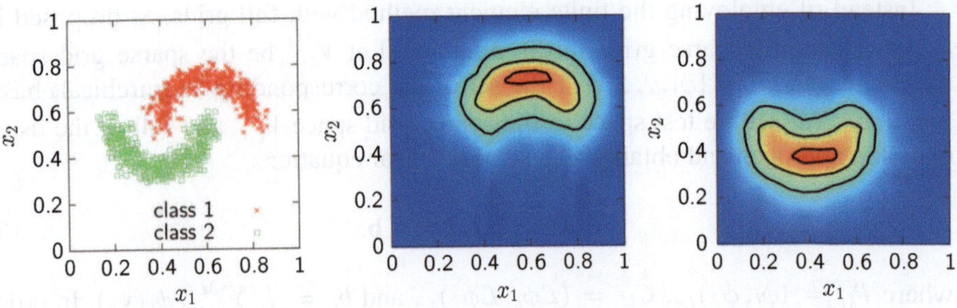

Fig. 2 The data points of the two-moons data set and the contours of the corresponding two density functions

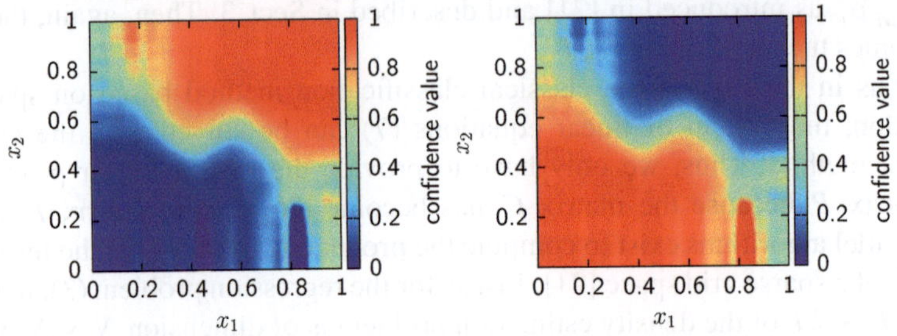

Fig. 3 The plot demonstrates the confidence value provided by our density-based classification method. The *left* plot corresponds to class 1 and the *right* one to class 2

$y_i \in \{1,\ldots,n\}$, see, e.g., Fig. 2 for an example of the two-dimensional two-moons data set with two classes. We now split the training data set S into n partitions $S^1,\ldots,S^n$ with

$$S^k = \{(\mathbf{x}_i, y_i) \in S \,|\, y_i = k\},$$

such that the set S^k contains all M_k pairs $(\mathbf{x}_i, y_i)$ with class label $y_i = k$. We then estimate the probability density functions $f^1,\ldots,f^n$ for each set $S^1,\ldots,S^n$ with the sparse grid method described in Sect. 4.

Figure 2 shows a contour map of the two estimated density functions corresponding to the two classes of the two-moons data set. Clearly, density function f^1 evaluates to greater values in the region where most data points with label 1 lie than in the rest of the domain. Just as density function f^2 yields greater values in the region of class 2. Furthermore, the contour lines of the two density functions approximate the shape of the boundary of the data point cluster with label 1 and label 2, respectively. Figure 3 demonstrates the confidence value we have discussed in Sect. 1. It shows the contour map of $f^i(x)/(f^1(x)+f^2(x))$ for $i = 1, 2$. Consider for example point $(0.4, 0.5)$. It lies between the two classes, thus it is very difficult to assign the correct class. This fact is well reflected by the two density functions.

Both functions evaluate to about the same value at $(0.4, 0.5)$, which confirms that we can only make an educated guess but not a profound statement about the class label of $(0.4, 0.5)$.

We can now summarize our proposed classification method based on sparse grid density estimation:

1. Split the training data set
$$S = \{(\mathbf{x}_i, y_i) \in \mathbb{R}^d \times \{1, \ldots, n\}\}_{i=1}^M$$
into its separate classes $S^1, \ldots, S^n$.
2. Estimate the density functions $f^1, \ldots, f^n : \mathbb{R}^d \to \mathbb{R}$ for the data sets $S^1, \ldots, S^n$ on a sparse grid with the minimization problem (4).
3. For a new point $\mathbf{x} \in \mathbb{R}^d$, evaluate the estimated density functions and assign class label k to $\mathbf{x}$ if function f^k yields the greatest value, i.e.,
$$y = \underset{k \in \{1,\ldots,n\}}{\arg\max}\, f^k(\mathbf{x}).$$

Let us again point out the close relationship to Bayesian classification methods [16, 17].

6 An Offline/Online Splitting for Classification

In the following, we describe our Offline/Online splitting of the computational procedure of the classification method based on sparse grid density estimation. We propose two matrix decompositions of the system matrix and discuss their properties in the context of the classification problem.

One advantage of our classification method is that the system matrix is truly decoupled from the number of the training data points M, cf. Sect. 4. However, the system matrix of the density estimation problem is even independent from the data points themselves. This allows us to introduce an Offline/Online splitting of the computational procedure. The Offline/Online splitting is only reasonable in settings where we want to have a very fast evaluation (Online phase) in favor of a quite costly pre-processing step (Offline phase). We can think of applications in the multi-query and real-time context. Another example is cross validation. There we want to construct a classifier for different regularization parameters λ with different training data sets and test them on a testing data set. Thus, we have to construct many classifiers where only the parameter λ as well as the training and testing data set change. Such an Offline/Online splitting of the computational procedure is very common in model order reduction where parameter-independent matrices, vectors and values are pre-computed and reused over and over again, cf. [18]. We follow this idea and want to pre-compute the system matrix $P + \lambda I$ of the density estimation problem (7). We would like to emphasize once more that this is not possible with the classical sparse grid classification method based on regression because there the system matrix depends on the data points.

Let $V_\ell^{(1)}$ be the sparse grid space of level ℓ and dimension d spanned by some (hierarchical) basis Φ. To explicitly form the matrix $P + \lambda I$ corresponding to $V_\ell^{(1)}$ we can employ the already available matrix-vector product procedures by multiplying with the unit vectors. Especially for higher dimensional data sets, this becomes rather expensive because one matrix-vector product with P is in $\mathcal{O}(2^d N)$ which is only linear in the number of grid points N but with the factor 2^d depending exponentially on the dimension d. Therefore, in our case, where we want to assemble the system matrix, it is better to follow the naive approach where we explicitly compute the L^2 dot products of all combinations of the basis functions. Even though this scales quadratically with the number of grid points N, we do not have the factor 2^d anymore. A key question is how to store the matrix during the Offline phase. We discuss here an LU decomposition and an eigendecomposition of the system matrix.

Since we want to solve a system of linear equations an LU decomposition is an obvious choice. Thus, in the Offline phase we compute

$$LU = P + \lambda I,$$

store the matrices L and U and can then solve the corresponding system of linear equations in the Online phase with backward and forward substitution which is in $\mathcal{O}(N^2)$ only. A severe drawback of the LU decomposition is that we have to fix the regularization parameter λ already in the Offline phase. This is indeed a disadvantage as cross validation with respect to the regularization parameter λ is in our context one key application of such an Offline/Online splitting. However, it has also been shown that the classification method is not very sensitive to different parameters λ for huge data sets and thus the LU decomposition might be used in cases where also the costs of the Offline phase is crucial.

If we store the eigendecomposition of the matrix P we are able to vary the parameter λ and the Online phase is still in $O(N^2)$. Let $P = USU^T$ be the eigendecomposition of P where U is an orthonormal matrix and S a diagonal matrix. Such an eigendecomposition exists because P is a Gram matrix and the basis functions in Φ are linearly independent. We store U and S in the Offline phase and can then construct a classifier, i.e., compute $(P + \lambda I)^{-1}\mathbf{b}$, in the Online phase as follows

$$(P + \lambda I)^{-1}\mathbf{b} = (USU^T + \lambda UU^T)^{-1}\mathbf{b} = U\,(S + \lambda I)^{-1}\,U^T\mathbf{b}.$$

Because U is orthonormal we have $U^{-1} = U^T$. The matrix $S + \lambda I$ is an $N \times N$ diagonal matrix which can be easily inverted in $O(N)$. Therefore, in order to solve the system (7) we only have to invert the diagonal matrix $S + \lambda I$ and perform three matrix-vector products with $N \times N$ matrices. Hence, the

Online phase is still in $O(N^2)$ and we can modify the parameter λ in the Online phase. Note that even though we invert the diagonal matrix $S + \lambda I$ we have not experienced any numerical instabilities for the numerical examples in Sect. 7. Computing the eigendecomposition in the Offline phases takes usually distinctly longer than computing the LU decomposition. Nevertheless, we will always employ the eigendecomposition in the following because it gives us greater flexibility with respect to the regularization parameter λ.

7 Examples with Artificial and Real-World Data Sets

In this section, we report on the performance of the proposed classification method based on sparse grid density estimation. Note that the reported accuracies do not depend on whether we use the Offline/Online splitting or not, of course. We compare with the classification approach based on sparse grid regression. We always employ sparse grids with linear hierarchical basis functions ("hat functions") and without basis functions at the boundary because we simply transform the data set in the $[0.1, 0.9]^d$ cube if necessary [20]. The selection of the regularization parameter λ was performed with k-fold cross validation. For more information about our parameter selection procedure and an in-depth discussion, we refer to [6]. To tackle the multi-class classification problem with the classical approach, we compute a sparse grid regression function for each class separately, as discussed in [21].

Let us first consider the results with density functions estimated on regular sparse grids, see Table 1. We compare the training and test accuracies of the classical approach [11, 21] and our proposed method for eight data sets. For all eight data sets we obtain competitive results. For four data sets we either obtained 100 % test accuracy or a better test accuracy than with the classical approach. For all other data sets (three spheres, svmguide1, olive oils, and shuttle) the classical approach is only slightly better, i.e., by around 1 %. However, the accuracies for all eight data sets clearly suggest that they can be learned by our proposed method. Hence, our method can cope with both artificial data sets (two-moons, three spheres, svmguide1) as well as real-world ones (old faithful, Iris flower, olive oils, shuttle and oil flow). Note that in rare cases our method can achieve slightly better results for the test data set than for the training set due to the optimization problem originating from density estimation which does not target explicitly the classification error.

Let us now come to the results for adaptively refined sparse grids, see Table 2. Note that adaptive sparse grid pose a challenge to the Offline/Online splitting, see the outlook in Sect. 8. We employ the standard sparse grid refinement criterion based on the absolute value of the hierarchical coefficients, see, e.g., [4, 21]. We only consider the data sets for which we did not obtain 100 % test accuracy in Table 1 and skipped the shuttle and oil flow data set because we will discuss them in the context of the Offline/Online splitting below. Except for the Iris flower data set,

Table 1 Percent of correctly classified data points for the classical and proposed classification method on regular sparse grids

	d	n	k	Grid points	Level	Regression-based Training	Regression-based Test	Density-based Training	Density-based Test
Two-moons [6]	2	2	10	17	3	100.00	100.00	100.00	100.00
Old faithful [3]	2	2	10	17	3	100.00	100.00	100.00	100.00
Three spheres [6]	3	3	10	351	5	100.00	99.85	99.92	99.71
Iris flower [5]	4	3	10	769	5	99.33	95.33	97.33	96.00
Svmguide1 [15]	4	2	–	769	5	97.05	96.75	95.21	95.85
Olive oils [1]	9	3	5	9439	5	100.00	99.77	99.77	99.08
Shuttle [11]	9	2	5	9439	5	99.59	99.57	97.54	97.66
Oil flow [3]	12	2	5	3249	4	86.11	65.64	86.11	84.57

We used k-fold cross validation to determine the test accuracy except for the svmguide1 data set where an extra test set is available. The table also includes references where to find more information about the data set, the dimension d and the number of classes n

Table 2 Results for the proposed method on adaptive sparse grids

	d	n	k	Grid points	Training	Test
Three spheres [6]	3	3	10	602	100.00	100.00
Iris flower [5]	4	3	10	209	96.37	93.33
Svmguide1 [15]	4	2	–	683	95.61	95.88
Olive oils [1]	9	3	5	732	100.00	100.00

Again the results have been obtained with k-fold cross validation

where overfitting occurs, we need distinctly fewer grid points if we employ adaptive refinement, i.e., if we adapt the grid to the data set.

In Fig. 4 we show a dimension-wise plot of the Iris flower data set, i.e., the four-dimensional data points are projected onto two dimensions each, as well as contour plots of the three density functions corresponding to the three classes of the data set. If we compare the dimension-wise plot of the data set with the dimension-wise plots of the density functions, we see that the density functions evaluate to higher values near the center of the corresponding class. This visualizes that it is reasonable to use the density value as a measure of confidence in the membership to its class.

Overall, we obtain either better or almost as good results as with the classical approach for all data sets. And, just as with the classical approach, we can drastically improve the results using adaptively refined sparse grids.

Finally, we want to report on the runtime of the proposed method with and without the Offline/Online splitting. We only consider the olive oils, shuttle and oil flow data set because only for those the classification took longer than one second. In Table 3 we show the runtime in seconds for the regression-based and density-based classification methods. In case of the density-based method we report the runtimes with and without the Offline/Online splitting. Without the Offline/Online splitting we did not assemble the system matrix but used the matrix-vector product procedures in combination with the CG method to solve the system of linear

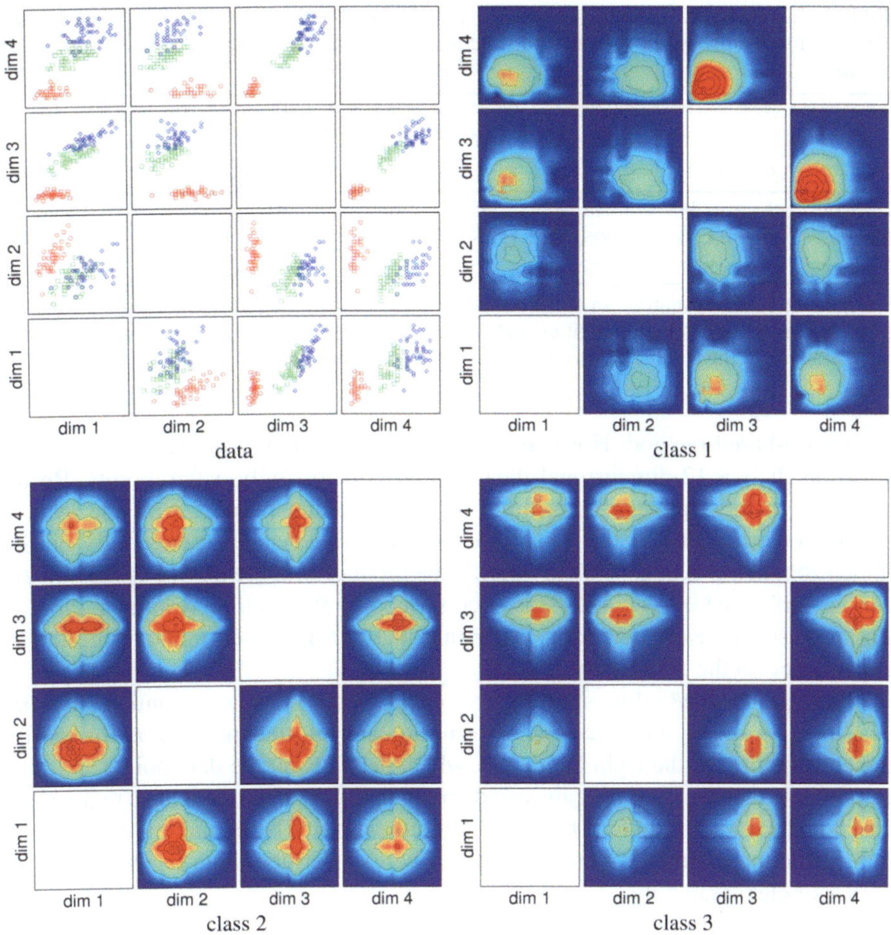

Fig. 4 Projection of the Iris flower data set to two dimensions each and the density functions corresponding to the three classes of the Iris flower data set. All three density functions evaluate to high values near the center of their corresponding class

equations. We stopped either after 50 CG iterations or when the norm of the residual was below 10^{-10}.

Let us first consider the runtimes of the regression-based method and the density-based method without the Offline/Online splitting. The olives oil data set is eight-dimensional and consists of 348 data points only. We do not see a distinct difference between the runtimes of the density-based and regression-based methods. The shuttle data set is also eight-dimensional but has 43,500 data points. For this large data set, we can clearly see the effect of the dependence of the matrix-vector product on the number of data points for the regression-based method. Our density-based method without the Offline/Online splitting is about five times faster than the

Table 3 Runtimes in seconds of the regression-based approach and the density-based classification method without (standard) and with (pre-computed) Offline/Online splitting are reported

			Regression-based		Density-based standard		Density-based Offline/Online	
	d	M	Solve (s)	Total (s)	Solve (s)	Total (s)	Solve (s)	Total (s)
Olive oil	8	348	55	55	60	61	<1	<1
Shuttle	8	43,500	1,006	1,018	215	242	10	36
Oil flow	12	1,318	25	25	567	567	<1	1

We split the (Online) runtime in the time spent to solve the system of linear equations and the total time including all reading, writing and pre-processing of the data. The table also includes the dimension d and the number of (training) data points M of the data sets

regression-based method. However, for the oil flow data set we have the opposite situation. It is a 12-dimensional data set and contains only a few points. Recall that, due to the UpDown algorithm, the matrix-vector product with the system matrix corresponding to the density-based method depends exponentially on the dimension, i.e., it is in $\mathcal{O}(2^d N)$, whereas the product with the matrix $\frac{1}{M}BB^T + \lambda I$ can be performed in $\mathcal{O}(NM)$. This is confirmed by the runtime results in Table 3. The regression-based method is faster than the density-based method due to the dependence on the dimension.

With the Offline/Online splitting we obtain by far the lowest runtimes for the classification of the data sets. For the shuttle data set, the speed up is limited by the construction of the right hand side, which depends on the data points and thus cannot be pre-computed. We gain speed ups of up to 100 compared to the regression-based method and of up to 500 compared to the density-based method without the Offline/Online splitting. The measurements were performed on an Intel Core i7 870 with a single thread only.

8 Conclusions

We presented a novel classification method based on probability density estimation with sparse grids. For each class of the training data set, we estimate a density function on a sparse grid. The computational complexity of the corresponding minimization problem scales only linearly with the number of training data points. Additionally, in contrast to the classification method based on sparse grid regression, the system matrix of the underlying system of linear equations is independent from the training data points. First, this makes the matrix-vector product with the system matrix truly independent from the number of data points which is especially advantageous for large data sets. Second, this allows an Offline/Online splitting of the computational procedure where we pre-compute the system matrix (Offline phase) and reuse it for different data sets (Online phase). This reduces the complexity of learning a data set from $\mathcal{O}(N^3)$ to only $\mathcal{O}(N^2)$. Furthermore, we have shown that if we store the eigendecomposition corresponding to the system matrix we can even change the regularization parameter λ in the Online phase.

The performance of the method has been demonstrated on numerous data sets of various dimensions and different numbers of classes. We obtained competitive results compared to the classical approach. Seven out of eight data sets were learned with an accuracy in test data well above 95 %. Furthermore, in 50 % of all cases the accuracies were either 100 % or better than the accuracies obtained with the classical method. And they were even improved by employing locally refined sparse grids. Finally, we have shown that for certain data sets we can reduce the time to solve the underlying system of linear equations by a factor of up to several hundred if we employ the Offline/Online splitting.

Future work includes the Offline/Online splitting for adaptively refine sparse grids. One option is to pre-compute the matrices corresponding to several refined sparse grids and to use a standard error indicator (e.g. the absolute value of the hierarchical coefficients) to select from them the most appropriate one.

References

1. A. Azzalini, N. Torelli, Clustering via nonparametric density estimation. Stat. Comput. **17**(1), 71–80 (2007)
2. B. Babcock, S. Babu, M. Datar, R. Motwani, J. Widom, Models and issues in data stream systems, in *Proceedings of the Twenty-First ACM SIGMOD-SIGACT-SIGART Symposium on Principles of Database Systems* (ACM, New York, 2002)
3. C. Bishop, *Pattern Recognition and Machine Learning* (Springer, Berlin, 2007)
4. H.-J. Bungartz, M. Griebel, Sparse grids. Acta Numer. **13**, 1–123 (2004)
5. R. Fisher, The use of multiple measurements in taxonomic problems. Ann. Hum. Genet. **7**(2), 179–188 (1936)
6. F. Franzelin, Classification with estimated densities on sparse grids. Master's thesis, Institut für Informatik, Technische Universität München, 2011
7. M. Gaber, A. Zaslavsky, S. Krishnaswamy, Mining data streams: a review. SIGMOD Rec. **34**(2), 18–26 (2005)
8. J. Gama, *Knowledge Discovery from Data Streams* (Chapman & Hall, London, 2010)
9. J. Garcke, Maschinelles Lernen durch Funktionsrekonstruktion mit verallgemeinerten dünnen Gittern. Doktorarbeit, Institut für Numerische Simulation, Universität Bonn, 2004
10. J. Garcke, M. Griebel, Classification with sparse grids using simplicial basis functions. Intell. Data Anal. **6**(6), 483–502 (2002)
11. J. Garcke, M. Griebel, M. Thess, Data mining with sparse grids. Computing **67**(3), 225–253 (2001)
12. T. Hastie, R. Tibshirani, J. Friedman, *The Elements of Statistical Learning* (Springer, Berlin, 2009)
13. M. Hegland, G. Hooker, S. Roberts, Finite element thin plate splines in density estimation. ANZIAM J. **42**, C712–C734 (2000)
14. A. Heinecke, B. Peherstorfer, D. Pflüger, Z. Song, Sparse grid classifiers as base learners for AdaBoost, in *2012 International Conference on High Performance Computing and Simulation (HPCS)*, Madrid (IEEE, New York, 2012)

15. C.-W. Hsu, C.-C. Chang, C.-J. Lin, A practical guide to support vector classification. Technical report, Department of Computer Science, National Taiwan University, 2003
16. T. Jebara, *Machine Learning. Discriminative and Generative* (Kluwer, Dordecht, 2004)
17. K. Murphy, *The Machine Learning: A Probabilistic Perspective* (MIT Press, Cambridge, 2012)
18. A. Patera, G. Rozza, *Reduced Basis Approximation and A Posteriori Error Estimation for Parametrized Partial Differential Equations*. MIT Pappalardo Graduate Monographs in Mechanical Engineering (MIT, 2007)
19. B. Peherstorfer, D. Pflüger, H.-J. Bungartz, Clustering based on density estimation with sparse grids, in *KI 2012: Advances in Artificial Intelligence*, ed. by B. Glimm, A. Krüger. Lecture Notes in Computer Science, vol. 7526 (Springer, Berlin, 2012), pp. 131–142
20. D. Pflüger, Data Mining mit Dünnen Gittern. Diplomarbeit, IPVS, Universität Stuttgart, 2005
21. D. Pflüger, *Spatially Adaptive Sparse Grids for High-Dimensional Problems* (Verlag Dr. Hut, München, 2010)
22. D. Pflüger, Spatially adaptive refinement, in *Sparse Grids and Applications*, ed. by J. Garcke, M. Griebel. Lecture Notes in Computational Science and Engineering, vol. 88 (Springer, Berlin, 2013)
23. D. Pflüger, B. Peherstorfer, H.-J. Bungartz, Spatially adaptive sparse grids for high-dimensional data-driven problems. J. Complex. **26**(5), 508–522 (2010)
24. A. Uzilov, J. Keegan, D. Mathews, Detection of non-coding RNAs on the basis of predicted secondary structure formation free energy change. BMC Bioinformatics **7**(1), 173 (2006)

Adjoint Error Estimation for Stochastic Collocation Methods

Bettina Schieche and Jens Lang

Abstract This paper deals with partial differential equations with random input data. An efficient way of solving such problems is adaptive stochastic collocation on sparse grids. For higher efficiency and a better understanding of the method, we derive adjoint error estimates for nonlinear stochastic solution functionals. The resulting adjoint problem also involves random parameters and can be treated by stochastic collocation as well. Only a few adjoint evaluations are required in order to estimate the deterministic error, while the stochastic error requires much more effort to be detected. To overcome these substantial additional costs, we suggest to replace the adjoint problem by a reduced model and demonstrate the applicability of the approach for nonlinear solution functionals and up to nine random dimensions.

1 Introduction

People in natural and engineering sciences have realized that simulations become more reliable when including uncertainties, such as natural fluctuations or deviations in manufacturing processes. The systematic analysis of output uncertainties with respect to input uncertainties is known as uncertainty quantification (UQ) [33].

B. Schieche
Numerical Analysis and Scientific Computing (Department of Mathematics), Technische Universität Darmstadt, Dolivostr. 15, 64293 Darmstadt, Germany
e-mail: schieche@mathematik.tu-darmstadt.de

J. Lang (✉)
Numerical Analysis and Scientific Computing (Department of Mathematics) & Graduate School of Computational Engineering, Technische Universität Darmstadt, Dolivostr. 15, 64293 Darmstadt, Germany
e-mail: lang@mathematik.tu-darmstadt.de

J. Garcke and D. Pflüger (eds.), *Sparse Grids and Applications - Munich 2012*, Lecture Notes in Computational Science and Engineering 97,
DOI 10.1007/978-3-319-04537-5_12,

One important branch of research deals with partial differential equations (PDEs) with correlated random parameters. Solution approaches are basically Monte Carlo methods, spectral methods of Galerkin type, and stochastic collocation on sparse grids. The first is for sure the best known sampling strategy and convinces with simplicity and generality. Low convergence rates can be overcome by the recently suggested multilevel Monte Carlo variance reduction technique [5, 15]. Spectral methods go back to [50] and are founded on stochastic, orthogonal basis polynomials, on whose span the PDE is projected, resulting in deterministic, coupled systems [20, 25, 34, 52] of certain structure [16, 17, 45, 46].

Stochastic collocation [51] came up as an alternative approach to spectral methods. The given random parameter space is here discretized into a number of collocation points that equal realizations of the random input. The corresponding deterministic solutions are then interpolated and integrated in order to extract statistical quantities of the solution. It is obvious that the number of collocation points is an important issue, especially for time-consuming underlying problems. So, a common approach is to choose the collocation points on structured, but sparse grids.

Stochastic collocation can be seen as an application of sparse grids. The name was proposed by Zenger [55], but the idea goes back to [48]. There are many works that study sparse grids as a tool for high-dimensional interpolation and integration in terms of convergence and polynomial exactness (see [6, 42] and the references therein). Adaptive versions can be found for example in [12, 24, 30]. A detailed overview is provided by [13]. In the context of stochastic collocation, important works addressing questions such as a priori error estimates are among others [4, 40, 41]. Last but not least, sparse grids have been discovered to be also useful to discretize deterministic equations [26–28].

We focus on stochastic collocation on dimension-adaptive sparse grids [24] with global Lagrange interpolation, because the approach combines the decoupling nature of Monte Carlo with fairly good convergence properties. The aim of this paper is to study the applicability of adjoint error estimation in the stochastic collocation setting. Conventional adaptivity criteria are usually based on relative changes of the solution or some solution functional, often stored as hierarchical surpluses when using hierarchical stochastic basis functions. Although the resulting error indicators provide reasonable results [21,38,54], it is by far not clear how much they over- or underestimate the true error in a particular application. Moreover, it is important to study the contribution of deterministic discretization errors and interpolation errors more closely in order to balance both error sources and avoid unnecessary computational costs.

So far, adjoint equations have been used to predict errors at randomly chosen points [14] or at the collocation points [2], leading to a random representation of the deterministic discretization error, which can be used to adapt spatial meshes. Our new contributions are a general derivation of stochastic, adjoint error estimates and extensive numerical studies for quite challenging problems, including several nonlinear solution functionals and rather high-dimensional random parameter spaces. Moreover, in contrast to [2], we address the question of deterministic and stochastic parts of the error, compare the results to traditional error indicators, and give a suggestion of how to overcome additional computational costs. Our numerical examples impressively show that adjoint error estimation is a powerful tool to gain understanding in the error behaviour of stochastic collocation methods.

The present paper is structured as follows: Sect. 2 formulates the PDE with random parameters including some mathematical preliminaries. Then, in Sect. 3 all ingredients for stochastic collocation are provided and Sect. 4 is dedicated to the general description of adjoint error estimates for different types of solution functionals. Numerical results are presented in Sect. 5, followed by a summary and conclusion in Sect. 6.

2 Problem Formulation

Throughout this work we consider a general PDE equipped with parameters such as material properties, forces, initial conditions, and boundary conditions. The particular application is not important to describe the general setting. In order to introduce uncertainties in the parameters, let $(\Omega, \Sigma, \mathcal{P})$ be a complete probability space, with Ω being the set of outcomes of a random experiment, the sigma algebra Σ of events in Ω, and a probability measure $\mathcal{P} : \Sigma \to \mathbb{R}$. By doing so, the uncertain parameters are modelled as random variables or random fields with dimension $\omega \in \Omega$ in addition to space and time. Without loss of generality, we describe the problem by the form

$$\left.\begin{aligned} \mathcal{A}(u, \mathbf{x}, t, \omega) &= f(\mathbf{x}, t, \omega) && \text{in } D \times (0, T_{\text{end}}] \times \Omega \\ \mathcal{B}(u, \mathbf{x}, t, \omega) &= g(\mathbf{x}, t, \omega) && \text{on } \partial D \times (0, T_{\text{end}}] \times \Omega \\ u(\mathbf{x}, 0, \omega) &= u_0(\mathbf{x}, \omega) && \text{in } D \times \Omega\,, \end{aligned}\right\} \tag{1}$$

with differential operator $\mathcal{A}$, boundary operator $\mathcal{B}$, and initial solution u_0. We seek for a random solution field $u(\mathbf{x}, t, \omega)$, that solves problem (1) almost surely. The spatial variable is denoted by $\mathbf{x} \in D \subset \mathbb{R}^d$ and the time t acts in the interval $[0, T_{\text{end}}]$. As a side note we want to mention that there are quite a few works available that deal with the analysis of problem (1) [3, 40]. Typically, deterministic results can be extended to the stochastic case, provided that the parameters satisfy some additional assumptions. Suitable solution spaces are Bochner spaces that contain L^p-integrable random variables taking values in deterministic solution spaces X:

$$L^p_{\mathcal{P}}(\Omega; X) = \{v : \Omega \to X \textit{ measurable} : \mathrm{E}[\|v\|_X^p] := \int_\Omega \|v\|_X^p \,\mathrm{d}\mathcal{P}(\omega) < \infty\}. \tag{2}$$

Note that $\mathrm{E}[\cdot]$ refers to the expected value operator. Furthermore, we will use the notation $\mathrm{Var}[\cdot]$ and $\mathrm{Cov}[\cdot, \cdot]$ for the variance and covariance, respectively.

We restrict our studies to correlated random fields, sometimes also known as colored noise in contrast to white noise problems. For numerical treatment, we follow the "Finite Noise Assumption", assuming that all random parameters can be expressed by finitely many random variables. This parametrization step is justified by the Karhunen-Loève expansion [25], for which we assume a general random parameter, here denoted by $\alpha(\mathbf{x}, \omega)$, with continuous covariance function $\mathrm{Cov}(\mathbf{x}, \tilde{\mathbf{x}})$, for simplicity time-independent. The covariance function defines a linear operator of integral type, whose spectrum is given by the integral equation

$$\int_D \mathrm{Cov}(\mathbf{x}, \tilde{\mathbf{x}})\, f_n(\tilde{\mathbf{x}})\, \mathrm{d}\tilde{\mathbf{x}} = \lambda_n f_n(\mathbf{x}), \tag{3}$$

with countable many normalized, orthogonal eigenfunctions f_n. The corresponding eigenvalues λ_n are positive, with 0 being the only limit point. The Karhunen-Loève expansion of the random field α is then given by

$$\alpha(\mathbf{x}, \omega) = \mathrm{E}[\alpha] + \sum_{n=1}^{\infty} f_n(\mathbf{x}) \sqrt{\lambda_n} \xi_n, \tag{4}$$

with uncorrelated random variables ξ_n, that can either be calculated from more accurate information about marginal distributions of α, or be part of the modelling. Usually, common types of distributions such as uniform, normal, or log-normal distributions are chosen. A further assumption that we are going to use in this work, is the stochastic independence of all arising random variables. Equation (4) allows for truncation after a finite number of terms. It can be shown that the truncation error in $L^2(D \times \Omega)$ is directly linked to the eigenvalues that correspond to the truncated terms. Hence, they can be used as a criterion to maintain a given amount of variance in the sequence. How many terms will be kept in practice depends on the decay of the eigenvalues λ_n, which in turn depends on the smoothness of the covariance kernel [10].

For convenience, let now M be the total number of random variables and collect them in the random vector $\boldsymbol{\xi} = (\xi_1(\omega), \ldots, \xi_M(\omega))$. Given that, the Doob-Dynkin-Lemma justifies that the unknown solution is a nonlinear functional of $\boldsymbol{\xi}$, namely $u(\mathbf{x}, t, \omega) = u(\mathbf{x}, t, \boldsymbol{\xi})$. A more intuitive way to look onto the resulting problem is provided by a change of the measure. To this end, we assume that the M random variables are strongly measurable with probability density functions $\rho_n(y_n)$. Due to the stochastic independence, the density function of $\boldsymbol{\xi}$ separates like $\rho(\mathbf{y}) = \prod_{n=1}^{M} \rho_n(y_n)$, with $\mathbf{y} = (y_1, \ldots, y_M)$. Let $\Gamma = \boldsymbol{\xi}(\Omega)$ be the image space of Ω. Then, we can change to the weighed Lebesgue measure $\rho(\mathbf{y})\,\mathrm{d}\mathbf{y}$ and finally arrive at the following counterpart of problem (1):

$$\left.\begin{aligned} \mathcal{A}(u, \mathbf{x}, t, \mathbf{y}) &= f(\mathbf{x}, t, \mathbf{y}) && \text{in } D \times (0, T_{\mathrm{end}}] \times \Gamma \\ \mathcal{B}(u, \mathbf{x}, t, \mathbf{y}) &= g(\mathbf{x}, t, \mathbf{y}) && \text{on } \partial D \times (0, T_{\mathrm{end}}] \times \Gamma \\ u(\mathbf{x}, 0, \mathbf{y}) &= u_0(x, \mathbf{y}) && \text{in } D \times \Gamma\,. \end{aligned}\right\} \tag{5}$$

Note that the hypercube Γ can be seen as a deterministic parameter space in $\mathbb{R}^M$. Having set up the general problem under consideration, the next section is devoted to the stochastic collocation approach to solve problem (5) numerically.

3 Stochastic Collocation on Sparse Grids

The idea of sparse grids in multiple dimensions is to start with a sequence of nodes for one dimension, given by $Y^i := \{y_j^i\}_{j=1}^{m_i}$, $i = 1, 2, \ldots$. Moreover, we claim $Y^i \subset Y^{i+1}$, resulting in a nested sequence of nodes. Using Lagrange interpolation, defined on these nodes, yields the approximation

$$U^i(u) = \sum_{j=1}^{m_i} u(y_j^i) L_j^i(y), \qquad (6)$$

for a function $u(y)$ in one dimension and nodal Lagrange polynomials $L_j^i(y)$. Let now $\Delta^i(u) := U^i(u) - U^{i-1}(u)$ be the difference between two consecutive interpolation rules, with $U_0 = 0$. Smolyak's formula for an interpolation rule in M dimensions is then given by

$$A(k, M)(u) := \sum_{|\mathbf{i}| \leq k+M} \left(\Delta^{i_1} \otimes \cdots \otimes \Delta^{i_M}\right)(u) \qquad (7)$$

and approximates a function $u(\mathbf{y})$ in the M variables $\mathbf{y} = (y_1, \ldots, y_M)$. The multi-index $\mathbf{i} := (i_1, \ldots, i_M)$ relates each dimension to an interpolation rule, and the sparsity property results from the restriction $|\mathbf{i}| := i_1 + \cdots + i_M \leq k + M$ with a natural number k, called the level of the algorithm.

A rigorous numbering of all points in the resulting sparse grid yields the set $\{\mathbf{y}^{(j)}\}_{j=1}^P$ with total number of points P. Let the corresponding nodal, multivariate Lagrange polynomials be denoted by $\mathcal{L}_j(\mathbf{y})$. Then we can write Eq. (7) in the short form

$$A(k, M)(u) = \sum_{j=1}^{P} u_j \mathcal{L}_j(\mathbf{y}), \quad u_j = u(\mathbf{y}^{(j)}). \qquad (8)$$

First note that sparse grids are again nested, which allows to reuse old points when adding new ones. This is especially interesting for adaptivity. Second, the collocation points should not be chosen equidistantly when applying Lagrange interpolation. And finally, the collocation points should also provide suitable quadrature nodes in order to calculate functionals of the solution. Examples of such functionals are the qth moments of the solution, which are given by

$$\mathrm{E}[A(k,M)(u)^q] = \sum_{j_1=1}^{P} \cdots \sum_{j_q=1}^{P} u_{j_1} \cdots u_{j_q} \mathrm{E}[\mathcal{L}_{j_1} \cdots \mathcal{L}_{j_q}], \tag{9}$$

and result in pre-computable expected values over products of multivariate Lagrange basis polynomials. However, it is convenient to approximate these expected values by means of the current set of collocation points, yielding

$$\mathrm{E}[\mathcal{L}_{j_1} \cdots \mathcal{L}_{j_q}] \approx \sum_{j=1}^{P} \mathcal{L}_{j_1}(\mathbf{y}^{(j)}) \cdots \mathcal{L}_{j_q}(\mathbf{y}^{(j)}) \underbrace{\mathrm{E}[\mathcal{L}_j]}_{=:w_j} . \tag{10}$$

Due to the fact that we have $\mathcal{L}_j(\mathbf{y}^{(i)}) = \delta_{ji}$, such a quadrature rule results in the following approximation of expression (9):

$$\mathrm{E}[A(k,M)(u)^q] \approx \sum_{j=1}^{P} w_j u_j^q . \tag{11}$$

To sum up, the collocation points in multiple dimensions are based on tensor products of one-dimensional quadrature nodes of increasing accuracy. These nodes should be nested, well-suited for Lagrange interpolation, and equal to quadrature nodes with respect to the considered probability measure. Basically, there are two classes of nodes that satisfy these conditions. The first are the Clenshaw-Curtis nodes, which equal the extrema of Chebyshev polynomials. The second are nested versions of Gauss quadrature rules, known as Gauss-Patterson nodes. The first has provided reasonable results in previous works [6, 19, 21]. However, convergence problems with sparse grids have been observed, especially in higher dimensions [37].

The question of nested extensions of Gauss quadrature was first addressed in [32]. Similar to the Gauss rule itself, the idea is to choose all degrees of freedom such that the highest possible polynomial exactness is obtained. If this procedure allows for a recursive extension, a nested sequence of nodes arises [43]. In the case of unweighted integrals, such extensions are known as Gauss-Patterson-Legendre rules and are linked to the uniform distribution. However, for arbitrary Gauss rules the derivation of extended formulas is not straightforward, especially for unbounded weighting functions, such as the density of the normal distribution [44]. One idea is to allow a lower degree of exactness, ending up with suboptimal extensions [8]. A sequence for the normal distribution can be found in [22, 29].

These rules together with sparse grids have, for example, been proposed in [23]. Recent research underlines that this choice of collocation points improves the accuracy of the integrals compared to Clenshaw-Curtis nodes [37]. Due to our own positive experiences with these nodes, we have decided to use Gauss-Patterson rules in this work. Figure 1 shows the sparse grids for $M = 2$ and levels $k = 0, \ldots, 4$ for the Gauss-Patterson-Legendre quadrature.

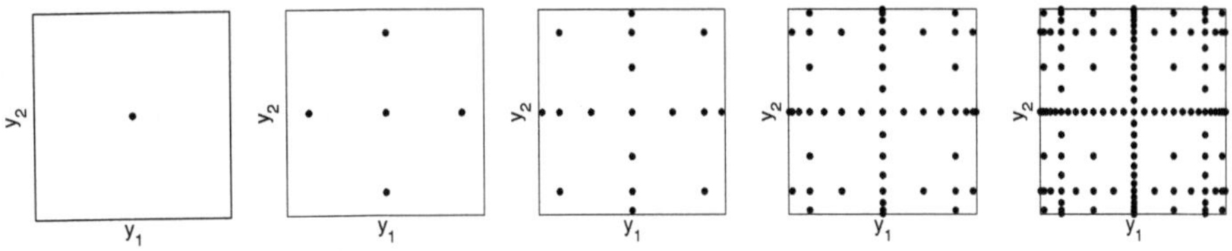

(a) $k = 0$ (b) $k = 1$ (c) $k = 2$ (d) $k = 3$ (e) $k = 4$

Fig. 1 Sparse grids with nested Gauss-Patterson-Legendre nodes in two dimensions for Smolyak levels $k = 0, \ldots, 4$

The appropriate Smolyak level is not known beforehand and has to be increased adaptively. One adaptivity concept is the dimension-adaptive, i.e. anisotropic, version of Smolyak's formula, which is well-suited for global Lagrange interpolation. We use this algorithm to generate sparse grids with more nodes in the dominating parameter ranges. This generalized adaptive Smolyak algorithm was introduced in [24], and we would like to sketch the idea briefly.

Compared to Eq. (7), the mentioned algorithm does not sum over all indices $|\mathbf{i}| \leq q$, but successively adds indices to the sum, whose nodes exhibit large error indicators. Note that an index $\mathbf{i}$ can only be part of the sum if the so-called father-indices $\mathbf{i} - \mathbf{e}_j$, $j = 1, \ldots, M$, are also used, where $\mathbf{e}_j$ denotes the jth unit vector. Otherwise the telescopic character of the sum is violated.

For each new index, a local error indicator is needed. The one with the largest contribution is then further refined. A global error indicator is given by the sum of all error indicators that refer to indices that were not yet refined, also called active indices in the spirit of [24]. In practice, we do not always interpolate the whole solution, but rather a functional like a space-time integral over the solution for example. The resulting interpolation then gives a random variable, for which we may be interested in the first or second moment. Hence, when adding a new index to the generalized Smolyak sum, a natural error indicator is given by this new contribution to the stochastic quantity of interest for the respective index. More precise, local error indicators refer to normalized changes in a stochastic solution functional throughout this paper.

The aim of this work is to analyze the global error behaviour and the quality of the error indicators very thoroughly. To this end, we make use of adjoint calculus, which is an important tool to construct error estimates for deterministic PDEs. There are some new issues that have to be addressed in the stochastic context, which will be covered by the next section.

4 Stochastic Adjoint Error Estimation

For deterministic problems, adjoint equations are already used to construct efficient a posteriori error estimates [7, 18, 35]. The approach is also known as dual-based or goal-oriented error estimation, because it controls the error in some quantity of interest. In this context the expression "primal" refers to the original problem.

Adjoint error estimates in the context of UQ are not that well understood. A perturbation-based approach can be found in [1], which is restricted to small input fluctuations. For spectral methods of Galerkin type, adjoint error estimates naturally arise from using a variational formulation for both deterministic and stochastic dimensions [39]. However, it is by far not clear how to approximate the adjoint solution such that the resulting error estimate captures both the deterministic and stochastic error contribution.

At this point we have to ask for the intentions, which may be of very different kind. A recent work [14] uses the adjoint framework to build up a representation of the error with the same number of terms as used for the original solution. The resulting surrogate for the error is then compared to the true errors point-wise in the random parameter space. Similarly, the adjoint problem can be solved in all collocation points [2] to provide error estimates.

However, the same collocation points for the adjoint problem hide information about the interpolation error, which vanishes in these points by construction. Our goal is to understand how the stochastic adjoint problem can be used to capture both deterministic and stochastic errors. This is not a trivial task and we will mainly study these effects numerically in Sect. 5. In this section we want to derive adjoint problems, discuss stochastic functionals, and present our ideas to gain a more efficient evaluation of the adjoint problem.

4.1 From Deterministic to Stochastic Adjoint Error Estimation

Let us first consider a deterministic solution functional $\mathcal{J}(u)$ of integral type, i.e.

$$\mathcal{J}(u) = \int \mathcal{N}(u), \tag{12}$$

with a possibly nonlinear functional $\mathcal{N}$. More precise, the integral is meant to cover the deterministic domain, which may be the spatial domain D, or, in the time-dependent case, $D \times (0, T_{\text{end}}]$. A typical expression for the error between the true value and the approximate value, denoted by $\mathcal{J}(u_h)$, is given by

$$\mathcal{J}(u) - \mathcal{J}(u_h) = \int \phi Res(u_h) + HOT, \tag{13}$$

with residual $Res(u_h)$, adjoint solution ϕ, and higher order terms (HOT) due to linearization in the case of nonlinearities [18]. Note that the index h refers to the whole deterministic discretization. For the extension to stochastic problems, we start with the parametrized PDE (5) derived in Sect. 2. For simplicity, we neglect the deterministic variables $\mathbf{x}$ and t as well as the boundary and initial conditions and consider

$$\mathcal{A}(u, \mathbf{y}) = f(\mathbf{y}), \qquad \mathbf{y} \in \Gamma. \tag{14}$$

We assume to have an approximate solution $u_{h\xi}$. By means of linearization about $u_{h\xi}$, we arrive at

$$\mathcal{A}(u, \mathbf{y}) = \mathcal{A}(u_{h\xi}, \mathbf{y}) + \mathcal{A}'(u_{h\xi}, \mathbf{y})(u - u_{h\xi}) + HOT. \tag{15}$$

In the following equations we will skip the HOT, because they are neglected in practice anyway. Recall that each value $\mathbf{y}$ refers to a realization of an underlying random vector $\boldsymbol{\xi}$. Hence, we can derive the adjoint system

$$\mathcal{A}^*(u_{h\xi}, \mathbf{y})\phi(\mathbf{y}) = \kappa(\mathbf{y}), \qquad \mathbf{y} \in \Gamma, \tag{16}$$

which apparently depends on the same parametrized random parameters as the primal problem, with solution $\phi(\mathbf{y})$, that can again be interpreted as a random field $\phi(\boldsymbol{\xi})$. The same holds for the not yet specified right hand side $\kappa(\mathbf{y})$ or $\kappa(\boldsymbol{\xi})$. In the following we will use the notation $\boldsymbol{\xi}$ when stressing the underlying random variables, and $\mathbf{y}$ when we are talking of realizations of $\boldsymbol{\xi}$.

For fixed $\mathbf{y} \in \Gamma$, the derivation of the adjoint operator $\mathcal{A}^*(u_{h\xi}, \mathbf{y})$ of the linearized operator $\mathcal{A}'(u_{h\xi}, \mathbf{y})$ is based on the usual duality condition given by

$$\int \mathcal{A}^*(u_{h\xi}, \mathbf{y})\phi(\mathbf{y})\, v(\mathbf{y}) \overset{!}{=} \int \phi(\mathbf{y})\, \mathcal{A}'(u_{h\xi}, \mathbf{y})v(\mathbf{y}) \tag{17}$$

for test functions v. Goal-oriented error estimation in this context now requires a stochastic quantity of interest. Many different kinds of such functionals are possible. The ones we would like to treat here are the following:

$$\mathcal{Q}_1(u) = \mathrm{E}[\mathcal{J}(u)^q], \qquad \mathcal{Q}_3(u) = \mathcal{J}(\mathrm{E}[u^q]),$$
$$\mathcal{Q}_2(u) = \mathrm{E}[\mathcal{J}(u)]^q, \qquad \mathcal{Q}_4(u) = \mathcal{J}(\mathrm{E}[u]^q).$$

These stochastic solution functionals involve a deterministic solution functional $\mathcal{J}$, now treated either as a random variable or evaluated in random fields, depending on the type of composition of $\mathcal{J}$ and $\mathrm{E}[\cdot]$. The number q allows higher moments to be part of the quantity of interest, but is not restricted to natural numbers. Note that these functionals can be arbitrarily combined, so that they also cover functionals that involve variances of the solution.

Likewise to the deterministic setting, we are interested in the difference $\mathcal{Q}_i(u) - \mathcal{Q}_i(u_{h\xi})$, $i = 1, \ldots, 4$. In order to find expressions for these errors, we use a first order Taylor expansion about $u_{h\xi}$ in all cases. For the first solution functional, this yields

$$\mathcal{Q}_1(u) - \mathcal{Q}_1(u_{h\xi}) \approx \mathrm{E}[\int \underbrace{q\mathcal{J}(u_{h\xi})^{q-1}\mathcal{N}'(u_{h\xi})}_{=:\kappa_1(\mathbf{y})}(u - u_{h\xi})]. \tag{18}$$

We set the right hand side of the adjoint system (16) equal to $\kappa_1(\mathbf{y})$ and call the corresponding unknown $\phi_1(\mathbf{y})$. Then the error allows for the following expression

$$\mathcal{Q}_1(u) - \mathcal{Q}_1(u_{h\xi}) \approx \mathrm{E}[\int \kappa_1(\mathbf{y})(u - u_{h\xi})] \tag{19}$$

$$= \mathrm{E}[\int \mathcal{A}^*(u_{h\xi}, \mathbf{y})\phi_1(\mathbf{y})(u - u_{h\xi})] \tag{20}$$

$$\stackrel{(17)}{=} \mathrm{E}[\int \phi_1(\mathbf{y})\mathcal{A}'(u_{h\xi}, \mathbf{y})(u - u_{h\xi})] \tag{21}$$

$$\stackrel{(15)}{\approx} \mathrm{E}[\int \phi_1(\mathbf{y})(\mathcal{A}(u, \mathbf{y}) - \mathcal{A}(u_{h\xi}, \mathbf{y}))] \tag{22}$$

$$\stackrel{(14)}{=} \mathrm{E}[\int \phi_1(\mathbf{y})\underbrace{(f(\mathbf{y}) - \mathcal{A}(u_{h\xi}, \mathbf{y}))}_{=:Res(u_{h\xi})}]. \tag{23}$$

We observe that the error has the same structure as the deterministic error (13), apart from the fact that the expected value arises, which is nothing else than an integral over the stochastic domain. For the other types of solution functionals a first order Taylor expansion about $u_{h\xi}$ gives

$$\mathcal{Q}_2(u) - \mathcal{Q}_2(u_{h\xi}) \approx \mathrm{E}[\int \underbrace{q\,\mathrm{E}[\mathcal{J}(u_{h\xi})]^{q-1}\mathrm{E}[\mathcal{N}'(u_{h\xi})]}_{=:\kappa_2(\mathbf{y})}(u - u_{h\xi})], \tag{24}$$

$$\mathcal{Q}_3(u) - \mathcal{Q}_3(u_{h\xi}) \approx \mathrm{E}[\int \underbrace{\mathcal{N}'(\mathrm{E}[u_{h\xi}^q])q u_{h\xi}^{q-1}}_{=:\kappa_3(\mathbf{y})}(u - u_{h\xi})], \quad \text{and} \tag{25}$$

$$\mathcal{Q}_4(u) - \mathcal{Q}_4(u_{h\xi}) \approx \mathrm{E}[\int \underbrace{\mathcal{N}'(\mathrm{E}[u_{h\xi}]^q)q\mathrm{E}[u_{h\xi}]^{q-1}}_{=:\kappa_4(\mathbf{y})}(u - u_{h\xi})]. \tag{26}$$

In order to see that the two latter approximations hold, note that the Taylor expansion yields $\mathrm{E}[u - u_{h\xi}]$ as part of the integrand. Drawing the expectation in front of the whole integral, reveals the given structure. We set the right hand side of the adjoint system (16) equal to $\kappa_i(\mathbf{y})$ and get the corresponding adjoint solutions $\phi_i(\mathbf{y})$, $i = 1, \ldots, 4$, yielding the error estimates

$$\mathcal{Q}_i(u) - \mathcal{Q}_i(u_{h\xi}) \approx \mathrm{E}[\int \phi_i(\mathbf{y})Res(u_{h\xi})], \qquad i = 1, \ldots, 4. \tag{27}$$

Note that it is also possible to consider boundary integrals or even function evaluations as underlying deterministic solution functional $\mathcal{J}$. For the latter, the resulting adjoint is also known as generalized Green's function [18]. We refer to [35] for details on how to include boundary and initial conditions into adjoint error estimates.

From now on, we will skip the index i, but assume a solution functional $\mathcal{Q}$ that allows for the general error expression

$$\mathcal{Q}(u) - \mathcal{Q}(u_{h\xi}) \approx \mathrm{E}[\underbrace{\int \phi(\boldsymbol{\xi}) Res(u_{h\xi})}_{=:\epsilon(\boldsymbol{\xi})}]. \tag{28}$$

4.2 Discussion About Deterministic and Stochastic Errors

In practice we have to approximate the adjoint random field ϕ. The brute-force approach that reuses the discretization of the primal solution usually does not provide acceptable results, because the interpolation error vanishes in all primal collocation points and, especially in the finite element context, the deterministic error vanishes as well due to the orthogonality of residual and adjoint function. This means that the adjoint has to be obtained in an enriched solution space [7].

From another point of view, different kinds of approximations of ϕ capture different information of the overall error. Let us consider the current P primal collocation points with quadrature weights $\{w_j\}_{j=1}^{P}$. We suggest to estimate the deterministic error according to

$$\mathcal{Q}(u_\xi) - \mathcal{Q}(u_{h\xi}) \approx \sum_{j=1}^{P} \epsilon_j w_j, \qquad \epsilon_j = \epsilon(\mathbf{y}^{(j)}), \tag{29}$$

because it hides all information about the interpolation error. This error expression means solving the adjoint problem for each primal collocation point. Depending on the required accuracy of the error, we claim that one can even get along with less adjoint computations to obtain quite accurate approximations of the deterministic error. If we now want to detect the interpolation error as well, it is obvious that an improved quadrature rule based on a refined sparse grid has to come into play. Let this number of adjoint collocation points be denoted by $\tilde{P}$ with weights $\{\tilde{w}_j\}_{j=1}^{\tilde{P}}$. So we propose to estimate the overall error by means of

$$\mathcal{Q}(u) - \mathcal{Q}(u_{h\xi}) \approx \sum_{j=1}^{\tilde{P}} \epsilon_j \tilde{w}_j, \tag{30}$$

with $\tilde{P}$ somehow large enough. Although formally not difficult, the choice of this fine quadrature rule is not a trivial task. To our best knowledge, no suggestions in literature can be found addressing this issue. In this paper we answer this question numerically for selected problems. Anyway, we are sure that the number of adjoint collocation points has to exceed the number of primal collocation points to detect the overall error. This is computationally demanding, which leads to the topic of the following subsection.

4.3 Application of Reduced Order Models

Reduced order models (ROMs) [9, 47, 49] play an important role in parametric studies or optimization. The approach is based on the observation that solutions of PDEs or corresponding quantities of interest often lie on low-dimensional manifolds of the solution space.

We do not want to go into details of ROMs, but sketch the typical procedure. In a first step, selected simulations are run to create a set of snapshots of the solution for either different points in time, different input parameter values, or both. Let us assume that the spatial discretization of the underlying PDE consists of N degrees of freedom. Hence, what we end up with is a set of vectors of length N. One way to find the most influencing modes of the solution is to perform a Proper Orthogonal Decomposition (POD) of the snapshot set. This technique basically consists in solving an eigenvalue problem. The resulting eigenvectors that correspond to the largest eigenvalues then serve as a reduced basis.

We assume now to have suitable reduced basis functions $\psi_1, \ldots, \psi_R$, with $R \ll N$. Just like for the truncated Karhunen-Loève expansion, one can, for example, choose R such that 95 % of the variance in the snapshots is maintained. We now want to approximate a solution u in terms of the reduced basis, i.e.

$$u(\mathbf{x}, t, \mathbf{y}) \approx u^R(\mathbf{x}, t, \mathbf{y}) := \sum_{r=1}^{R} a_r(t, \mathbf{y}) \psi_r(\mathbf{x}), \tag{31}$$

with unknown coefficient functions a_r, which depend on time t and the random parameters $\mathbf{y}$. Plugging expression (31) into the governing equations (5), followed by an $L^2(D)$-projection on the reduced basis, yields the conditions

$$\int_D Res(u^R(\mathbf{x}, t, \mathbf{y})) \psi_i(\mathbf{x}) \, d\mathbf{x} = 0, \qquad i = 1, \ldots, R, \tag{32}$$

where Res denotes the residual of problem (5). For linear problems, these conditions for the coefficient functions $a_r(t, \mathbf{y})$ can be simplified, so that all terms reduce to size R. The resulting systems of ordinary differential equations are supposed to be considerably smaller than the systems that would arise from the original finite element basis.

The reduced problem still involves the random parameter $\boldsymbol{\xi}$ and can be solved by stochastic collocation, where each collocation point requires a cheap equation to be solved.

To which extent we can benefit from ROMs is still an open issue, because their performance strongly depends on the kind of application, the pre-computed snapshots, and the dimension R of the reduced problem. Nevertheless, the approach can be important for problems that would hit their computational limits otherwise. We did not find much literature on reduced models in the context of stochastic collocation. However, the setting is similar to parametric studies, for which one may consult [11, 31, 36] for instance. Our aim is to replace full adjoint problems by ROMs to accelerate the error estimation.

5 Numerical Results

We present results for an elliptic problem in one spatial, but nine random dimensions, and a parabolic problem in two spatial and three random dimensions. So the first problem is more challenging from a stochastic point of view and the second from a deterministic point of view. We have considered several solution functionals for our numerical studies, of which we show here a representative selection. All results are obtained with MATLAB with linear finite elements in space, a build-in ODE solver, and the adaptive stochastic collocation procedure, outlined in Sect. 3. Since we only deal with uniformly distributed random variables, our results are obtained with Gauss-Patterson-Legendre quadrature nodes. Also recall from Sect. 3 that we use contributions of a new index set to a quantity of interest as natural local error indicators and the sum of contributions belonging to not yet refined collocation points serves as a natural global error indicator in comparison to adjoint error estimates.

5.1 Elliptic Problem: Stationary Diffusion Equation

The problem under consideration is the stationary, elliptic problem in one spatial dimension given by

$$\left.\begin{aligned} -\partial_x(\alpha(x,\omega)\partial_x u) &= 10, \quad && \text{for } x \in (-0.5, 0.5),\ \omega \in \Omega \\ u(x,\omega) &= 0, && \text{for } x = \pm 0.5, \omega \in \Omega\,, \end{aligned}\right\} \tag{33}$$

with correlated random diffusion coefficient α, parametrized by a truncated Karhunen-Loève expansion (see Eq. (4)) with constant mean $\mathrm{E}[\alpha] = 0.1$. More precise, we use an exponential covariance function given by

$$\mathrm{Cov}(x,\tilde{x}) = \sigma^2 \exp\left(-\frac{|x-\tilde{x}|}{c}\right), \tag{34}$$

with standard deviation $\sigma = 0.02$ and correlation length $c = 0.5$, which results in $M = 9$ random variables for a truncation error of less than 5 %. The random variables are assumed to be uniformly distributed. Note that the specific choice of the covariance function allows to solve the corresponding eigenvalue integral problem (3) analytically [25].

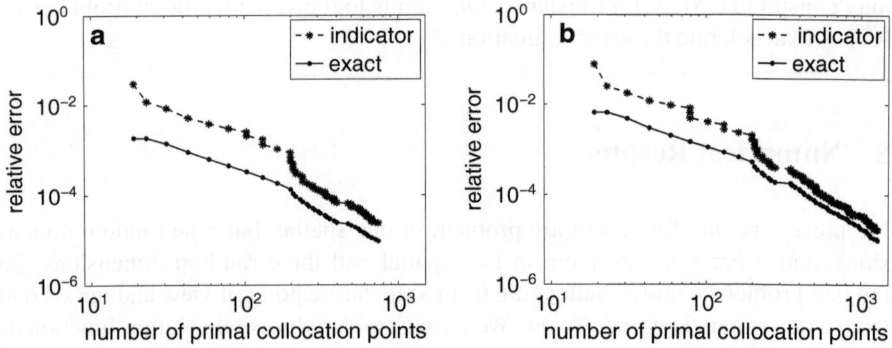

Fig. 2 Elliptic problem with random diffusion coefficient: we show the error indicators at each iteration for both quantities of interest, compared to the true difference to the reference solution, denoted as exact error. Both plots reveal monotonic convergence and over-estimation of the true error. (**a**) Convergence $\mathcal{Q}_1$. (**b**) Convergence $\mathcal{Q}_2$

For a goal-oriented error analysis, we want to consider the quantities of interest

$$\mathcal{Q}_1(u) = \int_{-0.5}^{0.5} \mathrm{E}[u]\,\mathrm{d}x \qquad \text{and} \qquad \mathcal{Q}_2(u) = \int_{-0.5}^{0.5} \mathrm{E}[u^2]\,\mathrm{d}x. \tag{35}$$

We aim to study the global interpolation error by means of adjoint calculus in each iteration of adaptive stochastic collocation. In order to exclude deterministic discretization errors from our studies, we use a rather fine spatial mesh size of 2^{-13} for all results shown here. The reference solution uses 2,007 collocation points and is the result of adaptive stochastic collocation with a tolerance of $2e-6$. We also checked the results for reference solutions based on more collocation points as well as different types of collocation points. So, we could make sure that our results are not biased by the choice of the reference solution. Figure 2 compares at each iteration the natural error indicators explained in Sect. 3 and the true errors. We notice that the error indicators drive the solution in the correct direction and interestingly exhibit a similar rate of convergence as the true error. However, they are around one magnitude larger than the actual error. So, we will now focus on adjoint error estimates for $\mathcal{Q}_1$ and $\mathcal{Q}_2$.

Estimating the error by means of adjoint problems means approximating the right-hand side of Eq. (28). Since we want to focus on the interpolation error, we use the same deterministic finite element mesh for both the primal and adjoint problem.

The first row of Fig. 3 shows the performance of the resulting adjoint error estimates based on Gauss-Patterson-Legendre nodes for both quantities of interest. We use sets of different sizes, namely 2,007, 1,103, 335, and 199 adjoint collocation points, and plot the effectivity index, i.e. the estimates divided by the true error. For the linear functional $\mathcal{Q}_1$ we observe that the curves start very close to the optimal value 1, but suffer from effectivity problems with increasing number of primal collocation points. The more adjoint collocation points we use, the better the

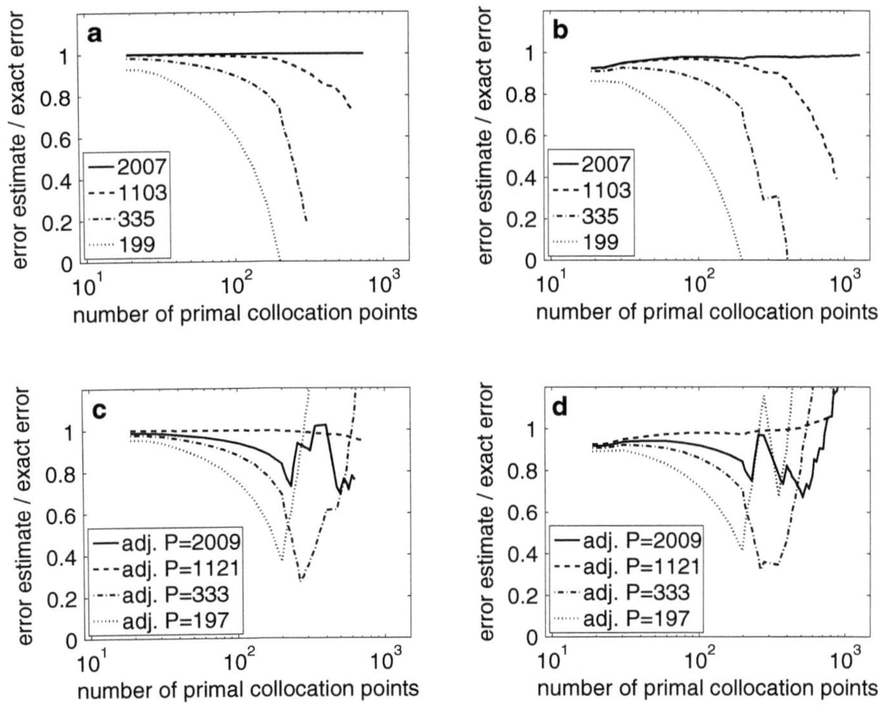

Fig. 3 Elliptic problem with random diffusion coefficient: we show the effectivity of adjoint error estimates with respect to the primal number of collocation points. The sparse grids for the estimates are based on several different numbers of Gauss-Patterson-Legendre and Clenshaw-Curtis nodes. (**a**) Estimates $\mathcal{Q}_1$ (Gauss-Patterson). (**b**) Estimates $\mathcal{Q}_2$ (Gauss-Patterson). (**c**) Estimates $\mathcal{Q}_1$ (Clenshaw-Curtis). (**d**) Estimates $\mathcal{Q}_2$ (Clenshaw-Curtis)

estimates are, and for the maximal used number of 2,007 points no quality break-off is observed at all. For the nonlinear functional $\mathcal{Q}_2$ the curves behave similarly, except that the estimates first improve due to decreasing HOT terms, before they also lack effectivity.

One could argue that the huge number of adjoint collocation points may only be caused by the fact that we use the same kind of quadrature rule for the primal and adjoint problem. That is why we want to approximate the error estimates now by means of a different type of nodes, namely Clenshaw-Curtis nodes, where we pre-constructed sparse grids of similar sizes (2,009, 1,121, 333, and 197). The second row of Fig. 3 shows the corresponding plots for both solution functionals. It is interesting to see that the set of 1,121 points gives the best performance, but is not as stable as the 2,007 Gauss-Patterson-Legendre points. Although the curves start with suitable error estimates, the estimates begin to stagger and become meaningless after some iterations. So, we can conclude that using a different rule does not improve the estimates. In this case the Clenshaw-Curtis nodes behave rather unpredictably and should not be used to obtain adjoint error estimates.

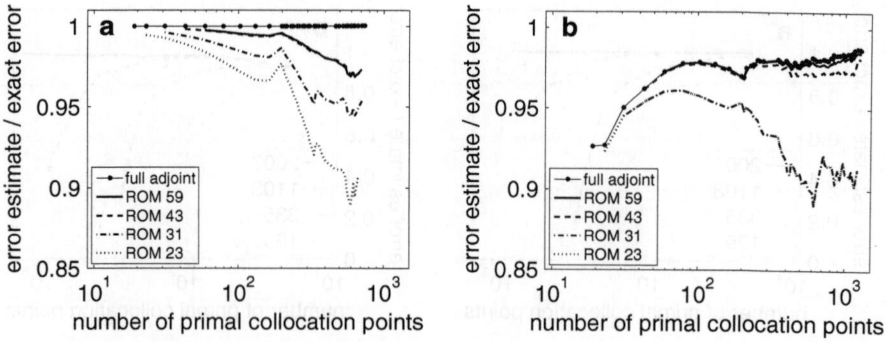

Fig. 4 Elliptic problem with random diffusion coefficient: we compare the effectivity of the full adjoint error estimator (2,007 adjoint collocation points) with adjoint error estimates based on reduced order models of the adjoint problem for increasing snapshot sets (see legend of the plots). (**a**) ROM estimates $\mathcal{Q}_1$. (**b**) ROM estimates $\mathcal{Q}_2$

We see that a suitable number of adjoint collocation points exceeds the number of primal collocation points by far. As mentioned in Sect. 4.3, we suggest to use a ROM for the adjoint problem. First we generate rather small snapshot sets consisting of only a few full solutions of the adjoint problem. By means of a standard POD-Galerkin method, we then generate a reduced basis and a reduced model for the adjoint problem, by which we replace the original problem to evaluate the huge number of adjoint collocation points. Figure 4 shows for both quantities of interest the results obtained with increasing size of the underlying snapshot set. Note that the maximal number of used snapshots is 59, which is still small compared to the 2,007 adjoint collocation points. What we see is that all estimates predict more than 90 % of the true error in nearly all cases. The estimates are thus indeed suitable alternatives to the full adjoint error estimates, because they are significantly cheaper without losing much accuracy. Note that the computational costs to evaluate the reduced model at the adjoint collocation points is negligible.

We conclude that adjoint calculus is able to predict the interpolation error with respect to some solution functional very accurately, provided that sufficiently many adjoint collocation points are used to approximate the error estimate. Hence, the estimates are very costly, which can be overcome to some extent by means of cheap ROMs.

5.2 Parabolic Problem: Heat Equation

The second numerical example we want to treat is similar to a parabolic problem suggested in [53], except of a slightly more challenging random input. The aim of this problem is to demonstrate adjoint error estimation also for a time-dependent problem in two spatial dimensions, which is technically more demanding than the elliptic problem in one dimension.

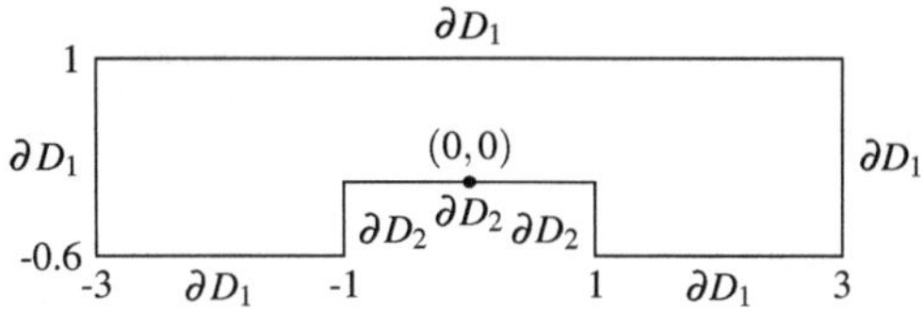

Fig. 5 Parabolic problem: computational domain D

The problem describes the heat conduction in an electronic chip with random heat conductivity α. The computational domain is depicted in Fig. 5, which we have discretized into triangles with 8465 degrees of freedom. At the cavity, a heat flux into the domain is assumed, while the remaining boundary is thermally isolated. The initial temperature is fixed by 0. The governing equation is the unsteady heat equation subject to the just described boundary conditions:

$$\left.\begin{aligned} \partial_t T - \nabla \cdot (\alpha(\mathbf{x},\omega)\nabla T) &= 0 && \text{in } D \times (0, T_{\text{end}}] \times \Omega \\ \alpha(\mathbf{x},\omega)\ \nabla T \cdot \mathbf{n} &= 0 && \text{in } \partial D_1 \times (0, T_{\text{end}}] \times \Omega \\ \alpha(\mathbf{x},\omega)\ \nabla T \cdot \mathbf{n} &= 1 && \text{in } \partial D_2 \times (0, T_{\text{end}}] \times \Omega \\ T(\mathbf{x},0,\omega) &= 0 && \text{in } D \times \Omega\,. \end{aligned}\right\} \tag{36}$$

The unknown temperature distribution is denoted by T and $\mathbf{n}$ refers to the outer normal vector. The temperature keeps growing over time due to the incoming flux together with adiabatic boundary conditions at the remaining walls. We set the final point in time to $T_{\text{end}} = 1$.

The conductivity parameter α is modelled as a spatially correlated random field with expected value $\mathrm{E}[\alpha] = 1$, correlation length $c = 4$, and standard deviation $\sigma = 0.2$. We use a covariance function, suitable for two-dimensional random fields,

$$C_\alpha(\mathbf{x},\tilde{\mathbf{x}}) = \sigma^2 \frac{\|\mathbf{x}-\tilde{\mathbf{x}}\|_2}{c} K_1\left(\frac{\|\mathbf{x}-\tilde{\mathbf{x}}\|_2}{c}\right), \tag{37}$$

where K_1 refers to the modified Bessel function of second kind and first order. The corresponding eigenfunctions for the Karhunen-Loéve expansion have to be calculated numerically (see [25] for a Galerkin approach). In order to maintain 95 % of the variance, we truncate the expansion of α after three terms and assume the resulting $M = 3$ random variables to be uniformly distributed.

In order to get a first impression of the solution, we present in Fig. 6 the expected value and standard deviation of the temperature at the final point in time $T_{\text{end}} = 1$. We observe that the solution exhibits most variance near the cavity. This gives rise to a quantity of interest that only acts in this region. Therefore, we consider the nonlinear stochastic solution functional

$$\mathcal{Q}(T) = \sqrt{\mathrm{Var}[\int_0^1 T|_{\mathbf{x}=(0,0)}\,\mathrm{d}t]}. \tag{38}$$

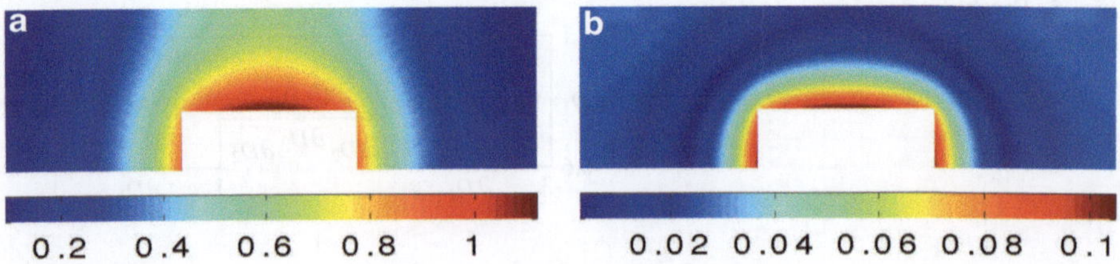

Fig. 6 Parabolic problem: on the *left* we show the **(a)** expected value of the temperature distribution at time $T_{\text{end}} = 1$, on the *right* the corresponding **(b)** standard deviation. Both plots reveal larger variations near the cavity of the domain

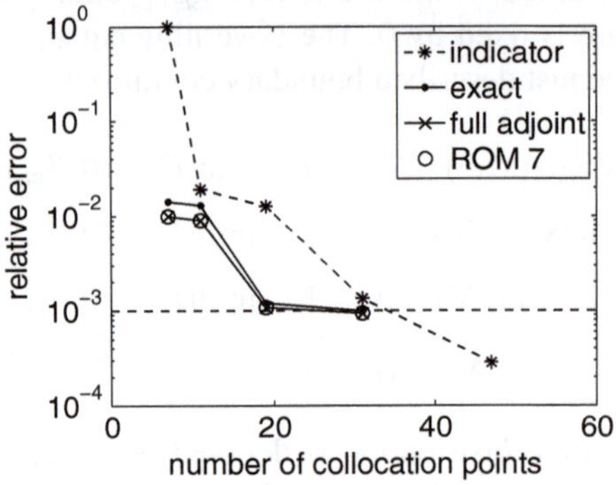

Fig. 7 Parabolic problem: the error indicators given by relative changes of the quantity of interest $\mathcal{Q}$ are compared to the exact errors. They appear to overestimate the exact error. Moreover, adjoint error estimates based on either the full or the reduced problem are applied at each iteration. They do not differ from each other and capture the exact error very well

The aim of this example is to use stochastic collocation with a tolerance of $1.0e-3$ for the relative error of $\mathcal{Q}$. We not only want to detect the interpolation error, but also deterministic error contributions. Therefore, we compute a reference solution with a triangulation of 530,049 degrees of freedom, a higher accuracy in time, and a stochastic tolerance of $1.0e-5$. Likewise, we can obtain semi-discrete reference solutions with respect to the deterministic and stochastic dimensions.

First, we apply the tolerance to the usual error indicators. Ignoring the curves for the error estimates for the moment, Fig. 7 shows the resulting error indicators at each iteration with respect to $\mathcal{Q}$, compared to the numerically exact error. We see that the indicators are satisfactory in the sense that the error is not underestimated. Instead, the error is slightly overestimated, so that 47 collocation points are demanded instead of the sufficient 31 collocation points. This behaviour is acceptable, but leads to the question whether an adjoint error estimator is able to detect the true error more accurately.

In order to answer this question, we take a closer look at the critical iteration, namely when reaching 31 collocation points. Here we evaluate the error estimate on different sparse grids and show the resulting effectivity for both the overall and

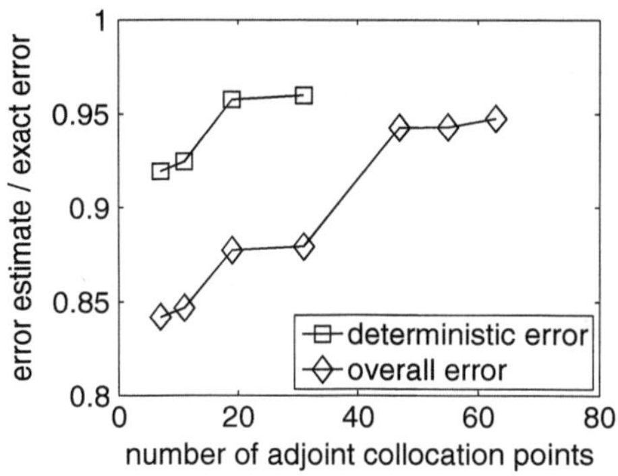

Fig. 8 Parabolic problem: we present the performance of adjoint error estimates for the deterministic and overall error contribution within the solution functional $\mathcal{Q}$ (recall Eqs. (29) and (30)). The plot refers to 31 primal collocation points, because these are needed to reach the tolerance of $1.0e-3$. Both curves increase up to an effectivity of about 0.95

deterministic error in Fig. 8. For the latter we compare the estimate to the semi-discrete reference solution. Moreover, all involved adjoint solutions are obtained on a refined spatial mesh. By doing so, each collocation point requires more computational costs for the adjoint than for the primal problem. However, the plot reveals that only a few adjoint collocation points are sufficient to detect the deterministic error contribution. On the contrary, the overall error really needs an extended set of adjoint collocation points to be captured accurately. Note that a suitable number of adjoint collocation points is not that large as shown for the elliptic case. This may be due to fact that the stochastic space is rather low-dimensional in the current setting and we do not focus on the pure interpolation error.

These results are very promising and give rise to reduced models, especially because of the fact that so few adjoint problems capture the deterministic error. We use the most simple sparse grid of $2M+1 = 7$ full adjoint solutions (compare $k = 1$ in Fig. 1) to build a reduced model of the adjoint problem. A POD of snapshots for these collocation points at several points in time is used to find the most important modes of the stochastic adjoint solution as a random field. Due to the fact that $\mathcal{Q}$ concentrates on one spatial point, the adjoint functions mainly give weight to this point. This is nicely reflected in the POD modes, depicted in Fig. 9. As expected, we observe a peak in the mentioned spatial point in the first modes, followed by the smaller scales of the adjoint solution.

We now use the resulting POD basis to build a reduced model of the adjoint problem, i.e. we have an approximation for the adjoint random field $\phi(\mathbf{x}, t, \boldsymbol{\xi})$ at hand that can be evaluated in an arbitrary number of adjoint collocation points at very low computational costs. The largest set of adjoint collocation points used in Fig. 8 consists of 63 points and is apparently sufficient to obtain accurate error estimates. Hence, the reduced model enables us to obtain cheap error estimates for the overall error of the solution functional $\mathcal{Q}$ at each iteration of the actual primal adaptive collocation procedure. Coming back to Fig. 7, we see this statement

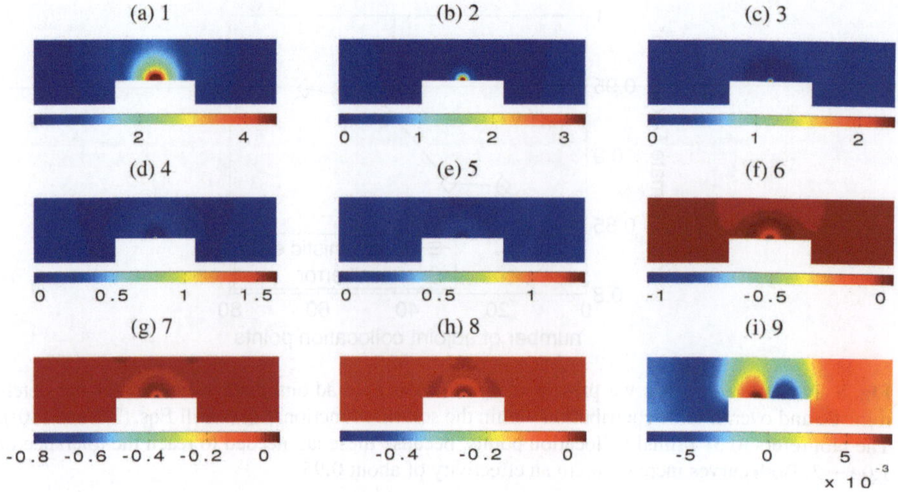

Fig. 9 Parabolic problem: we show the dominating nine POD modes of the stochastic adjoint solution. As expected most of the energy concentrates on the underlying spatial reference point $(0, 0)$

verified. The plot impressively shows that the effectivity of both the full and reduced adjoint error estimates tends to 1 with increasing number of iterations and lets the algorithm correctly terminate with the optimal number of 31 primal collocation points. Actually, there is no difference observable between the full and reduced estimates. We can conclude that the reduced model is very effective for the considered problem. A little drawback may be the expensive calculations of the first full adjoint problems. However, we are rewarded by a reduction of primal collocation points and an accurate statement about the final error.

6 Conclusions

We have extended adjoint error estimates to stochastic collocation methods for PDEs with random parameters. It turns out that solution functionals involving deterministic and stochastic integrals all lead to the same structure of error estimates by means of linearization of nonlinear terms. The resulting typical structure of an integral over the residual weighted by an adjoint solution is also known from deterministic adjoint error estimation.

The unknown adjoint function has to be approximated, where we use stochastic collocation as well. It is important to note that the choice of the adjoint collocation points strongly influences the kind of error we are able to capture. In this work we have shown numerically that the deterministic error only requires a small amount of adjoint collocation points, while the interpolation error may need up to magnitudes more adjoint than primal collocation points to be detected accurately. In order to overcome the substantial amount of additional costs, we suggest to replace the adjoint problem by a reduced model based on few expensive simulations. The resulting surrogate model is then very cheap and can be evaluated in arbitrary many adjoint collocation points.

Our results considerably contribute to the understanding of errors and estimates for stochastic collocation methods. A topic for future research could be a deeper analytical analysis of the arising estimates. Furthermore, the effective combination of reduced models for both primal and adjoint problems remains to be investigated. And finally, it would be interesting to apply the methodology also to more complex equations, such as they arise in fluid mechanics.

Acknowledgements This work was supported by the "Excellence Initiative" of the German Federal and State Governments and the Graduate School of Computational Engineering at Technische Universität Darmstadt.

References

1. A. Alekseev, I. Navon, M. Zelentsov, The estimation of functional uncertainty using polynomial chaos and adjoint equations. Int. J. Numer. Methods Fluids **67**(3), 328–341 (2011)
2. R. Almeida, J. Oden, Solution verification, goal-oriented adaptive methods for stochastic advection-diffusion problems. Comput. Methods Appl. Mech. Eng. **199**(37–40), 2472–2486 (2010)
3. I. Babuška, R. Tempone, G. Zouraris, Solving elliptic boundary value problems with uncertain coefficients by the finite element method: the stochastic formulation. Comput. Methods Appl. Mech. Eng. **194**(12–16), 1251–1294 (2005)
4. I. Babuška, F. Nobile, R. Tempone, A stochastic collocation method for elliptic partial differential equations with random input data. SIAM J. Numer. Anal. **45**(3), 1005–1034 (2007)
5. A. Barth, C. Schwab, N. Zollinger, Multi-level Monte Carlo finite element method for elliptic PDEs with stochastic coefficients. Numer. Math. **119**(1), 123–161 (2011)
6. V. Barthelmann, E. Novak, K. Ritter, High-dimensional polynomial interpolation on sparse grids. Adv. Comput. Math. **12**(4), 273–288 (2000)
7. R. Becker, R. Rannacher, An optimal control approach to a posteriori error estimation in finite element methods. Acta Numer. **10**, 1–102 (2001)
8. A. Begumisa, I. Robinson, Suboptimal Kronrod extension formulae for numerical quadrature. Numer. Math. **58**(8), 807–818 (1991)
9. G. Berkooz, P. Holmes, J. Lumley, The proper orthogonal decomposition in the analysis of turbulent flows. Annu. Rev. Fluid Dyn. **25**, 539–579 (1993)
10. M. Bieri, C. Schwab, Sparse high order FEM for elliptic SPDEs. Comput. Methods Appl. Mech. Eng. **198**(13–14), 1149–1170 (2009)
11. S. Boyaval, C. Le Bris, Y. Maday, N. Nguyen, A. Patera, A reduced basis approach for variational problems with stochastic parameters: application to heat conduction with variable Robin coefficient. Comput. Methods Appl. Mech. Eng. **198**(41–44), 3187–3206 (2009)
12. H. Bungartz, S. Dirnstorfer, Multivariate quadrature on adaptive sparse grids. Computing **71**(1), 89–114 (2003)
13. H. Bungartz, M. Griebel, Sparse grids. Acta Numer. **13**, 147–269 (2004)
14. T. Butler, C. Dawson, T. Wildey, A posteriori error analysis of stochastic differential equations using polynomial chaos expansions. SIAM J. Sci. Comput. **33**(3), 1267–1291 (2011)

15. K. Cliffe, M. Giles, R. Scheichl, A. Teckentrup, Multilevel Monte Carlo methods and applications to elliptic PDEs with random coefficients. Comput. Vis. Sci. **14**(1), 3–15 (2011)
16. H. Elman, O. Ernst, D. O'Leary, M. Stewart, Efficient iterative algorithms for the stochastic finite element method with application to acoustic scattering. Comput. Methods Appl. Mech. Eng. **194**(9–11), 1037–1055 (2005)
17. O. Ernst, E. Ullmann, Stochastic Galerkin matrices. SIAM J. Matrix Anal. Appl. **31**(4), 1848–1872 (2010)
18. D. Estep, A short course on duality, adjoint operators, Green's functions, and a posteriori error analysis. Technical report, Department of Mathematics, Colorado State University, Fort Collins, 2004. Available on http://www.stat.colostate.edu/~estep/
19. J. Foo, X. Wan, G. Karniadakis, The multi-element probabilistic collocation method (ME-PCM): error analysis and applications. J. Comput. Phys. **227**(22), 9572–9595 (2008)
20. P. Frauenfelder, C. Schwab, R. Todor, Finite elements for elliptic problems with stochastic coefficients. Comput. Methods Appl. Mech. Eng. **194**(2–5), 205–228 (2005)
21. B. Ganapathysubramanian, N. Zabaras, Sparse grid collocation schemes for stochastic natural convection problems. J. Comput. Phys. **225**(1), 652–685 (2007)
22. A. Genz, B. Keister, Fully symmetric interpolatory rules for multiple integrals over infinite regions with Gaussian weight. J. Comput. Appl. Math. **71**(2), 299–309 (1996)
23. T. Gerstner, M. Griebel, Numerical integration using sparse grids. Numer. Algorithms **18**(3–4), 209–232 (1998)
24. T. Gerstner, M. Griebel, Dimension-adaptive tensor-product quadrature. Computing **71**(1), 65–87 (2003)
25. R. Ghanem, P. Spanos, *Stochastic Finite Elements: A Spectral Approach* (Springer, New York, 1991)
26. M. Griebel, J. Hamaekers, Sparse grids for the Schrödinger equation. Math. Model. Numer. Anal. **41**(2), 215–247 (2007)
27. M. Griebel, S. Knapek, Optimized general sparse grid approximation spaces for operator equations. Math. Comput. **78**(268), 2223–2257 (2009)
28. M. Hegland, A. Hellander, P. Lötstedt, Sparse grids and hybrid methods for the chemical master equation. BIT Numer. Math. **48**, 265–283 (2008)
29. F. Heiss, V. Winschel, Quadrature on sparse grids: code to generate and readily evaluated nodes and weights, 2007. Available on http://sparse-grids.de/
30. A. Klimke, B. Wohlmuth, Algorithm 847: Spinterp: piecewise multilinear hierarchical sparse grid interpolation in MATLAB. ACM Trans. Math. Softw. **31**(4), 561–579 (2005)
31. D. Knezevic, N. Nguyen, A. Patera, Reduced basis approximation and a posteriori error estimation for the parametrized unsteady Boussinesq equations. Math. Models Methods Appl. Sci. **21**(7), 1415–1442 (2011)
32. A. Kronrod, *Nodes and Weights of Quadrature Formulas. Sixteen-Place Tables. Authorized Translation from the Russian*, 2nd edn. (Consultants, New York, 1966)
33. O. Le Maître, O. Knio, *Spectral Methods for Uncertainty Quantification. With Applications to Computational Fluid Dynamics* (Springer, Dordrecht, 2010)
34. O. Le Maître, M. Reagan, H. Najm, G. Ghanem, O. Knio, A stochastic projection method for fluid flow. II: random process. J. Comput. Phys. **181**(1), 9–44 (2002)
35. S. Li, L. Petzold, Adjoint sensitivity analysis for time-dependent partial differential equations with adaptive mesh refinement. J. Comput. Phys. **198**(1), 310–325 (2004)
36. C. Lieberman, K. Willcox, O. Ghattas, Parameter and state model reduction for large-scale statistical inverse problems. SIAM J. Sci. Comput. **32**(5), 2523–2542 (2010)
37. M. Liu, Z. Gao, J. Hesthaven, Adaptive sparse grid algorithms with applications to electromagnetic scattering under uncertainty. Appl. Numer. Math. **61**(1), 24–37 (2011)
38. X. Ma, N. Zabaras, An adaptive hierarchical sparse grid collocation algorithm for the solution of stochastic differential equations. J. Comput. Phys. **228**(8), 3084–3113 (2009)
39. L. Mathelin, O. Le Maître, Dual-based a posteriori error estimate for stochastic finite element methods. Commun. Appl. Math. Comput. Sci. **2**, 83–115 (2007)

40. F. Nobile, R. Tempone, Analysis and implementation issues for the numerical approximation of parabolic equations with random coefficients. Int. J. Numer. Methods Eng. **80**(6–7), 979–1006 (2009)
41. F. Nobile, R. Tempone, C. Webster, An anisotropic sparse grid stochastic collocation method for partial differential equations with random input data. SIAM J. Numer. Anal. **46**(5), 2411–2442 (2008)
42. E. Novak, K. Ritter, Simple cubature formulas with high polynomial exactness. Constr. Approx. **15**(4), 499–522 (1999)
43. T. Patterson, The optimum addition of points to quadrature formulae. Math. Comput. **22**, 847–856 (1968)
44. T. Patterson, Modified optimal quadrature extensions. Numer. Math. **64**(4), 511–520 (1993)
45. C. Powell, H. Elman, Block-diagonal preconditioning for spectral stochastic finite-element systems. IMA J. Numer. Anal. **29**(2), 350–375 (2009)
46. E. Rosseel, S. Vandewalle, Iterative solvers for the stochastic finite element method. SIAM J. Sci. Comput. **32**(1), 372–397 (2010)
47. L. Sirovich, Turbulence and the dynamics of coherent structures I: coherent structures. II: symmetries and transformations. III: dynamics and scaling. Q. Appl. Math. **45**, 561–590 (1987)
48. S. Smolyak, Quadrature and interpolation formulas for tensor products of certain classes of functions. Dokl. Akad. Nauk SSSR **148**, 1042–1045 (1963)
49. S. Volkwein, Model reduction using proper orthogonal decomposition, in *Script in English language* (2011), 43 pp. Available on http://www.math.uni-konstanz.de/numerik/personen/volkwein/teaching/scripts.php
50. N. Wiener, The homogeneous chaos. Am. J. Math. **60**, 897–936 (1938)
51. D. Xiu, J. Hesthaven, High-order collocation methods for differential equations with random inputs. SIAM J. Sci. Comput. **27**(3), 1118–1139 (2005)
52. D. Xiu, G. Karniadakis, The Wiener-Askey polynomial chaos for stochastic differential equations. SIAM J. Sci. Comput. **24**(2), 619–644 (2002)
53. D. Xiu, G. Karniadakis, A new stochastic approach to transient heat conduction modeling with uncertainty. Int. J. Heat Mass Transf. **46**(24), 4681–4693 (2003)
54. N. Zabaras, B. Ganapathysubramanian, A scalable framework for the solution of stochastic inverse problems using a sparse grid collocation approach. J. Comput. Phys. **227**(9), 4697–4735 (2008)
55. C. Zenger, Sparse grids, in *Parallel Algorithms for Partial Differential Equations, Proceedings of the 6th GAMM-Seminar*, Kiel, 1990, ed. by W. Hackbusch. Notes on Numerical Fluid Mechanics, vol. 31 (Vieweg, Braunschweig, 1991), pp. 241–251

POD-Galerkin Modeling and Sparse-Grid Collocation for a Natural Convection Problem with Stochastic Boundary Conditions

Sebastian Ullmann and Jens Lang

Abstract The computationally most expensive part of the stochastic collocation method are usually the numerical solutions of a deterministic equation at the collocation points. We propose a way to reduce the total computation time by replacing the deterministic model with its Galerkin projection on the space spanned by a small number of basis functions. The proper orthogonal decomposition (POD) is used to compute the basis functions from the solutions of the deterministic model at a few collocation points. We consider the computation of the statistics of the Nusselt number for a two-dimensional stationary natural convection problem with a stochastic temperature boundary condition. It turns out that for the estimation of the mean and the probability density, the number of finite element simulations can be significantly reduced by the help of the POD-Galerkin reduced-order model.

1 Introduction and Problem Formulation

We study the convective flow in the unit square $D = (0,1) \times (0,1)$ with stochastic temperature Dirichlet conditions [8]. The governing equations are the dimensionless incompressible Navier-Stokes equations with the Boussinesq approximation for the

S. Ullmann
Department of Mathematics, Technische Universität Darmstadt, Dolivostraße 15, 64293 Darmstadt, Germany
e-mail: ullmann@mathematik.tu-darmstadt.de

J. Lang (✉)
Department of Mathematics, Technische Universität Darmstadt, Dolivostraße 15, 64293 Darmstadt, Germany

Graduate School of Computational Engineering, Technische Universität Darmstadt, Dolivostraße 15, 64293 Darmstadt, Germany
e-mail: lang@mathematik.tu-darmstadt.de

J. Garcke and D. Pflüger (eds.), *Sparse Grids and Applications - Munich 2012*, Lecture Notes in Computational Science and Engineering 97, DOI 10.1007/978-3-319-04537-5_13,

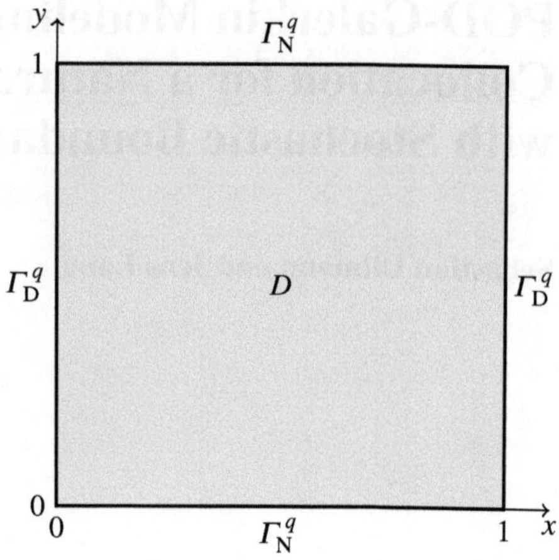

Fig. 1 Sketch of the domain D with the locations of the temperature boundaries Γ_D^q and Γ_N^q

temperature forcing term. We only consider cases for which the solution tends to a unique stationary solution independently of the initial data, so that the stationary version of the equations is applicable. We fix all parameters and incorporate them explicitly in the equations, which are consequently given as

$$\mathbf{u} \cdot \nabla \mathbf{u} + \nabla p - \nabla \cdot (\nabla \mathbf{u} + (\nabla \mathbf{u})^T) + 5000\mathbf{g}q = \mathbf{0} \quad \text{in } D, \tag{1}$$

$$\mathbf{u} \cdot \nabla q - \nabla \cdot \nabla q = 0 \quad \text{in } D, \tag{2}$$

$$\nabla \cdot \mathbf{u} = 0 \quad \text{in } D. \tag{3}$$

The unknown field variables are the velocity $\mathbf{u} = (u, v)^T$, the pressure p and the temperature q. The direction of gravity is given by $\mathbf{g} = (0, -1)^T$. A general vector in the physical space will be denoted by $\mathbf{x} = (x, y)$.

The boundary Γ of the domain D is split into the parts Γ_D^q and Γ_N^q, as sketched in Fig. 1. The following boundary conditions are imposed:

$$\mathbf{u} = \mathbf{0} \quad \text{on } \Gamma, \tag{4}$$

$$q = q_D \quad \text{on } \Gamma_D^q, \tag{5}$$

$$\nabla q \cdot \mathbf{n} = 0 \quad \text{on } \Gamma_N^q. \tag{6}$$

Here, $\mathbf{n}$ denotes the boundary unit normal vector pointing outward of the domain. The temperature Dirichlet boundary data are given by

$$q_D(\mathbf{x}, \alpha(y, \theta)) = \begin{cases} 0.5 & \text{if } x = 0, \\ -0.5 + \alpha(y, \theta) & \text{if } x = 1, \end{cases} \tag{7}$$

where $\alpha(y, \theta)$ is a centered stochastic process and θ is a member of the space of random events Ω. The process $\alpha(y, \theta)$ is spatially correlated with the exponential covariance function

$$\mathcal{K}(y_1, y_2) = e^{-|y_1 - y_2|}.$$

The Nusselt number Nu at the left boundary is chosen as the output of the simulation:

$$Nu = -\int_0^1 \frac{\partial q}{\partial x}\bigg|_{x=0} \mathrm{d}y. \tag{8}$$

The goal of this paper is to demonstrate how a reduced-order formulation of this Boussinesq problem can be used to compute accurate statistics of the Nusselt number at a low computational cost. While the problem serves as a prototype for related scenarios, we limit our focus to sparse-grid collocation [27] for the discretization of the stochastic dimensions and POD-Galerkin modeling [23] for the reduced-order treatment of the deterministic sub-problems. We note that the stochastic collocation method has been combined with reduced-order models based on the reduced basis method [21] in the context of elliptic boundary value problems with uncertain coefficients [3, 4, 19] and uncertain boundary conditions [1]. The reduced basis method has also been applied to steady incompressible Navier-Stokes problems with uncertainty in the viscosity and in the boundary conditions [7] and to the unsteady Boussinesq equations in a deterministic setting [17].

2 Numerical Solution

To be able to compute the statistics of the Nusselt number for the given stochastic Boussinesq problem we have to perform a number of modeling steps. In each of these steps we introduce a parameter with which we can control the accuracy of the modeling. At first, we transform the infinite-dimensional stochastic process to a low-dimensional one using a Karhunen-Loève expansion. The number of terms in the expansion will be denoted by K. Then, we discretize the resulting stochastic dimensions of the problem using collocation on a sparse grid with a sparse grid refinement level of L. The collocation method requires a set of P deterministic problems to be solved independently, where P is depending on the dimension K and the level L. We employ the finite element method for the discretization of the physical dimensions of the deterministic problems and denote the finite element grid refinement level as M.

2.1 *Karhunen-Loève Expansion*

A truncated Karhunen-Loève expansion of the stochastic process that determines the temperature boundary condition is given by

$$\alpha(y,\theta) \approx \alpha(y,\boldsymbol{\xi}^\star(\theta)) = \sum_{i=1}^{K} \sqrt{\lambda_i}\xi_i^\star(\theta) f_i(y), \tag{9}$$

where $\boldsymbol{\xi}^\star(\theta) = (\xi_1^\star(\theta),\ldots,\xi_K^\star(\theta))^T$ is a random vector. For the case of exponential covariance it is possible to derive the eigenvalues $\lambda_1,\ldots,\lambda_K$ and the eigenfunctions $f_1,\ldots,f_K$ analytically [10]. It holds that $E[\xi_i^\star(\theta)] = 0$ and $E[\xi_i^\star(\theta)\xi_j^\star(\theta)] = \delta_{ij}$ for all $i, j = 1,\ldots,K$. An explicit representation of the elements of the random vector is given by

$$\xi_i^\star(\theta) = \frac{1}{\sqrt{\lambda_i}} \int_0^1 \alpha(y,\theta) f_i(y)\,\mathrm{d}y, \qquad i = 1,\ldots,K.$$

Since we do not possess full knowledge about the stochastic process, we replace $\boldsymbol{\xi}^\star(\theta)$ by the modeled random vector $\boldsymbol{\xi} = (\xi_1,\ldots,\xi_K)^T$. As modeling assumptions, we impose that $\xi_1,\ldots,\xi_K$ are mutually independent and that $\xi_i \sim \mathcal{U}[-\sqrt{3},\sqrt{3}]$ for $i = 1,\ldots,K$, so that each random variable has unit variance and is contained in a closed domain. To compute a realization of the stochastic boundary condition, we use

$$q_{\mathrm{D}}(\mathbf{x},\boldsymbol{\xi}) = \begin{cases} 0.5 & \text{if } x = 0, \\ -0.5 + \sum_{i=1}^K \sqrt{\lambda_i}\xi_i f_i(y) & \text{if } x = 1, \end{cases} \tag{10}$$

where a realization of the random vector can be computed with a suitable pseudo-random number generator. Figure 2 illustrates the spatial components of the decomposition and presents typical realizations of the modeled process.

2.2 *Stochastic Collocation*

We have described how the stochastic boundary condition can be modeled as a function of K independent, uniformly distributed random variables. For a given realization of $\boldsymbol{\xi}$ we can compute an approximate Nu by the following steps:

1. Define the temperature boundary condition q_{D} via (10).
2. Solve the deterministic problem (1)–(6) numerically.
3. Evaluate the boundary integral (8).

If the Nusselt number is to be computed for a large number of realizations, or if integrations over the stochastic space are to be performed, it is beneficial to replace the steps 1–3 by a simpler model, as provided by the sparse grid collocation method [2, 9]. The cost of the method is determined by the numerical solutions of a decoupled set of deterministic problems at predefined collocation points in the stochastic domain.

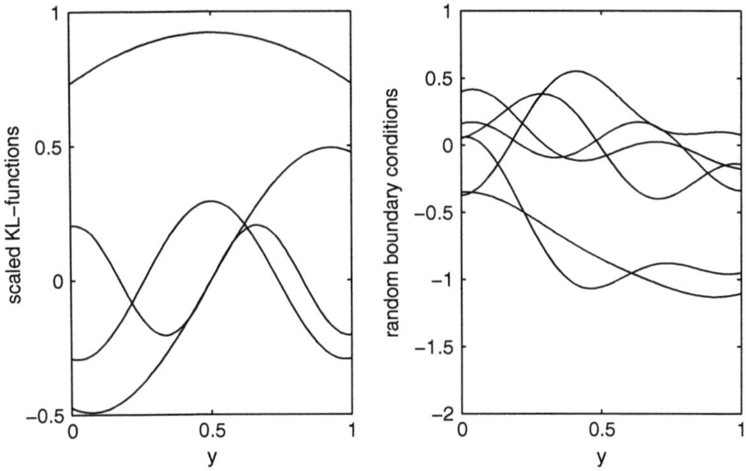

Fig. 2 Scaled Karhunen-Loève functions $\sqrt{\lambda_i} f_i(y)$ for $i = 1, \ldots, 4$ (*left*) and six random realizations of the boundary condition q_D using $K = 4$ (*right*)

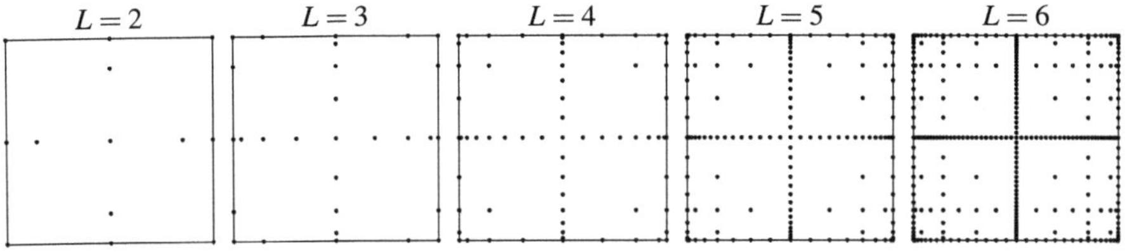

Fig. 3 Sparse grids of dimension $K = 2$ based on a nested sequence of Chebyshev points for different Smolyak levels L

For the stochastic collocation we use Smolyak's algorithm [24] with an underlying nested sequence of Chebyshev points in the interval $[-\sqrt{3}, \sqrt{3}]$. The resulting sparse grids for $K = 2$ and $L = 2, \ldots, 6$ are presented in Fig. 3.

The collocation points for a fixed stochastic dimension K and Smolyak level L are given as $\boldsymbol{\xi}_1, \ldots, \boldsymbol{\xi}_P$. The respective multivariate Lagrange polynomials are denoted as $\mathcal{L}_1(\boldsymbol{\xi}), \ldots, \mathcal{L}_P(\boldsymbol{\xi})$. It holds that $\mathcal{L}_i(\boldsymbol{\xi}_j) = \delta_{ij}$ for all $i, j = 1, \ldots, P$. The sparse grid approximation of the temperature field depending on the random vector is given by

$$q(\mathbf{x}, \boldsymbol{\xi}) \approx \sum_{p=1}^{P} \mathcal{L}_p(\boldsymbol{\xi}) q(\mathbf{x}, \boldsymbol{\xi}_p), \tag{11}$$

It follows from (8) and (11) that the Nusselt number computed from the sparse grid approximation of the temperature field is given by

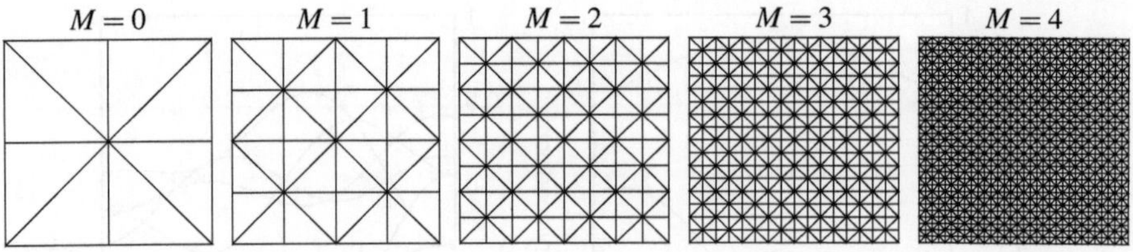

Fig. 4 Uniform mesh refinement

$$Nu(\boldsymbol{\xi}) \approx \sum_{p=1}^{P} \mathcal{L}_p(\boldsymbol{\xi}) Nu(\boldsymbol{\xi}_p). \tag{12}$$

To set up the sparse grid interpolation, we have to perform the steps 1–3 for each collocation point and store the respective Nusselt numbers. For any other realization of $\boldsymbol{\xi}$, the Nusselt number can then be evaluated quickly by using the sparse interpolation (12). Furthermore, integrals over the stochastic domain can be evaluated using exact integrations of the Lagrange polynomials.

2.3 Finite Element Discretization

As the stochastic collocation in our case relies on finite element simulations, a sufficiently fine mesh is crucial for the accuracy of the stochastic simulation. For the numerical simulation we use a simple triangular mesh with uniform refinement, where M denotes the refinement level. The mesh refinement is sketched in Fig. 4.

Following the usual finite element modeling steps [12], we first derive a weak form of the Boussinesq problem: Find $p \in L^2$, $\mathbf{u} \in \mathbf{H}_0^1$, $q - q_{\mathrm{D}} \in H_0^1$, for which

$$(\mathbf{u} \cdot \nabla \mathbf{u}, \boldsymbol{\psi}^{\mathbf{u}}) + (\nabla \mathbf{u} + (\nabla \mathbf{u})^T, \nabla \boldsymbol{\psi}^{\mathbf{u}}) - (p, \nabla \cdot \boldsymbol{\psi}^{\mathbf{u}}) + (5000 \mathbf{g} q, \boldsymbol{\psi}^{\mathbf{u}}) = 0 \quad \forall \boldsymbol{\psi}^{\mathbf{u}} \in \mathbf{H}_0^1, \tag{13}$$

$$(\mathbf{u} \cdot \nabla q, \psi^q) + (\nabla q, \nabla \psi^q) = 0 \quad \forall \psi^q \in H_0^1, \tag{14}$$

$$(\nabla \cdot \mathbf{u}, \psi^p) = 0 \quad \forall \psi^p \in L^2. \tag{15}$$

We approximate the velocity and temperature fields in each mesh triangle with piecewise quadratic polynomials and the pressure field with piecewise linear polynomials. This choice of elements, also known as Taylor-Hood finite elements [25], fulfills an inf-sup stability condition [13] and could be called a standard approach for incompressible flow calculations. We introduce the finite element spaces $L_h^2 \subset L^2$ and $\mathbf{H}_{0h}^1 \subset \mathbf{H}_0^1$, which are spanned by the pressure and velocity finite element basis functions, respectively, as well as $H_h^1 \subset H^1$ and $H_{0h}^1 \subset H_0^1$, which are spanned by the temperature basis functions.

Fig. 5 Temperature fields resulting from a deterministic simulation with $K = 0$ (*upper left*) and from simulations with $K = 4$ using random realizations of the temperature boundary condition (other plots). All plots use the same color scaling. A mesh refinement level of $M = 4$ was used for the computations

We define a continuous extension $q_{\mathrm{D}}^h \in H_h^1$ of the Dirichlet data into the computational domain D. We require that q_{D}^h is equal to the Dirichlet condition q_{D} of the problem for all mesh nodes on the Dirichlet boundary and equal to zero at all other mesh nodes. Based on the Karhunen-Loève expansion we can define suitable functions $f_0^h, \ldots, f_K^h \in H_h^1$ so that

$$q_{\mathrm{D}}^h(\mathbf{x}, \boldsymbol{\xi}) = f_0^h(\mathbf{x}) + \sum_{i=1}^{K} \sqrt{\lambda_i} \xi_i f_i^h(\mathbf{x}). \tag{16}$$

By using the finite element spaces as test spaces in (13)–(15) and replacing the unknowns by their finite element representations, we obtain the following discrete weak form: Find $p^h \in L_h^2, \mathbf{u}^h \in \mathbf{H}_{0h}^1, q^h - q_{\mathrm{D}}^h \in H_{0h}^1$, for which

$$(\mathbf{u}^h \cdot \nabla \mathbf{u}^h, \boldsymbol{\psi}^{\mathbf{u}}) + (\nabla \mathbf{u}^h + (\nabla \mathbf{u}^h)^T, \nabla \boldsymbol{\psi}^{\mathbf{u}})$$

$$-(p^h, \nabla \cdot \boldsymbol{\psi}^{\mathbf{u}}) + (5000 \mathbf{g} q^h, \boldsymbol{\psi}^{\mathbf{u}}) = 0 \qquad \forall \boldsymbol{\psi}^{\mathbf{u}} \in \mathbf{H}_{0h}^1, \tag{17}$$

$$(\mathbf{u}^h \cdot \nabla q^h, \psi^q) + (\nabla q^h, \nabla \psi^q) = 0 \qquad \forall \psi^q \in H_{0h}^1, \tag{18}$$

$$(\nabla \cdot \mathbf{u}^h, \psi^p) = 0 \qquad \forall \psi^p \in L_h^2. \tag{19}$$

After testing against the finite element basis functions in turn and evaluating the inner products, we obtain a system of non-linear algebraic equations. We employ the trust-region dogleg method [20] of the Matlab® R2012a Optimization Toolbox to solve the system. A few pseudo-time-steps with the backward Euler method are sufficient to obtain good initial values for the solver. The cost of the finite element solution is determined by the LU-decomposition of the Jacobi matrix.

The temperature solution of the deterministic finite element simulation with $K = 0$ and a few solutions with random boundary conditions and $K = 4$ are shown in Fig. 5. Notice the absence of thin boundary layers, which justifies the choice of a uniform grid.

3 Reduced-Order Modeling

A considerable amount of computational work may be needed for the setup of the sparse grid interpolation if the discretizations of the stochastic and the physical space are sufficiently fine. As a possible remedy we propose the use of reduced-order models that are obtained by a Galerkin projection of the deterministic equations on the space spanned by a set of reduced basis functions. To compute the reduced basis functions we use a proper orthogonal decomposition (POD) of a set of representative solution snapshots. The POD is very closely related to the Karhunen-Loève decomposition. Just like the sparse grid collocation method, the considered reduced-order models provide a computationally inexpensive mapping from a realization of the random vector to the respective Nusselt number. While the sparse grid collocation is based purely on polynomial interpolations, however, POD-Galerkin models include features of the governing equations. The computational work as well as the accuracy of the reduced-order models depend on the parameter R, which is equal to the number of POD basis functions that are used to build the model.

In the following we first introduce the POD in a general context. Then we describe the POD approximation of the velocity field as a special case, which inherits the discrete divergence-freeness and the fulfillment of the Dirichlet conditions from the finite element simulation. The POD approximation of the temperature field involves the approximation of the stochastic process at the temperature Dirichlet boundary. Having the reduced-order approximations of the velocity and temperature available, we are able to formulate the reduced-order model and describe how it can be used to reduce the computational cost of the sparse-grid collocation method.

3.1 Proper Orthogonal Decomposition

We briefly summarize a few computational aspects of proper orthogonal decomposition that are necessary for the development of the reduced-order model. For details we refer to the literature [15].

Let a set of snapshots $u_1, \ldots, u_N \in L^2$ be given. The extension to snapshots which are vector-valued functions is straightforward and will not be detailed, here. We would like to find a set of basis functions $\phi_1, \ldots, \phi_R \in L^2$ which are mutually orthonormal, $(\phi_i, \phi_j) = \delta_{ij}$ for $i, j = 1, \ldots, R$, and which minimize the differences between the snapshots and their orthogonal projection on the space spanned by the basis functions. In particular, the basis functions are supposed to solve the constrained minimization problem

$$\min_{\phi_1,\ldots,\phi_R} \sum_{n=1}^{N} \left\| u_n - \sum_{r=1}^{R} (u_n, \phi_r)\phi_r \right\|^2 \quad \text{subject to} \quad (\phi_i, \phi_j) = \delta_{ij} \quad \text{for} \quad i, j = 1, \ldots, R.$$

To solve the minimization problem we use the "method of snapshots" [23]. The Gramian matrix $\mathcal{G} \in \mathbb{R}^{N \times N}$ of the snapshots consists of the mutual inner products (u_i, u_j). We compute its eigenvalue decomposition

$$\mathcal{G}\mathcal{V} = \mathcal{V}\Lambda,$$

where $\mathcal{V} \in \mathbb{R}^{N \times N}$ is an orthogonal matrix that contains the eigenvectors of $\mathcal{G}$ as columns and $\Lambda \in \mathbb{R}^{N \times N}$ is a diagonal matrix with the eigenvalues $\lambda_1, \ldots, \lambda_N$ sorted decreasingly on the diagonal. The POD basis functions are then given as

$$\phi_j = \sum_{i=1}^{N} \frac{v_{ij}}{\sqrt{\lambda_j}} u_i, \qquad j = 1, \ldots, R, \tag{20}$$

where v_{ij} are the entries of $\mathcal{V}$. Alternatively, the POD can be computed by singular value decomposition [18].

3.2 Reduced-Order Approximation of the Velocity Field

Suppose that for a set of collocation points $\boldsymbol{\xi}_1, \ldots, \boldsymbol{\xi}_N$ we have obtained respective finite element velocity snapshots $\mathbf{u}_1^h, \ldots, \mathbf{u}_N^h \in \mathbf{H}_{0h}^1$. We perform a POD of the snapshots, which provides us with POD basis functions $\boldsymbol{\phi}_1^{\mathbf{u}}, \ldots, \boldsymbol{\phi}_R^{\mathbf{u}} \in \mathbf{H}_{0h}^1$. A POD representation $\mathbf{u}^R \in \mathbf{H}_{0R}^1$ of the velocity field is given as

$$\mathbf{u}^R(\mathbf{x}, \boldsymbol{\xi}) = \sum_{r=1}^{R} a_r(\boldsymbol{\xi}) \boldsymbol{\phi}_r^{\mathbf{u}}(\mathbf{x}), \tag{21}$$

where $\mathbf{H}_{0R}^1 \subset \mathbf{H}_{0h}^1$ is the space spanned by the first R velocity POD basis functions.

By applying (20) to the present case of vector-valued functions and assuming that the snapshots fulfill the discretized weak form (17)–(19), one can see that the POD basis functions fulfill homogeneous Dirichlet conditions. By (21) this property carries over to the POD representation of the velocity field. In a similar way, it can be shown that the POD basis functions are discretely divergence-free by construction,

$$(\nabla \cdot \boldsymbol{\phi}_r^{\mathbf{u}}, \psi^p) = 0 \quad \forall\, \psi^p \in L_h^2, \qquad r = 1, \ldots, R. \tag{22}$$

and that this property is inherited by the POD representation of the velocity,

$$(\nabla \cdot \mathbf{u}^R, \psi^p) = 0 \quad \forall\, \psi^p \in L_h^2. \tag{23}$$

3.3 Reduced-Order Approximation of the Temperature Field

The reduced-order approximation of the temperature field is performed in a way that a priori incorporates the Dirichlet conditions, similar to the approach taken in the finite element discretization. The procedure is a variant of the control function method [11] for multiple parameters [14]. Let the finite element temperature snapshots for a set of collocation points $\boldsymbol{\xi}_1, \ldots, \boldsymbol{\xi}_N$ be given as $q_1^h, \ldots, q_N^h$. We define modified temperature snapshots $q_1', \ldots, q_N' \in H_{0h}^1$ as

$$q_i'(\mathbf{x}) = q_i^h(\mathbf{x}) - q_{\mathrm{D}}^h(\mathbf{x}, \boldsymbol{\xi}_i), \qquad i = 1, \ldots, N,$$

where $q_{\mathrm{D}}^h(\mathbf{x}, \boldsymbol{\xi})$ is a continuous extension of the Dirichlet data into the domain, see (16). A POD of rank R of the modified snapshots provides us with POD basis functions $\phi_1^q, \ldots, \phi_R^q \in H_{0h}^1$. The POD basis functions fulfill homogeneous Dirichlet conditions because, by (20), they are linear combinations of the modified snapshots. We introduce the POD representation of the temperature as

$$q^R(\mathbf{x}, \boldsymbol{\xi}) = q_{\mathrm{D}}^h(\mathbf{x}, \boldsymbol{\xi}) + \sum_{r=1}^{R} b_r(\boldsymbol{\xi}) \phi_r^q(\mathbf{x}). \tag{24}$$

It holds that $q^R - q_{\mathrm{D}}^h \in H_{0R}^1$, where $H_{0R}^1 \subset H_{0h}^1$ is the space spanned by the first R temperature POD basis functions. After substituting the definition (16) of q_{D}^h in (24) and declaring

$$\begin{aligned} \phi_{R+1}^q(\mathbf{x}) &= f_0^h(\mathbf{x}), & b_{R+1}(\boldsymbol{\xi}) &= 1, & \\ \phi_{R+1+k}^q(\mathbf{x}) &= \sqrt{\lambda_k}\, f_k^h(\mathbf{x}), & b_{R+1+k}(\boldsymbol{\xi}) &= \xi_k, & k &= 1, \ldots, K, \end{aligned}$$

we obtain

$$q^R(\mathbf{x}, \boldsymbol{\xi}) = \sum_{r=1}^{R'} b_r(\boldsymbol{\xi}) \phi_r^q(\mathbf{x}) \tag{25}$$

with $R' = R + 1 + K$. It turned out that it was beneficial for the computation of the Nusselt number to replace $f_0^h(\mathbf{x}), \ldots, f_K^h(\mathbf{x})$ with functions which are linear in x throughout the domain. This approach, which uses different continuous extensions of the boundary data as compared to the finite element model, will be used for all reduced-order computations.

Figure 6 shows the functions $\phi_{R+1}^q, \ldots, \phi_{R+5}^q$, which are required to implement the Dirichlet conditions for $K = 4$, as well as the first five basis functions of the temperature POD approximation using snapshots with $M = 4$ at the collocation points of $K = 4$ and $L = 6$.

Fig. 6 Dirichlet functions $\phi^q_{R+1}, \ldots, \phi^q_{R+5}$ (*top row*) and POD basis functions $\phi^q_1, \ldots, \phi^q_5$ (*bottom row*) of the temperature snapshots at the collocation points of $K = 4$ and $L = 6$. The color scales are the same within the *top row* and within the *bottom row*

3.4 POD-Galerkin Reduced-Order Model

The discretized weak form (17)–(19) is the starting point for the derivation of the reduced-order model. At first, we substitute $\mathbf{u}^h$ by $\mathbf{u}^R$ and q^h by q^R. Due to (23), the discrete continuity equation is automatically fulfilled for $\mathbf{u}^R$ and, thus, can be neglected. We choose the POD spaces $\mathbf{H}^1_{0\,R}$ and $H^1_{0\,R}$ as test spaces for the remaining equations. Due to (22), the pressure term becomes zero for all discrete pressures $p^h \in L^2_h$. This leaves us with the following reduced-order weak formulation: Find $\mathbf{u}^R \in \mathbf{H}^1_{0\,R}$ and $q^R - q^h_{\mathrm{D}} \in H^1_{0\,R}$, for which

$$
\begin{aligned}
(\mathbf{u}^R \cdot \nabla \mathbf{u}^R, \boldsymbol{\phi}^{\mathbf{u}}) + (\nabla \mathbf{u}^R + (\nabla \mathbf{u}^R)^T, \nabla \boldsymbol{\phi}^{\mathbf{u}}) \\
+ (5000\mathbf{g} q^R, \boldsymbol{\phi}^{\mathbf{u}}) = 0 \quad \forall \boldsymbol{\phi}^{\mathbf{u}} \in \mathbf{H}^1_{0\,R}, \qquad (26)
\end{aligned}
$$

$$
(\mathbf{u}^R \cdot \nabla q^R, \phi^q) + (\nabla q^R, \nabla \phi^q) = 0 \quad \forall \phi^q \in H^1_{0\,R}, \qquad (27)
$$

After testing against the POD basis functions in turn, the reduced-order model can be written as

$$
\begin{aligned}
&\sum_{i=1}^{R} \sum_{j=1}^{R} (\boldsymbol{\phi}^{\mathbf{u}}_i \cdot \nabla \boldsymbol{\phi}^{\mathbf{u}}_j, \boldsymbol{\phi}^{\mathbf{u}}_r) a_i a_j + \sum_{i=1}^{R} (\nabla \boldsymbol{\phi}^{\mathbf{u}}_i + (\nabla \boldsymbol{\phi}^{\mathbf{u}}_i)^T, \nabla \boldsymbol{\phi}^{\mathbf{u}}_r) a_i \\
&\quad + \sum_{i=1}^{R'} (5000\mathbf{g} \phi^q_i, \boldsymbol{\phi}^{\mathbf{u}}_r) b_i = 0, \qquad r = 1, \ldots, R, \qquad (28)
\end{aligned}
$$

$$
\sum_{i=1}^{R} \sum_{j=1}^{R'} (\boldsymbol{\phi}^{\mathbf{u}}_i \cdot \nabla \phi^q_j, \phi^q_r) a_i b_j + \sum_{i=1}^{R'} (\nabla \phi^q_i, \nabla \phi^q_r) b_i = 0, \qquad r = 1, \ldots, R. \qquad (29)
$$

The inner products can be evaluated via the finite element representations of the POD basis functions, which leads to a set of algebraic equations with quadratic non-linearities. The unknowns are the POD coefficients $a_1, \ldots, a_R$ and $b_1, \ldots, b_R$, while $b_{R+1} = 1$ is fixed and $b_{R+1+k} = \xi_k$ for $k = 1, \ldots, K$ is the input data. The reduced-order model is solved by the same non-linear solver as the finite element model. The cost of performing one iterative step is determined by the evaluation of the quadratic terms and by the LU-decomposition of the Jacobi-matrix.

To obtain a reduced-order equation for the Nusselt number we substitute (25) in (8) to see that

$$Nu(q^R) = \sum_{r=1}^{R'} b_r Nu(\phi_r^q),$$

where $Nu(\phi_r^q)$ can be evaluated and stored once for each $r = 1, \ldots, R$, so that the Nusselt number can be directly computed from any solution of the reduced-order model, without forming the approximation of the temperature field.

In the following we will evaluate the reduced-order model for input vectors $\boldsymbol{\xi}$ whose length can be incompatible with the dimensions R' of the reduced-order model. This case occurs, for example, when we evaluate one model for different stochastic dimensions. We remove the last entries of $\boldsymbol{\xi}$ when it is too large and we fill it with zeros when it is too small. For instance, suppose we have a model with a stochastic dimension of $K = 2$, so that $R' = R + 3$. If we want to evaluate this model at a collocation point of the stochastic dimension $K^\star = 4$, we are faced with a random vector $(\xi_1, \xi_2, \xi_3, \xi_4)$. In this case, we evaluate the model at (ξ_1, ξ_2).

3.5 *Application to Stochastic Collocation*

For a stochastic collocation with dimension $K^\star$ and Smolyak level $L^\star$, we propose to replace the finite element simulations with reduced-order simulations. This results in the following POD-aided stochastic collocation method:

1. Choose K and L for the computation of the POD snapshots.
2. Create temperature boundary conditions for the collocation points of K and L, perform the respective finite element simulations and store the numerical solutions.
3. Create a reduced-order model from the numerical solutions. The model takes a realization of $\boldsymbol{\xi}$ as input and provides the respective Nu as output.
4. Evaluate the reduced-order model at the collocation points of $K^\star$ and $L^\star$, and store the resulting values of Nu.
5. Create a sparse grid interpolation of $K^\star$ and $L^\star$ using the data generated with the reduced-order model.

This procedure provides us with a sparse grid interpolation of dimension $K^\star$ and level $L^\star$ at the cost of the finite element simulations necessary for dimension K and level L plus the computational overhead from building and solving the reduced-order model. If we can obtain an accurate reduced-order model with snapshots of dimension $K < K^{\star}$ or level $L < L^{\star}$ and with R much smaller than the number of finite element mesh nodes, then we can expect savings of computation time.

4 Results

To compare the POD-aided stochastic collocation method with the standard stochastic collocation, we consider three aspects. Firstly, we estimate the approximation properties of the respective sparse-grid interpolation of Nu throughout the parameter domain. Secondly, we assess the accuracy of the probability density function that is computed via a Monte-Carlo simulation using random evaluations of the sparse-grid interpolation. And, finally, we investigate the computation of the mean and variance of the Nusselt number using an exact integration of the underlying Lagrange polynomials.

To enable a quantitative comparison of the methods, in Table 1 we show the numbers of collocation points for the different stochastic dimensions and Smolyak levels. In Table 2 we present total and average computation times of the finite element model (FEM), the reduced-order model (ROM) and of the sparse-grid interpolation (SG) as well as the times needed to create the models. The computation time of the sparse-grid was tested in one instance for the approximation of the probability density function via Monte Carlo sampling with 1,000,000 evaluations of the sparse-grid interpolation at random points (see Sect. 4.2). In another instance, the computation of the mean via an exact integration of the multivariate Lagrange polynomial was considered (see Sect. 4.3). All computations were performed with Matlab®. The timings were measured on a single computational thread on an Intel® Xeon® E5-2670 CPU. The Monte Carlo sampling as well as the independent finite element and reduced-order simulations are well suited for parallel processing, though.

4.1 Approximation of the Nusselt Number

The goal of this section is a sparse grid representation of the Nusselt number that is valid throughout the domain of $\boldsymbol{\xi}$. We compare the results from a standard stochastic collocation with the results from a POD-aided stochastic collocation.

Suppose we have computed the numerical solutions with a finite element mesh refinement level of M at the collocation points of the stochastic dimension K and the Smolyak level L. At first, we create a sparse grid interpolation from the available numerical solutions (see Sect. 2.2) using the standard stochastic collocation. Then we use the available numerical solutions to build a POD-Galerkin reduced-order model with R basis functions. We evaluate the reduced-order model

Table 1 Number of collocations points P depending on the stochastic dimension K and the sparse grid refinement level L

	$L=0$	$L=1$	$L=2$	$L=3$	$L=4$	$L=5$	$L=6$	$L=7$	$L=8$
$K=0$	1	1	1	1	1	1	1	1	1
$K=1$	1	3	5	9	17	33	65	129	257
$K=2$	1	5	13	29	65	145	321	705	1,537
$K=3$	1	7	25	69	177	441	1,073	2,561	6,017
$K=4$	1	9	41	137	401	1,105	2,929	7,537	18,945
$K=5$	1	11	61	241	801	2,433	6,993	19,313	51,713
$K=6$	1	13	85	389	1,457	4,865	15,121	44,689	127,105

Table 2 Total and average computation times and relevant parameters

	K	L	P	M	R	Total (s)	Average (s)
FEM setup				4		0.46	
FEM simulation	4	6	2,929	4		13,245.52	4.5222
FEM simulation	4	3	137	4		602.46	4.3975
ROM setup	4	3	137	4	30	1.67	
ROM simulation	4	6	2,929	4	30	63.57	0.0217
SG Nusselt density	4	6	2,929			3,053.72	0.0031
SG Nusselt mean	4	6	2,929			0.27	

at the collocation points that are necessary to build a sparse grid interpolation of stochastic dimension $K^\star = 4$ and sparse grid refinement level $L^\star = 6$.

We want to compare the standard collocation and the POD-aided collocation, for varying discretization parameters. We vary the finite element mesh refinement level M from 1 to 4, the stochastic dimension K from 0 to 4, the Smolyak level L from 0 to 6 and, for the reduced order model, we compare the numbers of POD basis functions $R = 5, 10, 15, 20, 25, 30$. To estimate the errors of the Nusselt number, we use a reference solution which is obtained by finite element computations with $M = 5$ at the collocation points of $K = 6$ and $L = 8$. For each of these collocation points we can compute the difference between the reference Nusselt number and interpolated Nusselt number. The maximum absolute difference is our error estimate. A comparison of the estimated errors of the standard and the POD-aided stochastic collocation is given in Fig. 7. In the following, we comment on each of the plots.

The first plot of Fig. 7 shows the dependence of the error on the stochastic dimension K. Note that $K = 0$ is equivalent to a fully deterministic simulation. As we would expect, the error in the Nusselt number decreases when the number of Karhunen-Loève terms is increased. Furthermore, it turns out that both methods lead to results of similar accuracy, which shows that the POD-Galerkin modeling does not introduce any significant error. The results suggest, however, that the number of finite element simulations can not be reduced by using a POD of snapshots of a lower stochastic dimension and that one should use $K = K^\star$.

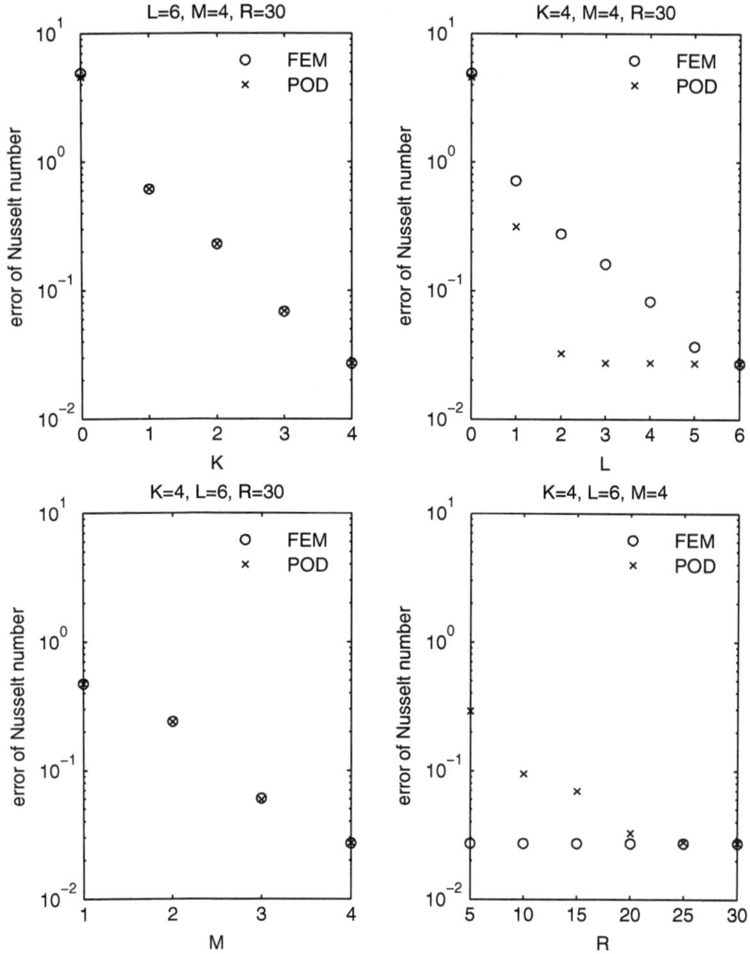

Fig. 7 Estimates of the maximum absolute error of the Nusselt number throughout the stochastic domain for the sparse grid interpolation (FEM) and the POD-aided sparse grid interpolation (POD), both relying on finite element solutions of mesh level M at the collocation points of stochastic dimension K and Smolyak level L. In the POD case, a reduced-order model with R basis functions was evaluated at the collocation points of dimension $K^{\star} = 4$ and level $L^{\star} = 6$ and a respective sparse grid interpolation was formed

In the second plot of Fig. 7 the error is presented for varying Smolyak level L. The case $L = 0$ is equivalent to a fully deterministic simulation. We can see that the error of the POD-aided collocation method drops much faster than the error of the standard method when the level is increased. Therefore, by using finite element snapshots at the collocation points of a low level, e.g. $L = 2$, it is possible via the reduced-order model to create a sparse-grid interpolation that is about as accurate as a standard collocation with $L = 5$ or $L = 6$.

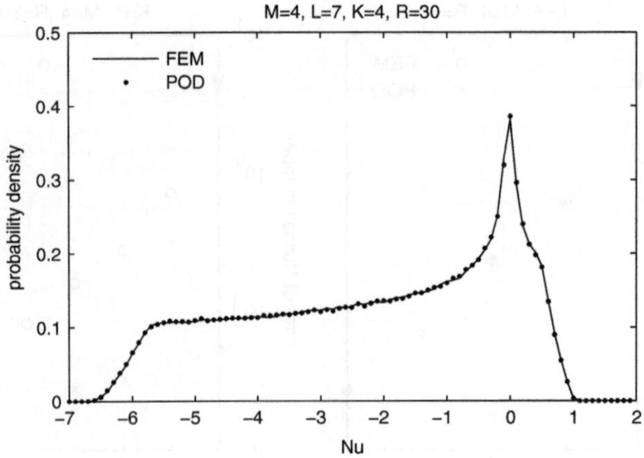

Fig. 8 Probability density function of the Nusselt number obtained from a Monte Carlo simulation. The *line* represents the results obtained directly from the finite element simulations with $K = 4$ and $L = 6$ while the *dots* present the results obtained from a reduced-order model with $R = 30$ that was created from finite element solutions at the collocation points of $K = 4$ and $L = 2$ and evaluated at $K^\star = 4$ and $L^\star = 6$

In Fig. 7, the third plot demonstrates the accuracy with respect to the mesh refinement level M. With this plot we want to ensure that the resolution of the finite element mesh is adequate. When comparing computation times, an over-resolving finite element mesh would give an advantage to the reduced-order model. One can see that by decreasing the mesh resolution from $M = 4$ to $M = 3$ the error of the Nusselt increases significantly, so that $M = 4$ is indeed an appropriate choice.

The last plot of Fig. 7 shows the dependence of the POD-aided stochastic collocation on the number of basis functions. The results of the standard method, which do not depend on R, are shown for comparison. We see that a reduced-order model with 25–30 basis functions is sufficient to build a sparse grid whose error is very close to the error of the sparse grid obtained directly from the finite element snapshots.

We can conclude that the refinement parameters $K = 4$, $L = 6$ and $M = 4$ are a choice that balances the errors introduced by the Karhunen-Loève expansion, the sparse-grid interpolation and the finite element simulation, respectively. The error introduced by a reduced-order model with 30 basis functions is small enough as to be dominated by the other error components.

4.2 *Probability Density*

We want to assess the applicability of the POD-aided collocation method to the computation of the probability density function of the Nusselt number via the Monte Carlo method. Figure 8 shows the probability density functions approximated with a Monte Carlo simulation using 1,000,000 samples and a bin width of 0.1. For the POD-aided collocation we used finite element solutions of $K = 4$ and $L = 2$

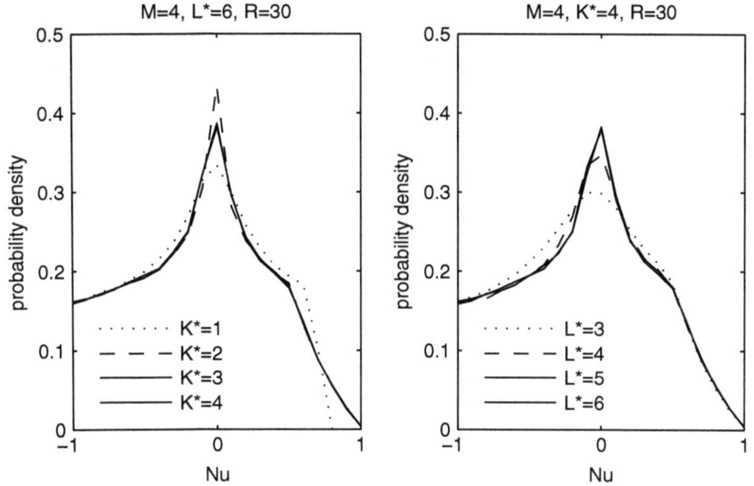

Fig. 9 Probability density functions of the Nusselt number obtained from a Monte Carlo simulation using a POD-aided stochastic interpolation of varying $K^\star$ and $L^\star$ built from snapshots at the collocation points of $K = 4$ and $L = 2$

to create a reduced-order model with 30 POD basis functions. We evaluated the reduced-order model at $K^\star = 4$ and $L^\star = 6$ to create a respective sparse grid interpolation. For the standard collocation we created an interpolation directly from the finite element solutions of $K = 4$ and $L = 6$. All finite element computations were performed on a mesh with $M = 4$. We observe that the POD-aided collocation produces a probability density function that is visually indistinguishable from the one obtained with the standard method.

The computation time that is required to evaluate the stochastic interpolation at a given point in the stochastic domain is dependent on the number of collocation points. To see whether a smaller number of collocation points is sufficient for an accurate computation of the probability density function, we compare the results of the POD-aided stochastic collocation method for different stochastic dimensions $K^\star$ and Smolyak levels $L^\star$ in Fig. 9. We used snapshots of $M = 4$, $K = 4$ and $L = 2$ to build a reduced-order model with $R = 30$ and created sparse grid interpolations of varying $L^\star$ and $K^\star$. For the case of varying stochastic dimension, we see that at least $K^\star = 2$ is needed to capture the correct slope at $Nu > 0.5$ and at least $K^\star = 3$ is needed to capture the peak. For the case of varying Smolyak level we see that at least $L^\star = 5$ is needed to approximate the peak correctly.

4.3 *Mean and Variance*

Often one is not interested in the detailed probability density function of a stochastic quantity, but rather in simple statistics like the mean and variance.

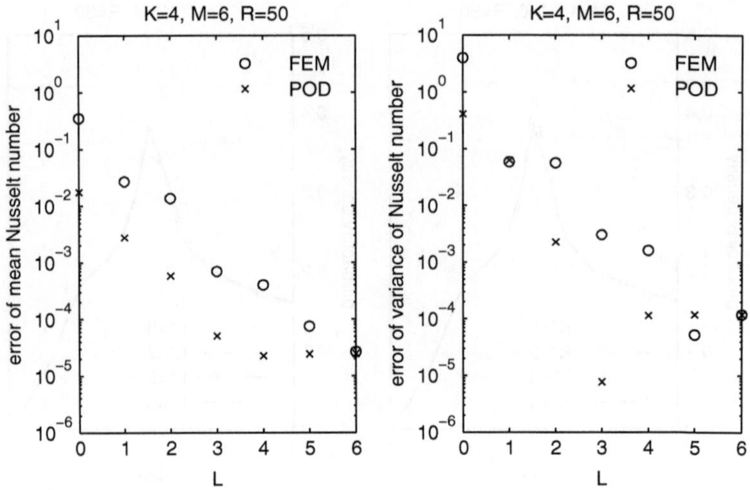

Fig. 10 Absolute errors of the mean (*left*) and variance (*right*) of the Nusselt number depending on the Smolyak level L. The errors were obtained with respect to reference solutions with $K = 4$, $L = 6$ and $M = 7$

To compute the moments of the Nusselt number we use exact integration of the multivariate Lagrangian polynomials. The computation of the integrals is based on one-dimensional Clenshaw-Curtis quadrature weights [5], which can be efficiently computed with a routine based on an inverse FFT [26]. To obtain the second moments we compute the square of the Nusselt number at the collocation points and subsequently use the same quadrature weights as in the case of the mean Nusselt number.

The resulting plots of the mean and variance of the Nusselt number depending on the Smolyak level L are presented in Fig. 10 for the standard collocation and the POD-aided collocation. The other parameters were set to $K = 4$, $M = 6$ and $R = 50$. Snapshots of level L were used to build a reduced-order model, which was subsequently evaluated at $K^{\star} = 4$ and $L^{\star} = 6$ to build the sparse-grid interpolation. To estimate the errors we used reference values of the mean and variance obtained with $K = 4$, $L = 6$ and $M = 7$. Note that with this choice at $L = 6$ the error estimator does not capture the effects of K and L anymore. The second plot of Fig. 10 contains two values which are significantly lower than the finite element error. They were most probably a result of errors that canceled by chance, and are therefore not representative. For the computation of the mean Nusselt number we observe that the sparse grid refinement level can be reduced by 1 or 2 when POD-Galerkin modeling is employed.

5 Concluding Remarks

In our study we have assessed the applicability of POD-Galerkin reduced-order modeling to the context of stochastic collocation on sparse grids. For the test case of the computation of the Nusselt number in a natural convective flow setting, we have studied the behavior of the different error contributions in a systematic way with the help of a reference solution. By employing a reduced-order model we were able to significantly decrease the number of finite element simulations necessary to create an accurate sparse grid interpolation. For example (see Fig. 7), by building a reduced-order model from the finite element solutions at the 137 collocation points of Smolyak level $L = 3$ and evaluating the model at the 2,929 collocation points of Smolyak level $L^{\star} = 6$, we were able to obtain a sparse grid interpolation with a similar accuracy as an interpolation created directly from 2,929 finite element solutions of level $L = 6$.

To further assess the efficiency of the method we comment on the computation times (see Table 2): It took about 3 h 41 min to perform the 2,929 finite element computations of level $L = 6$ and about 10 min to perform the 137 finite element computations of level $L = 3$. In comparison, 2,929 evaluations of the reduced-order model took only 64 s. The additional computational overhead of the reduced-order model, i.e. the creation of the basis functions from the snapshots and the assembly of the system matrices, was less than 2 s. Therefore, by using reduced-order modeling the total computation time could be decreased by almost a factor of 20. The resulting sparse grids had the same number of collocation points, independently of whether they were created using finite element solutions or reduced-order solutions, and so their evaluation times were identical.

We have limited our scope to the context of a sparse grid collocation method based on the standard Smolyak formula with underlying Clenshaw-Curtis nodes. Possible enhancements on the side of the employed sparse grids include the incorporation of adaptivity [16] and error estimation [22]. To create the reduced-order model, we have used a proper orthogonal decomposition of snapshots at the collocation points of a lower Smolyak level. In principle, one could place the snapshots freely in the parameter domain, which opens the possibility of further enhancements on the side of the reduced-order model. In particular, the reduced basis method [21] provides automatic positioning of the snapshots in the parameter domain with the help of an error estimator. Further, alternative choices of the continuous extension of the Dirichlet data could be explored [6].

References

1. S. Boyaval, C. Le Bris, Y. Maday, N. Nguyen, A. Patera, A reduced basis approach for variational problems with stochastic parameters: application to heat conduction with variable Robin coefficient. Comput. Methods Appl. Mech. Eng. **198**, 3187–3206 (2009)
2. H.-J. Bungartz, M. Griebel, Sparse grids. Acta Numer. **13**, 147–269 (2004)

3. P. Chen, A. Quarteroni, G. Rozza, Comparison between reduced basis and stochastic collocation methods for elliptic problems. MATHICSE Report 34.2012, Mathematics Institute of Computational Science and Engineering, École Polytechnique Fédérale de Lausanne, 2012
4. P. Chen, A. Quarteroni, G. Rozza, A weighted reduced basis method for elliptic partial differential equations with random input data. MATHICSE Report 04.2013, Mathematics Institute of Computational Science and Engineering, École Polytechnique Fédérale de Lausanne, 2013
5. C. Clenshaw, A. Curtis, A method for numerical integration on an automatic computer. Numer. Math. **2**, 197–205 (1960)
6. J.L. Eftang, E.M. Rønquist, Evaluation of flux integral outputs for the reduced basis method. Math. Models Methods Appl. Sci. **20**, 351–374 (2010)
7. H.C. Elman, Q. Liao, Reduced basis collocation methods for partial differential equations with random coefficients. Technical Report CS-TR-5011, Computer Science Department, University of Maryland, 2012
8. B. Ganapathysubramanian, N. Zabaras, Sparse grid collocation schemes for stochastic natural convection problems. J. Comput. Phys. **225**, 652–685 (2007)
9. J. Garke, Sparse grids in a nutshell, in *Sparse Grids and Applications*, ed. by J. Garke, M. Griebel (Springer, Heidelberg, 2013), pp. 57–80
10. R. Ghanem, P. Spanos, *Stochastic Finite Elements: A Spectral Approach* (Dover publications, New York, 1991)
11. W.R. Graham, J. Peraire, K.Y. Tang, Optimal control of vortex shedding using low-order models. Part I – open-loop model development. Int. J. Numer. Methods Eng. **44**, 945–972 (1999)
12. P.M. Gresho, R.L. Sani, *Incompressible Flow and the Finite Element method* (Wiley, Chichester, 2000)
13. M.D. Gunzburger, *Finite Element Methods for Viscous Incompressible Flows*. Computer Science and Scientific Computing (Academic, Boston, 1989)
14. M.D. Gunzburger, J.S. Peterson, J.N. Shadid, Reduced-order modeling of time-dependent PDEs with multiple parameters in the boundary data. Comput. Methods Appl. Mech. Eng. **196**, 1030–1047 (2007)
15. P. Holmes, J. Lumley, G. Berkooz, *Turbulence, Coherent Structures, Dynamical Systems and Symmetry* (Cambridge University Press, Cambridge, 1996)
16. J.D. Jakeman, S.G. Roberts, Local and dimension adaptive stochastic collocation for uncertainty quantification, in *Sparse Grids and Applications*, ed. by J. Garke, M. Griebel (Springer, Heidelberg, 2013), pp. 181–204
17. D. Knezevic, N. Nguyen, A.T. Patera, Reduced basis approximation and a posteriori error estimation for the parametrized unsteady boussinesq equation. Math. Models Methods Appl. Sci. **21**, 1415–1442 (2011)
18. K. Kunisch, S. Volkwein, Control of the Burgers equation by a reduced-order approach using proper orthogonal decomposition. J. Optim. Theory Appl. **102**, 345–371 (1999)
19. B. Peherstorfer, S. Zimmer, H.-J. Bungartz, Model reduction with the reduced basis method and sparse grids, in *Sparse Grids and Applications*, ed. by J. Garke, M. Griebel (Springer, Heidelberg, 2013), pp. 223–242
20. M. Powell, A Fortran subroutine for solving systems of nonlinear algebraic equations, in *Numerical Methods for Nonlinear Algebraic Equations*, Chap. 7, ed. by P. Rabinowitz (Gordon and Breach, London, 1970), pp. 115–161
21. G. Rozza, D.B.P. Huynh, A.T. Patera, Reduced basis approximation and a posteriori error estimation for affinely parametrized elliptic coercive partial differential equations: application to transport and continuum mechanics. Arch. Comput. Methods Eng. **15**, 229–275 (2008)
22. B. Schieche, J. Lang, Uncertainty quantification for thermo-convective Poiseuille flow using stochastic collocation. Int. J. Comput. Sci. Eng. (2014, in press)
23. L. Sirovich, Turbulence and the dynamics of coherent structures. Parts I, II and III. Q. Appl. Math. **45**, 561–571 (1987)
24. S. Smolyak, Quadrature and interpolation formulas for tensor products of certain classes of functions. Sov. Math. Dokl. **4**, 240–243 (1963)

25. C. Taylor, P. Hood, A numerical solution of the Navier-Stokes equations using the finite element technique. Int. J. Comput. Fluids **1**, 73–100 (1973)
26. J. Waldvogel, Fast construction of the Fejér and Clenshaw-Curtis quadrature rules. BIT Numer. Math. **46**, 195–202 (2006)
27. D. Xiu, J.S. Hesthaven, High-order collocation methods for differential equations with random inputs. SIAM J. Sci. Comput. **27**, 1118–1139 (2005)

Opticom and the Iterative Combination Technique for Convex Minimisation

Matthias Wong and Markus Hegland

Abstract Since "A combination technique for the solution of sparse grid problems" Griebel et al. (1992), the sparse grid combination technique has been successfully employed to approximate sparse grid solutions of multi-dimensional problems. In this paper we study the technique for a minimisation problem coming from statistics. Our methods can be applied to other convex minimisation problems. We improve the combination technique by adapting the "Opticom" method developed in Hegland et al. (Linear Algebra Appl 420:249–275, 2007). We also suggest how the Opticom method can be extended to other numerical problems. Furthermore, we develop a new technique of using the combination technique iteratively. We prove this method yields the true sparse grid solution rather than an approximation. We also present numerical results which illustrate our theory.

1 Introduction

The full grid discretisation, though accurate, suffers fully from the curse of dimensionality. At the sixth GAMM Seminar in 1990, Zenger presented the sparse grid discretisation for the solution of partial differential equations [21]. The curse can be partially overcome through sparse grids. Since the seminal paper by Zenger, sparse grids have been used successfully in other areas such as interpolation, integral equations, data mining and other areas. The helpful Sparse Grid Tutorial is a good entry to understanding sparse grids [9].

The sparse grid combination technique is a way of approximating the sparse grid solution through a combination of low-resolution approximations [11]. This avoids using the hierarchal basis intrinsic to a direct sparse grid approach. Moreover, the

M. Wong (✉) • M. Hegland
The Australian National University, Canberra, ACT, Australia
e-mail: matthias.wong@anu.edu.au; markus.hegland@anu.edu.au

J. Garcke and D. Pflüger (eds.), *Sparse Grids and Applications - Munich 2012*, Lecture Notes in Computational Science and Engineering 97,
DOI 10.1007/978-3-319-04537-5_14,

combination technique adds an extra level of parallelism. The technique has been widely applied [4, 7, 15, 18].

There are studies to improve the accuracy of the combination technique [4, 15]. Most notably the "Opticom" method was developed for orthogonal projection problems [13] and was successfully applied to regression and eigenvalue problems [6, 8].

Interestingly, there are few studies which compare the combined solution, the Opticom solution *and* the true sparse grid solution together. Most studies compare either pair of these or with the full grid solution. However, a quick reflection reveals a rich set of possibilities. For example, even if the combined solution is not close to the full grid solution, it does not mean it is not close to the true sparse grid solution. This is obvious, but the same applies to comparing with the Opticom solution. Even if the combined solution is not close to the true sparse grid solution, it does not tell us whether there are better approximations or whether it is already optimal. This reflection warrants study and is one motivation of the paper. We hope these questions may lead to a deeper understanding of the combination technique.

In the next section we define the numerical problem of interest. We also explain the statistical problem which will be the subject of our numerical experiments. In the same section we introduce the concept of Bregman projection as a conceptual tool. In the third section, we give a brief overview of the combination technique and present our numerical results. We use those results later for comparison. In the fourth section we introduce the method of Opticom for minimisation. For these we present numerical results. We also sketch how the Opticom method can be extended to more general numerical problems. In the fifth section we explain a new way of using the combination technique iteratively. Combined with Opticom, we can recover the true sparse grid solution. Numerical results support the theory and raise new questions at the same time. Finally, we give a summary of the methods and present our conclusions.

2 The Minimisation Problem

In this section we define our numerical problem of interest. We also introduce the application problem on which we will test the techniques.

2.1 Theoretical Formulation

We are interested in a standard minimisation problem over a space of functions. Let $T = [0, 1]^d$ and H is a Hilbert space of functions $u : T \to \mathbb{R}$. Furthermore, let $j : H \to \mathbb{R}$ to be a differentiable and strictly convex functional with a unique minimum. We are interested in the minimisation problem

$$u_{min} = \arg\min_{u \in H} j(u).$$

We require j to have zero derivative at the unique minimum. In particular

$$\langle \nabla j(u_{min}), v \rangle = 0 \tag{1}$$

for all $v \in H$, where ∇ denotes the Gâteaux derivative. In practice, we cannot solve our minimisation problem over the infinite dimensional space H and must restrict ourselves to discretised subspaces $V_h \subset H$. Let u_h denote the minimiser of j on V_h. To find calculate this $u_h \in V_h$, we require

$$\langle \nabla j(u_h), v \rangle = 0 \tag{2}$$

whenever $v \in V_h$. This is the only property we need for our subsequent analysis.

Our application problem comes from [10]. The functional j is given by

$$j(u) = \frac{1}{2}\|u\|_H^2 + \log \int_T e^u \, d\mu - \frac{1}{n}\sum_{k=1}^{n} u(t_k) \tag{3}$$

where $t_1, \ldots, t_n \in T = [0,1]^d$ are n i.i.d. data samples. If we write

$$f(t|u) = \frac{e^{u(t)}}{\int_T e^{u(t)} \, d\mu(t)},$$

then $f(t|u_{min})$ is an estimator for the underlying probability density function (pdf).

Equation (2) is satisfied if H is a reproducing kernel Hilbert space continuously embedded into $\mathcal{C}(T)$. The continuous embedding implies

$$\|u\|_\infty \leq C_H \|u\|_H$$

for some finite positive constant C_H. The H-norm is implicitly parameterised by two statistical parameters α and β. The parameter α is the usual regularisation parameter used in regression [6, 17]. It has been absorbed into the H-norm for notational convenience. The constant β is the structural parameter determining the kernel of H. Readers may refer to [10] for other statistical aspects of the problem, including a more detailed description of α and β.

2.2 Implementation

For implementation, we choose a discretised space $V_h \subset H$ with basis functions $\phi_1, \ldots, \phi_m$. At the moment we are not concerned with the specific choice of basis functions. The flexibility in choosing any basis will become important later on.

The infinite dimensional problem on H is reduced to an m-dimensional problem. If we define $J : \mathbb{R}^m \to \mathbb{R}$ by

$$J(\mathbf{c}) = J(c_1, \ldots, c_m) = j\left(\sum_{i=1}^{m} c_i \phi_i\right),$$

our minimisation problem can be recast into

$$\mathbf{c_{min}} = \arg\min_{\mathbf{c} \in \mathbb{R}^m} J(\mathbf{c}).$$

At this point one can choose a suitable minimisation technique. In our case, the finite dimensional problem was studied in another submitted paper [20]. We use Newton's method to solve Eq. (2). Explicitly, we would like to find the roots of

$$\nabla J(\mathbf{c}) = 0.$$

Using Newton's method, we start with guess $\mathbf{c_0}$. At the k-th step we have $\mathbf{c^k}$ and therefore $\nabla J(\mathbf{c^k})$ and $\nabla^2 J(\mathbf{c^k})$. With these we can solve the linear system

$$\nabla^2 J(\mathbf{c^k}) \delta c^k = \nabla J(\mathbf{c^k})$$

for the update step δc^k. Next, we find a step size λ^k and update using the formula

$$\mathbf{c^{k+1}} = \mathbf{c^k} + \lambda^k \delta c^k.$$

The step size should be chosen appropriately for convergence [10]. One way to choose λ^k is through the Armijo criterion [14].

2.3 Minimisation as Bregman Projection

Every differentiable strictly convex functional induces a Bregman divergence. Roughly speaking, the Bregman divergence induced by j is a distance function "with respect to" j. The Bregman divergence of j can be used as a notion of distance between two functions and therefore give us an error measure. They enjoy desirable properties which we will explain. Readers may find a fuller treatment of Bregman divergences in [1].

Assume $j : H \to \mathbb{R}$ is differentiable and strictly convex. The Bregman divergence of j, denoted $d_j : H \times H \to \mathbb{R}^+$ is defined by

$$d_j(u, v) = j(u) - j(v) - \langle \nabla j(v), u - v \rangle.$$

The function $d_j(u, v)$ is non-negative, but it may not be symmetric nor satisfy the triangle inequality. A Bregman divergence is therefore not a true metric.

In the following subsection we show how Bregman divergences leads to an interpretation of our convex minimisation problem as a Bregman projection problem. The concept is interesting in itself but is also useful for understanding Opticom.

2.3.1 Bregman Divergence for Convex Minimisation

Minimising j turns out to be equivalent to minimising the Bregman divergence of u to the true solution u_{min}. In fact, we have a stronger statement.

Proposition 1. *Let u_h denote the minimiser of the functional j on a subspace $V_h \subset H$ and let $v \in V_h$, the Bregman divergence of v from u_h is*

$$d_j(v, u_h) = j(v) - j(u_h).$$

Proof. Using Eq. (2), we have $\langle \nabla j(u_h), v \rangle = 0$ for all $v \in V_h$. Since v, u_h are both in V_h, their difference is in V_h. The linear term from the definition of Bregman divergence drops out and we have the formula.

In other words, the minimiser of j on V_h is a *Bregman projection* of the true minimum u_{min} onto the space V_h. Orthogonal projection is an instance of the Bregman projection with $j(u) = \|u\|^2$.

The other two benefits for using Bregman divergences are specific to our statistical functional j. We omit the proofs. The first one is a type of norm equivalence.

Proposition 2. *If $u, v \in H$ are functions, then*

$$\frac{1}{2}\|u - v\|_H^2 \le d_j(u, v) \le \frac{1 + C_H^2}{2}\|u - v\|_H^2.$$

Recall C_H is the embedding constant such that $\|u\|_\infty \le C_H \|u\|_H$ for all $u \in H$.

Finally, the Bregman divergence of j can be interpreted statistically using the Kullback–Leibler (KL) divergence important in statistically learning.

Proposition 3. *If $u, v \in H$ and if $KL(u, v)$ denotes their KL-divergence, then*

$$d_j(u, v) = \frac{1}{2}\|u - v\|_H^2 + KL(v, u)$$

2.4 *Numerical Examples*

For our numerical experiments, we sampled 50,000 data points from two known probability distributions. A large sample size was used to reduce influence from statistical error. This allows us to focus on the numerical errors themselves. The first sample is a strongly correlated, two dimensional, multivariate normal distribution. We picked correlated datasets because they tend to cause problems for the combination technique. Using this sample of 50,000 data points, we performed two numerical experiments with different statistical parameters. We chose different regularisation parameters to show how the smoothness of u_{min} affects the combination technique. To demonstrate the effect of dimensionality, we again conduct a third experiment using a strongly correlated, multivariate normal distribution—but in three dimensions. At the moment, the quadrature of $\int_T e^u \, d\mu$ limits us to three dimensions. This is not the limitation of the combination technique (or our modifications thereof) but of our statistical method. The problematic quadrature is still a subject of research.

In the following we wish to measure two different accuracies. The first compares the solution accuracy between different refinement levels. The second compares the solution accuracy across our different methods.

For refinement accuracy we calculate the error through the Bregman divergence $d_j(u, u_{min})$. We do not have $j(u_{min})$, but for artificial examples we can calculate high resolution solutions until they converge. We use this value to approximate the true solution.

Next, we are interested in comparing the combination technique, Opticom and the iterative method in their ability to approximate the full grid solution u_h of the same refinement level h. For this, we use the same error measure as [20] and calculate the percentage error in j using the formula

$$e_h(u) = \frac{j(u) - j(u_h)}{j(u_h)}.$$

In the following we present our numerical results for three datasets. All the plots of numerical results are gathered near the end in Figs. 1, 2, and 3 of Sect. 5.2.

Example 1 (2D Dataset A). The 2D dataset A is a two dimensional, strongly correlated dataset sampled from a multivariate normal distribution. We set $\alpha = 2.0$ and $\beta = 5.0$ to impose a heavier regularisation and therefore a more stringent smoothness for u_{min}. As an approximation for the true minimum value, we use $j(u_{min}) = -2.3320$. The results are presented in Table 1.

As predicted by the theory in [10], the error decreases as refinement increases. The decrease is exponential before slowing down for convergence, Sect. 5.2, Fig. 1.

Example 2 (2D Dataset B). 2D dataset B is the same as before but with smaller parameters $\alpha = 0.2$ and $\beta = 1.0$. The true solution u_{min} is therefore less smooth. We use the approximation $j(u_{min}) = -3.2560$. The results are presented in Table 2.

The rate of convergence is not significantly affected by decreasing regularisation.

Example 3 (3D Dataset). The third dataset is the correlated data set in three dimensions. Computational resources restricted us to a lower refinement level. Running the combination technique to very high levels give us an estimate $j(u_{min}) = -4.2630$. The results are in Table 3.

This dataset present the most difficulties to the combination technique.

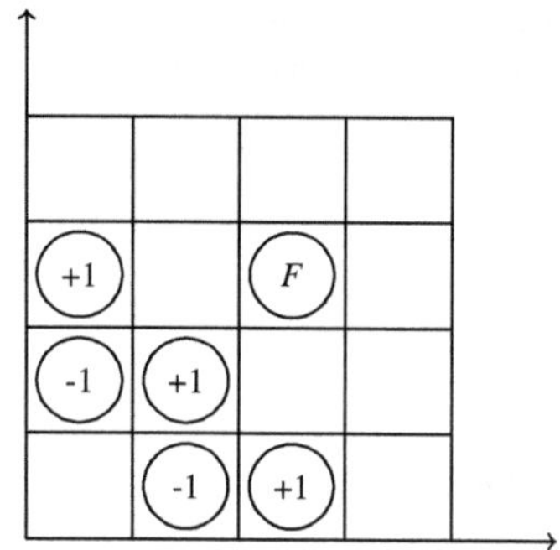

Fig. 1 A combination technique illustrated in the grid of grids—the ij-th cell corresponds to a grid with discretisation (i, j)

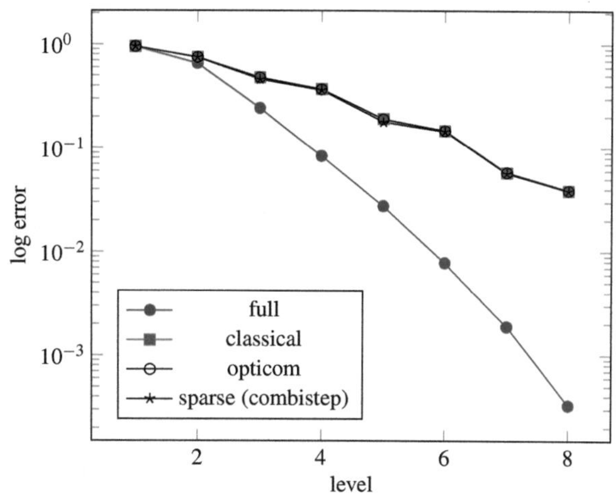

Fig. 2 2D dataset, strong regularisation

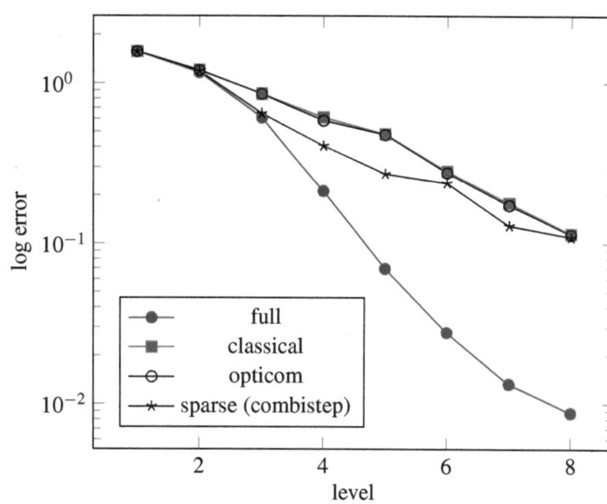

Fig. 3 2D dataset, weak regularisation

Table 1 Full grids: 2D dataset, strong regularisation

Level	Functional $j(u)$	Error $d_j(u, u_{min})$
1	−1.3827	0.949
2	−1.6757	0.656
3	−2.0897	0.242
4	−2.2485	0.084
5	−2.3045	0.028
6	−2.3242	0.008
7	−2.3301	0.002

Table 2 Full grids: 2D dataset, weak regularisation

Level	Functional $j(u)$	Error $d_j(u, u_{min})$
1	−1.6913	1.569
2	−2.0854	1.165
3	−2.6517	0.608
4	−3.0485	0.212
5	−3.1904	0.070
6	−3.2322	0.028
7	−3.2467	0.013

Table 3 Full grids: 3D dataset

Level	Functional $j(u)$	Error $d_j(u, u_{min})$
1	−2.4702	1.793
2	−3.1282	1.135
3	−3.9000	0.363
4	−4.1674	0.096
5	−4.2391	0.024

3 The Classical Combination Technique

We call the combination technique developed in [11] the *classical* combination technique. This distinguishes it from the techniques later on.

Let $(l_1, \dots l_d)$ denote the resolution levels of a coarse grid so, for example, $(2, 4)$ would denote a $2^2 \times 2^4$ grid with $(2^2 + 1) \times (2^4 + 1)$ nodes.

In the combination technique, the user specifies a desired level and the technique defines the set of coarse grids and their coefficients. For the level 3 case, the combination technique specifies levels $(3, 1), (2, 2), (1, 3), (2, 1)$ and $(1, 2)$ for solutions $u_1, \dots, u_5$, and piece them together to form a solution

$$u_c = u_1 + u_2 + u_3 - u_4 - u_5 = \sum_{i=1}^{5} c_i u_i,$$

where the coefficients c_i are respectively $1, 1, 1, -1, -1$, see Fig. 1.

The combination scheme illustrated by Fig. 1 approximates the more expensive full grid solution $(3, 3)$ indicated by an F. This is the classical combination technique. The classical coefficients are selected using the inclusion/exclusion principle and is independent of the numerical problem [12].

Table 4 Classical combination: 2D dataset, strong regularisation

Level	Functional $j(u)$	Error $d(u, u_{min})$	% error $e_h(u)$
1	−1.3827	0.949	–
2	−1.5838	0.748	5.485
3	−1.8549	0.477	11.238
4	−1.9628	0.369	12.708
5	−2.1410	0.191	7.096
6	−2.1849	0.147	5.992
7	−2.2740	0.058	2.410

Table 5 Classical combination: 2D dataset, weak regularisation

Level	Functional $j(u)$	Error $d(u, u_{min})$	% error $e_h(u)$
1	−1.6913	1.569	–
2	−2.0588	1.201	1.748
3	−2.4047	0.855	9.315
4	−2.6498	0.610	13.080
5	−2.7800	0.480	12.864
6	−2.9779	0.282	7.869
7	−3.0818	0.178	5.081

The combined solution u_c does not live in any of the coarse grid spaces but is a member of the *sparse grid space* with the corresponding mesh level. However, the technique only approximates the sparse grid solution. As a result, we cannot use error bounds developed for the sparse grid discretisation. Error bounds for the classical combination technique have been found for model problems [2, 16, 19]. These analysis assume specific error forms which is not often available. In general, there are no theoretical bounds on the approximation error. Indeed, the technique has been known to fail [5].

3.1 Numerical Examples

Example 4 (2D Dataset A).

As the refinement level increases, the functional values in Table 4 drops. The errors $e_h(u)$ suggest the combination technique is often much inferior to the full grid solution. However, this may be because the sparse grid solution itself is inferior. As a result, a comparison between the combined solution and the full grid solution alone is insufficient to reveal the accuracy of the combined solution. We will see later that, in fact, the combined solution actually approximates the sparse grid solution very well with this dataset.

Example 5 (2D Dataset B).

The results of Table 5 is similar to 2D dataset A. The errors in $e_h(u)$ drops similarly. The similarity is superficial, as we will see later that the true sparse grid solution is much better, Fig. 2. We found that in general, a weaker regularisation

Table 6 Classical combination: 3D dataset

Level	Functional $j(u)$	Error $d(u, u_{min})$	% error $e_h(u)$
1	−2.4702	1.793	–
2	−2.7253	1.538	12.879
3	−2.9048	1.358	25.519
4	−3.4801	0.783	16.490
5	−3.6490	0.614	12.879

(that is, when the solution is less smooth) causes more problem for the combination technique.

Example 6 (3D Dataset).

Table 6 shows the classical technique failing completely. The discrepancy from the full grid solution is significant. For some levels it was actually better not to combine but to pick the best of the coarse grid solutions.

4 The Opticom Method

The method of Opticom was developed in [13] to improve the classical combination technique for orthogonal projection problems. In this section we explain how Opticom can be adapted to more general minimisation problem.

4.1 The Idea of Opticom

Consider the orthogonal projection of $u \in H$ onto the Hilbert subspace $V_h \subset H$ with respect to the space norm $\|\cdot\|_H$. Algorithmically, we seek $P_h u \in V_h$ such that for all $v \in V_h$

$$(u, v)_H = (P_h u, v)_H.$$

Suppose $V_1, \ldots, V_m$ are subspaces of V_s and V_s is itself a subspace of H. Here we are thinking of $V_1, \ldots, V_m$ as the coarse grid subspaces of the combination technique and V_s stands for the *sparse grid* function space. Let $u \in H$, $u_i = P_i u$ be the orthogonal projections of u onto the spaces V_i and $u_s = P_s u$ be the orthogonal projection of u onto V_s. Instead of computing the sparse solution u_s directly, we calculate u_i's and approximate u_s through a linear combination of u_i's. We wish to minimise the approximation error. Precisely, we seek c_i's such that

$$F(c_1, \ldots, c_m) = \left\| u_s - \sum_{i=1}^{m} c_i u_i \right\|_H^2 \tag{4}$$

is minimised.

The minimiser of F can be obtained by setting the derivative to zero. We obtain the equation

$$\begin{pmatrix} (u_1,u_1) & (u_1,u_2) & \dots & (u_1,u_{m-1}) & (u_1,u_m) \\ \vdots & \ddots & \ddots & \ddots & \vdots \\ \vdots & \ddots & \ddots & \ddots & \vdots \\ (u_m,u_1) & (u_m,u_2) & \dots & (u_m,u_{m-1}) & (u_m,u_m) \end{pmatrix} \cdot \begin{pmatrix} c_1 \\ \vdots \\ \vdots \\ c_m \end{pmatrix} = \begin{pmatrix} (u_1,u_s) \\ \vdots \\ \vdots \\ (u_m,u_s) \end{pmatrix}.$$

We do not have u_s but we know $(u_i, u_s) = (u_i, u_i)$ in the right hand side. This allows us to calculate optimal coefficients c_i's. This method proposed by [13].

The Opticom method can be applied for numerical methods which are orthogonal projections in nature. One example is Galerkin projection for elliptic PDEs. Another example is for regression. Indeed, the Opticom method was applied to improve the combination technique for regression [6].

The Opticom method cannot be applied for other problems. While we can always write down Eq. (4) and differentiate, we cannot solve the linear system because we do not have (u_i, u_s). The idea of Opticom, however, can be adapted to minimisation problems in a straightforward way.

4.2 *Opticom for Minimisation Problems*

Consider the coarse grid subspaces $V_1, \dots, V_m \subset H$ designated by the combination technique. Let $u_i = \arg\min_{v \in V_i} j(v)$. Naturally, we can choose c_i's such that the combined solution minimises the functional j. In other words, we seek coefficients such that

$$J(\mathbf{c}) = J(c_1, \dots, c_m) = j\left(\sum_{i=1}^{m} c_i u_i\right)$$

is minimised. The minimising c_i's can be found by solving $\nabla J(\mathbf{c}) = 0$. We are in fact minimising j on the space spanned by the coarse grid solutions u_i:

$$V_c = \operatorname{span}\{u_1, \dots, u_m\} \subset V_s.$$

The subscript c stands for the space of all *combined* solutions.

By Proposition 1, the Opticom solution minimises the Bregman divergence

$$d_j\left(\sum_{i=1}^{m} c_i u_i, u_s\right).$$

In other words, Opticom is a Bregman projection onto the space V_c.

Table 7 Opticom: 2D dataset, strong regularisation

	Functional $j(u)$		Error $d(u, u_{min})$		% error $e_h(u)$
Level	Opticom	Classical	Opticom	Classical	Opticom
1	−1.3827	0.949	0.949	–	–
2	−1.5844	0.748	0.748	5.485	5.446
3	−1.8555	0.477	0.477	11.238	11.211
4	−1.9639	0.369	0.368	12.708	12.656
5	−2.1420	0.191	0.190	7.096	7.050
6	−2.1855	0.147	0.147	5.992	5.967
7	−2.2741	0.058	0.058	2.410	2.403

This is another way of viewing the Opticom technique for convex minimisation problems—instead of orthogonal projections using the space norm, we use Bregman projections with the Bregman divergence of j. Since $\| \cdot \|_H^2$ is a convex functional, our way of looking at Opticom generalises the idea from [13].

This is exactly the same as the original minimisation problem we were solving. The only difference is that we are using the coarse grid solutions rather than hat functions as the basis set. The implementation of the Opticom algorithm is therefore similar to the original problem. Much of the original code can be reused. If one is using an object oriented approach, the class structure of the Opticom is algorithm is almost identical. Moreover, the new problem has size equal to the number of coarse grid subspaces. The system is therefore very small.

In summary, the Opticom solution is a Bregman projection on to the space V_c. This attains the best possible combination of coarse grid solutions to minimise j. Methods of improving the combination technique by truncating some coarse grids is equivalent to setting some of the coefficients to zero and adjusting the coefficients of the other grids using the inclusion/exclusion principle. Opticom is by definition superior in accuracy.

Opticom still may not be the true sparse grid solution because $V_c \neq V_s$, nor do we have an error bound. In fact, our numerical examples reflect a diversity of phenomenon.

4.3 Numerical Examples

Example 7 (2D Dataset A).

Example 8 (2D Dataset B).

The results in Tables 7 and 8 are not significantly better than the ones in the classical technique. We expect this for the strongly regularised experiment. This is because, as remarked, the classical solution is already close to the true sparse

Table 8 Opticom: 2D dataset, weak regularisation

	Functional $j(u)$		Error $d(u, u_{min})$		% error $e_h(u)$
Level	Opticom	Classical	Opticom	Classical	Opticom
1	−1.6913	1.569	1.569	–	–
2	−2.0610	1.201	1.199	1.748	1.642
3	−2.4066	0.855	0.853	9.315	9.241
4	−2.6774	0.610	0.583	13.080	12.174
5	−2.7836	0.480	0.476	12.864	12.751
6	−2.9842	0.282	0.276	7.869	7.672
7	−3.0874	0.178	0.173	5.081	4.908

Table 9 Opticom: 3D dataset

	Functional $j(u)$		Error $d(u, u_{min})$		% error $e_h(u)$
Level	Opticom	Classical	Opticom	Classical	Opticom
1	−2.4702	1.793	1.793	–	–
2	−2.8991	1.538	1.364	12.879	7.322
3	−3.2974	1.358	0.966	25.519	5.453
4	−3.6411	0.783	0.622	16.490	12.627
5	−3.8898	0.614	0.373	12.879	8.238

grid solution. Opticom, however, did not offer any significant improvement for the weakly regularised set. This stands in contrasts with the next set of results.

Example 9 (3D Dataset).

For the three dimensional problem we see in Table 9 that Opticom improved on the classical technique. Both the errors measured are significantly less.

4.4 Opticom for General Problems

Before proceeding further, let us consider how insights from this section generalise the Opticom method to other problems beyond [13].

A numerical solution u_h on the discretised space $V_h \subset H$ can be understood as an approximation of the true solution in $u \in H$. In general, we may write the numerical solution u_h on the space V_h as

$$u_h = P_h u$$

where $P_h : H \to V_h$ is some operator. The operator may or may not be known explicitly. In minimising the H-norm error, P_h is the orthogonal projection operator. In convex minimisation problems, P_h is the Bregman projection operator. The Finite Element Method for elliptic PDE exploits the Galerkin projection.

We do not know u itself for $P_h u$. The operator P_h is merely an abstract way of thinking about the numerical method as *mapping* u to its discretised approximations.

Definition 1 (Opticom). Let $u \in H$ denote the true solution. Let $P_s : H \to V_s$ be an operator where V_s is a sparse grid space. Denote $u_s = P_s u$. Let $P_i : H \to V_i$ be operators where $V_1, \ldots, V_m$ are coarse grid spaces. The combination space is defined as

$$V_c = \operatorname{span}\{u_1, \ldots, u_m\}.$$

If $P_c : H \to V_c$ is defined, then the *Opticom solution* is given by $u_c = P_c u$.

This defines Opticom for a large class of problems. They include any numerical scheme which is defined for a generic basis set (as opposed to being able to use only specific ones like hat functions). In this case the operator P_h seeks coefficients for some basis functions. There is no reason why we cannot use the coarse grid solutions $u_1, \ldots, u_m$ as a basis set and solve the problem again. This replaces the classical coefficients and gives the Opticom solution just proposed.

In general, what is the relationship between u_c and u_s? What are the conditions for an appropriate error bound? What is the relationship between u_c and the classical solution? What are the conditions for an appropriate error bound?

This definition of Opticom raises theoretical questions about the combination technique. We now have three solutions to consider: the classical solution, the Opticom solution and the sparse grid solution. The Opticom solution serves as a "bridge" between the classical technique and the sparse grid solution. It allows us to ask questions concerning the combination technique in a new way. We hope to pursue these questions further in the future. A concrete place to start may be to focus on the case where P_h is the Bregman projection onto spaces V_h. In this case the Opticom solution is necessarily defined.

5 An Iterative Combination Technique

The coarse grid solutions do not span the whole sparse grid space V_s. Therefore the Opticom solution which minimises j on V_c may not be the true sparse grid solution which minimises j on V_s. The first motivation for our new technique is the desire to keep the parallelism and simplicity of the combination technique but improve the accuracy by finding a better solution in the space V_s.

The second motivation comes from a completely different perspective. In the future of petascale computing and beyond, fault-tolerance and robustness of our numerical methods become a concern [3]. This is particularly the case for "silent" errors. These are the unreported errors committed by the computation devices. In these cases, iterative algorithms are advantageous because they are intrinsically robust. Iterations can automatically correct errors which have crept in.

These are two possible motivations for the new computation technique we will describe. It is a slight modification of the combination technique which allows us to parallelise while giving us an iterative method.

5.1 The Method of Combistep

As the name suggests, Combistep applies the combination technique to approximate each sparse grid Newton step. Consider the Newton's method on the sparse grid space V_s. The algorithm can be summarised as follows.

1. Start with guess $u_0 \in V_s$
2. At the k-th step, find Newton step Δu^k and step size λ^k
3. Update using

$$u^{k+1} = u^k + \lambda^k \Delta u^k$$

4. Repeat until convergence

Instead of running a full Newton's method on each coarse grid until convergence *and then* combining the solutions, we run only one Newton step on each coarse grid before we combine them for a sparse grid step. Algorithmically,

1. Start with guess $u_0 \in V_s$
2. At the k-th step, for each coarse grid space V_i, find Newton step Δu_i^k
3. Combine the Newton steps for a sparse grid Newton update

$$u^{k+1} = u^k + \Delta u^k = u^k + \sum_{i=1}^{m} c_i^k \Delta u_i^k$$

4. Repeat until convergence

We call this method "Combistep". The combination is made for each step rather than for the solution. An iterative scheme is thereby retained.

Note that Combistep retains the parallelism of the classical combination technique. Just like the classical combination technique, each component grid can be computed in parallel. Also like the combination technique, there is a need to address load balancing so we do not waste time waiting for the final component grid at each synchronisation step.

Does the method converge to the true sparse grid solution or to somewhere else? In fact, how can we even ensure j goes down with each iteration? This can be solved by exploiting the Opticom idea. We choose coefficients for the coarse grid steps which minimises j. It turns out the Opticom choice of the coefficients is what we need.

Theorem 1. *Assume exact arithmetic. If we choose step coefficients to minimise functional j, then the Combistep iteration will terminate only at the sparse grid minimum.*

Proof. Let V_s denote a sparse grid function space, $V_1, \ldots, V_m \subset V_s$ be coarse grid spaces such that $V_s = \sum_{i=1}^m V_i$. Furthermore let $u^k \in V_s$ and u_s be the minimiser of j in V_s. We claim that if $u^k \neq u_s$, then there are coarse grid spaces with non-zero updates. Furthermore, there must be at least one choice of step coefficients which will decrease j.

Recall the minimiser of j on V_h is characterised by $\nabla j(u_h) = 0$. If there are no updates in any of the coarse grids $V_1, \ldots, V_m$ then $\nabla j(u^k)$ is the zero map on each of the coarse grid spaces. In particular, for each i and for all $v \in V_i$

$$\langle \nabla j(u^k), v \rangle = 0.$$

Since $V_s = \sum_{i=1}^m V_i$, we can write, for any $v \in V_s$ $v = \sum_{i=1}^m v_i$ for some choice of v_i's. Since $\nabla j(u^k)$ is a linear we have

$$\langle \nabla j(u^k), v \rangle = \sum_{i=1}^m \langle \nabla j(u^k), v_i \rangle.$$

All the summation terms evaluate to zero by assumption. This is to say $\nabla j(u^k)$ is the zero map on V_s, contrary to the assumption u^k is not the minimum.

Therefore at the k-th step, at least one of the coarse grid spaces V_i will have a non-zero update Δu_i^k and step size λ_i^k such that $j(u^k + \lambda_i^k \Delta u_i^k) < j(u^k)$. There is therefore a choice of coefficients c_i^k's such that the update $u^{k+1} = u^k + \sum_{i=1}^m c_i^k \Delta u_i^k$ will give $j(u^{k+1}) < j(u^k)$. This will continue as long as $u^k \neq u_s$.

5.2 Numerical Examples

In theory the results of Combistep should give the true sparse grid solution. This will allow us to see all the previous results in a new light. When does the classical technique approximate the true sparse grid solution? When does it not? When does Opticom offer an improvement?

Example 10 (2D Dataset A).

The results of Table 10 are close to the original classical solutions. Even though the classical technique had a large discrepancy from the full grid solution for some levels, the results of Table 10 show it was not the problem of the combination technique but the limitations of the sparse grid discretisation.

Table 10 Combistep: 2D dataset, strong regularisation

	Functional $j(u)$		Error $d(u, u_{min})$		% error $e_h(u)$
Level	Combistep	Classical	Combistep	Classical	Combistep
1	−1.3827	0.949	0.949	–	–
2	−1.5852	0.748	0.747	5.485	5.400
3	−1.8744	0.477	0.458	11.238	10.305
4	−1.9717	0.369	0.360	12.708	12.310
5	−2.1545	0.191	0.178	7.096	6.510
6	−2.1880	0.147	0.144	5.992	5.858
7	−2.2753	0.058	0.057	2.410	2.352

Table 11 Combistep: 2D dataset, weak regularisation

	Functional $j(u)$		Error $d(u, u_{min})$		% error $e_h(u)$
Level	Combistep	Classical	Combistep	Classical	Combistep
1	−1.6913	1.569	1.569	–	–
2	−2.0780	1.201	1.182	1.748	0.830
3	−2.6153	0.855	0.645	9.315	1.374
4	−2.8536	0.610	0.406	13.080	6.394
5	−2.9886	0.480	0.271	12.864	6.325
6	−3.0222	0.282	0.238	7.869	6.499
7	−3.1312	0.178	0.129	5.081	3.560

Table 12 Combistep: 3D dataset

	Functional $j(u)$		Error $d(u, u_{min})$		% error $e_h(u)$
Level	Combistep	Classical	Combistep	Classical	Combistep
1	−2.4702	1.793	1.793	–	–
2	−2.9194	1.538	1.344	12.879	6.676
3	−3.5980	1.358	0.665	25.519	7.744
4	−3.8417	0.783	0.421	16.490	7.815
5	−4.0606	0.614	0.202	12.879	4.210

Example 11 (2D Dataset B).

Table 11 shows a different scenario. The sparse grid solution is significantly better than for both Opticom and the classical solution. This is confirmed vividly for the 3D set.

Example 12 (3D Dataset).

The results in Table 12 show the results for the 3D dataset. The errors in the Bregman divergence drop as for the two dimensional datasets. Moreover, the discrepancy from the full grid solution drops by two to three times. The numerical results of all the datasets are summarised in Figs. 2, 3, and 4.

In conclusion, the numerical results of this paper illustrate the theory. Denote the classical combined solution $u_{classical}$. Recall V_c denotes the span of the coarse grid

Fig. 4 3D dataset

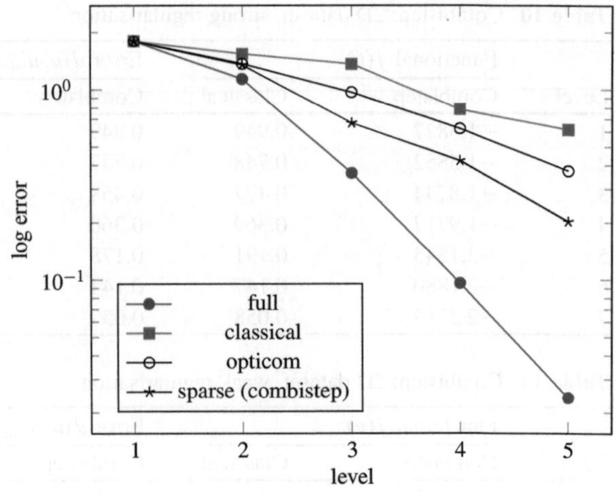

solutions and V_s denotes the sparse grid space. In general we have

$$u_{classical} \in V_c \subsetneq V_s.$$

Opticom minimises j over V_c while Combistep provides the true sparse grid solution. As a result we expect the methods, ordered by increasing accuracy, to be (1) the classical combination technique, (2) Opticom and (3) the iterative combination technique. This is confirmed by all the numerical results we presented.

We do not, however, have specific error bounds and expect a range of phenomenons. The classical technique are at times close to the sparse grid solution and sometimes not. We found that the combination technique tends to work better when the true solution is smoother. The Opticom solution may or may not greatly improve the classical coefficients. This too, is evident throughout the numerical experiments. The results of the 2D dataset with weak regularisation is particularly interesting. We see the classical technique does not approximate well the sparse grid solution though it does approximate the Opticom solution. This is perhaps surprising. We see it is a different question to ask whether the classical combination technique is optimal and whether it is a good approximation for the sparse grid solution.

6 Conclusion: Applications and Further Research

In this paper we presented two new combination technique methods for convex minimisation. The first one is a generalised Opticom method and the second is Combistep, an iterative combination technique. Numerical results confirm our theory. The solution from the Combistep is better than the Opticom solution which is in turn better than the classical solution. Moreover, all of them are inferior to the full grid solution.

The Opticom method developed here generalises the work of [13] through Bregman projections. The Opticom method seeks the optimal combination of the coarse grid solutions by using the coarse grid solutions as a new set of basis functions. The implementation is simple programming-wise because the class structures can be maintained.

The Combistep method is a robust, iterative use of the combination technique. The method is based on using the combination technique to approximate each iterative step rather than the final solution. When combined with Opticom, the method converges to the true sparse grid solution. The method can also be used to refine a classical combined solution. In our paper we focused on the theoretical aspects and not yet on a numerically efficient implementation. For now, our experience is that the method scales like the combination technique.

Most importantly, this paper raise research questions concerning the combination technique. We compared all the relevant solutions—the combined solution, the Opticom solution and the true sparse grid solution. The numerical results have been revealing. Furthermore, we proposed a framework for understanding Opticom in a general setting. This allows us to pose theoretical questions about the combination technique in a new way. We hope research into this area could lead to a deeper understanding of the combination technique. A less ambitious undertaking would be to concentrate on the combination technique for convex minimisation problems.

References

1. A. Banerjee, S. Merugu, I.S. Dhillon, J. Ghosh, Clustering with Bregman divergences. J. Mach. Learn. Res. **6**, 1705–1749 (2005)
2. H. Bungartz, M. Griebel, D. Röschke, C. Zenger, Pointwise convergence of the combination technique for laplace's equation. East-West J. Numer. Math. **2**, 21–45 (1994)
3. F. Cappello, Fault tolerance in petascale/exascale systems: current knowledge, challenges and research opportunities. Int. J. High Perform. Comput. Appl. **23**(3), 212–226 (2009)
4. C. Chiarella, B. Kang, The evaluation of American compound option prices under stochastic volatility using the sparse grid approach. Research Paper Series 245, Quantitative Finance Research Centre, University of Technology, Sydney, February 2009
5. J. Garcke, Maschinelles Lernen durch Funktionsrekonstruktion mit verallgemeinerten dünnen Gittern. Doktorarbeit, Institut für Numerische Simulation, Universität Bonn, 2004
6. J. Garcke, Regression with the optimised combination technique, in *Proceedings of the 23rd International Conference on Machine learning*, ICML '06 (ACM, New York, 2006), pp. 321–328
7. J. Garcke, A dimension adaptive sparse grid combination technique for machine learning. ANZIAM J. **48**(C), C725–C740 (2007)
8. J. Garcke, An optimised sparse grid combination technique for eigenproblems. PAMM **7**(1), 1022301–1022302 (2007)
9. J. Garcke, Sparse grid tutorial. TU Berlin **479**(7371), 1–25 (2007)
10. M. Griebel, M. Hegland, A finite element method for density estimation with Gaussian process priors. SIAM J. Numer. Anal. **47**(6), 4759–4792 (2010)

11. M. Griebel, M. Schneider, C. Zenger, A combination technique for the solution of sparse grid problems, in *Iterative Methods in Linear Algebra*, Brussels, 1991 (North-Holland, Amsterdam, 1992), pp. 263–281
12. M. Hegland, Adaptive sparse grids. ANZIAM J. **44**(C), C335–C353 (2002)
13. M. Hegland, J. Garcke, V. Challis, The combination technique and some generalisations. Linear Algebra Appl. **420**(2–3), 249–275 (2007)
14. C.T. Kelley, *Solving Nonlinear Equations with Newton's Method*. Fundamentals of Algorithms (Society for Industrial and Applied Mathematics, Philadelphia, 2003)
15. J. Kraus, Option pricing using the sparse grid combination technique. Master's thesis, University of Waterloo and Technical University of Munich, Munich, January 2007
16. C. Pflaum, A. Zhou, Error analysis of the combination technique. Numer. Math. **84**, 327–350 (1999)
17. D. Pflüger, *Spatially Adaptive Sparse Grids for High-Dimensional Problems* (Verlag Dr. Hut, München, 2010)
18. C. Reisinger, Numerische Methoden für hochdimensionale parabolische Gleichungen am Beispiel von Optionspreisaufgaben. Master's thesis, Universität Heidelberg, 2004
19. C. Reisinger, Analysis of linear difference schemes in the sparse grid combination technique. IMA J. Numer. Anal. (2013)
20. M. Wong, M. Hegland, Maximum a posteriori density estimation and the sparse grid combination technique. ANZIAM J. **54** (2013)
21. C. Zenger, Sparse grids, in *Parallel Algorithms for Partial Differential Equations, Proceedings of the Sixth GAMM-Seminar*, vol. 31, 1990

Editorial Policy

1. Volumes in the following three categories will be published in LNCSE:

i) Research monographs
ii) Tutorials
iii) Conference proceedings

Those considering a book which might be suitable for the series are strongly advised to contact the publisher or the series editors at an early stage.

2. Categories i) and ii). Tutorials are lecture notes typically arising via summer schools or similar events, which are used to teach graduate students. These categories will be emphasized by Lecture Notes in Computational Science and Engineering. **Submissions by interdisciplinary teams of authors are encouraged.** The goal is to report new developments – quickly, informally, and in a way that will make them accessible to non-specialists. In the evaluation of submissions timeliness of the work is an important criterion. Texts should be well-rounded, well-written and reasonably self-contained. In most cases the work will contain results of others as well as those of the author(s). In each case the author(s) should provide sufficient motivation, examples, and applications. In this respect, Ph.D. theses will usually be deemed unsuitable for the Lecture Notes series. Proposals for volumes in these categories should be submitted either to one of the series editors or to Springer-Verlag, Heidelberg, and will be refereed. A provisional judgement on the acceptability of a project can be based on partial information about the work: a detailed outline describing the contents of each chapter, the estimated length, a bibliography, and one or two sample chapters – or a first draft. A final decision whether to accept will rest on an evaluation of the completed work which should include

- at least 100 pages of text;
- a table of contents;
- an informative introduction perhaps with some historical remarks which should be accessible to readers unfamiliar with the topic treated;
- a subject index.

3. Category iii). Conference proceedings will be considered for publication provided that they are both of exceptional interest and devoted to a single topic. One (or more) expert participants will act as the scientific editor(s) of the volume. They select the papers which are suitable for inclusion and have them individually refereed as for a journal. Papers not closely related to the central topic are to be excluded. Organizers should contact the Editor for CSE at Springer at the planning stage, see *Addresses* below.

In exceptional cases some other multi-author-volumes may be considered in this category.

4. Only works in English will be considered. For evaluation purposes, manuscripts may be submitted in print or electronic form, in the latter case, preferably as pdf- or zipped ps-files. Authors are requested to use the LaTeX style files available from Springer at http://www. springer.com/authors/book+authors?SGWID=0-154102-12-417900-0.

For categories ii) and iii) we strongly recommend that all contributions in a volume be written in the same LaTeX version, preferably LaTeX2e. Electronic material can be included if appropriate. Please contact the publisher.

Careful preparation of the manuscripts will help keep production time short besides ensuring satisfactory appearance of the finished book in print and online.

5. The following terms and conditions hold. Categories i), ii) and iii):

Authors receive 50 free copies of their book. No royalty is paid.
Volume editors receive a total of 50 free copies of their volume to be shared with authors, but no royalties.

Authors and volume editors are entitled to a discount of 33.3 % on the price of Springer books purchased for their personal use, if ordering directly from Springer.

6. Commitment to publish is made by letter of intent rather than by signing a formal contract. Springer-Verlag secures the copyright for each volume.

Addresses:

Timothy J. Barth
NASA Ames Research Center
NAS Division
Moffett Field, CA 94035, USA
barth@nas.nasa.gov

Michael Griebel
Institut für Numerische Simulation
der Universität Bonn
Wegelerstr. 6
53115 Bonn, Germany
griebel@ins.uni-bonn.de

David E. Keyes
Mathematical and Computer Sciences
and Engineering
King Abdullah University of Science
and Technology
P.O. Box 55455
Jeddah 21534, Saudi Arabia
david.keyes@kaust.edu.sa

and

Department of Applied Physics
and Applied Mathematics
Columbia University
500 W. 120 th Street
New York, NY 10027, USA
kd2112@columbia.edu

Risto M. Nieminen
Department of Applied Physics
Aalto University School of Science
and Technology
00076 Aalto, Finland
risto.nieminen@tkk.fi

Dirk Roose
Department of Computer Science
Katholieke Universiteit Leuven
Celestijnenlaan 200A
3001 Leuven-Heverlee, Belgium
dirk.roose@cs.kuleuven.be

Tamar Schlick
Department of Chemistry
and Courant Institute
of Mathematical Sciences
New York University
251 Mercer Street
New York, NY 10012, USA
schlick@nyu.edu

Editor for Computational Science
and Engineering at Springer:
Martin Peters
Springer-Verlag
Mathematics Editorial IV
Tiergartenstrasse 17
69121 Heidelberg, Germany
martin.peters@springer.com

Lecture Notes in Computational Science and Engineering

1. D. Funaro, *Spectral Elements for Transport-Dominated Equations.*
2. H.P. Langtangen, *Computational Partial Differential Equations.* Numerical Methods and Diffpack Programming.
3. W. Hackbusch, G. Wittum (eds.), *Multigrid Methods V.*
4. P. Deuflhard, J. Hermans, B. Leimkuhler, A.E. Mark, S. Reich, R.D. Skeel (eds.), *Computational Molecular Dynamics: Challenges, Methods, Ideas.*
5. D. Kröner, M. Ohlberger, C. Rohde (eds.), *An Introduction to Recent Developments in Theory and Numerics for Conservation Laws.*
6. S. Turek, *Efficient Solvers for Incompressible Flow Problems.* An Algorithmic and Computational Approach.
7. R. von Schwerin, ***Multi Body System SIM**ulation.* Numerical Methods, Algorithms, and Software.
8. H.-J. Bungartz, F. Durst, C. Zenger (eds.), *High Performance Scientific and Engineering Computing.*
9. T.J. Barth, H. Deconinck (eds.), *High-Order Methods for Computational Physics.*
10. H.P. Langtangen, A.M. Bruaset, E. Quak (eds.), *Advances in Software Tools for Scientific Computing.*
11. B. Cockburn, G.E. Karniadakis, C.-W. Shu (eds.), *Discontinuous Galerkin Methods.* Theory, Computation and Applications.
12. U. van Rienen, *Numerical Methods in Computational Electrodynamics.* Linear Systems in Practical Applications.
13. B. Engquist, L. Johnsson, M. Hammill, F. Short (eds.), *Simulation and Visualization on the Grid.*
14. E. Dick, K. Riemslagh, J. Vierendeels (eds.), *Multigrid Methods VI.*
15. A. Frommer, T. Lippert, B. Medeke, K. Schilling (eds.), *Numerical Challenges in Lattice Quantum Chromodynamics.*
16. J. Lang, *Adaptive Multilevel Solution of Nonlinear Parabolic PDE Systems.* Theory, Algorithm, and Applications.
17. B.I. Wohlmuth, *Discretization Methods and Iterative Solvers Based on Domain Decomposition.*
18. U. van Rienen, M. Günther, D. Hecht (eds.), *Scientific Computing in Electrical Engineering.*
19. I. Babuška, P.G. Ciarlet, T. Miyoshi (eds.), *Mathematical Modeling and Numerical Simulation in Continuum Mechanics.*
20. T.J. Barth, T. Chan, R. Haimes (eds.), *Multiscale and Multiresolution Methods.* Theory and Applications.
21. M. Breuer, F. Durst, C. Zenger (eds.), *High Performance Scientific and Engineering Computing.*
22. K. Urban, *Wavelets in Numerical Simulation.* Problem Adapted Construction and Applications.

23. L.F. Pavarino, A. Toselli (eds.), *Recent Developments in Domain Decomposition Methods.*

24. T. Schlick, H.H. Gan (eds.), *Computational Methods for Macromolecules: Challenges and Applications.*

25. T.J. Barth, H. Deconinck (eds.), *Error Estimation and Adaptive Discretization Methods in Computational Fluid Dynamics.*

26. M. Griebel, M.A. Schweitzer (eds.), *Meshfree Methods for Partial Differential Equations.*

27. S. Müller, *Adaptive Multiscale Schemes for Conservation Laws.*

28. C. Carstensen, S. Funken, W. Hackbusch, R.H.W. Hoppe, P. Monk (eds.), *Computational Electromagnetics.*

29. M.A. Schweitzer, *A Parallel Multilevel Partition of Unity Method for Elliptic Partial Differential Equations.*

30. T. Biegler, O. Ghattas, M. Heinkenschloss, B. van Bloemen Waanders (eds.), *Large-Scale PDE-Constrained Optimization.*

31. M. Ainsworth, P. Davies, D. Duncan, P. Martin, B. Rynne (eds.), *Topics in Computational Wave Propagation.* Direct and Inverse Problems.

32. H. Emmerich, B. Nestler, M. Schreckenberg (eds.), *Interface and Transport Dynamics.* Computational Modelling.

33. H.P. Langtangen, A. Tveito (eds.), *Advanced Topics in Computational Partial Differential Equations.* Numerical Methods and Diffpack Programming.

34. V. John, *Large Eddy Simulation of Turbulent Incompressible Flows.* Analytical and Numerical Results for a Class of LES Models.

35. E. Bänsch (ed.), *Challenges in Scientific Computing - CISC 2002.*

36. B.N. Khoromskij, G. Wittum, *Numerical Solution of Elliptic Differential Equations by Reduction to the Interface.*

37. A. Iske, *Multiresolution Methods in Scattered Data Modelling.*

38. S.-I. Niculescu, K. Gu (eds.), *Advances in Time-Delay Systems.*

39. S. Attinger, P. Koumoutsakos (eds.), *Multiscale Modelling and Simulation.*

40. R. Kornhuber, R. Hoppe, J. Périaux, O. Pironneau, O. Wildlund, J. Xu (eds.), *Domain Decomposition Methods in Science and Engineering.*

41. T. Plewa, T. Linde, V.G. Weirs (eds.), *Adaptive Mesh Refinement – Theory and Applications.*

42. A. Schmidt, K.G. Siebert, *Design of Adaptive Finite Element Software.* The Finite Element Toolbox ALBERTA.

43. M. Griebel, M.A. Schweitzer (eds.), *Meshfree Methods for Partial Differential Equations II.*

44. B. Engquist, P. Lötstedt, O. Runborg (eds.), *Multiscale Methods in Science and Engineering.*

45. P. Benner, V. Mehrmann, D.C. Sorensen (eds.), *Dimension Reduction of Large-Scale Systems.*

46. D. Kressner, *Numerical Methods for General and Structured Eigenvalue Problems.*

47. A. Boriçi, A. Frommer, B. Joó, A. Kennedy, B. Pendleton (eds.), *QCD and Numerical Analysis III.*

48. F. Graziani (ed.), *Computational Methods in Transport.*

49. B. Leimkuhler, C. Chipot, R. Elber, A. Laaksonen, A. Mark, T. Schlick, C. Schütte, R. Skeel (eds.), *New Algorithms for Macromolecular Simulation.*

50. M. Bücker, G. Corliss, P. Hovland, U. Naumann, B. Norris (eds.), *Automatic Differentiation: Applications, Theory, and Implementations.*

51. A.M. Bruaset, A. Tveito (eds.), *Numerical Solution of Partial Differential Equations on Parallel Computers.*

52. K.H. Hoffmann, A. Meyer (eds.), *Parallel Algorithms and Cluster Computing.*

53. H.-J. Bungartz, M. Schäfer (eds.), *Fluid-Structure Interaction.*

54. J. Behrens, *Adaptive Atmospheric Modeling.*

55. O. Widlund, D. Keyes (eds.), *Domain Decomposition Methods in Science and Engineering XVI.*

56. S. Kassinos, C. Langer, G. Iaccarino, P. Moin (eds.), *Complex Effects in Large Eddy Simulations.*

57. M. Griebel, M.A Schweitzer (eds.), *Meshfree Methods for Partial Differential Equations III.*

58. A.N. Gorban, B. Kégl, D.C. Wunsch, A. Zinovyev (eds.), *Principal Manifolds for Data Visualization and Dimension Reduction.*

59. H. Ammari (ed.), *Modeling and Computations in Electromagnetics: A Volume Dedicated to Jean-Claude Nédélec.*

60. U. Langer, M. Discacciati, D. Keyes, O. Widlund, W. Zulehner (eds.), *Domain Decomposition Methods in Science and Engineering XVII.*

61. T. Mathew, *Domain Decomposition Methods for the Numerical Solution of Partial Differential Equations.*

62. F. Graziani (ed.), *Computational Methods in Transport: Verification and Validation.*

63. M. Bebendorf, *Hierarchical Matrices.* A Means to Efficiently Solve Elliptic Boundary Value Problems.

64. C.H. Bischof, H.M. Bücker, P. Hovland, U. Naumann, J. Utke (eds.), *Advances in Automatic Differentiation.*

65. M. Griebel, M.A. Schweitzer (eds.), *Meshfree Methods for Partial Differential Equations IV.*

66. B. Engquist, P. Lötstedt, O. Runborg (eds.), *Multiscale Modeling and Simulation in Science.*

67. I.H. Tuncer, Ü. Gülcat, D.R. Emerson, K. Matsuno (eds.), *Parallel Computational Fluid Dynamics 2007.*

68. S. Yip, T. Diaz de la Rubia (eds.), *Scientific Modeling and Simulations.*

69. A. Hegarty, N. Kopteva, E. O'Riordan, M. Stynes (eds.), *BAIL* 2008 – *Boundary and Interior Layers.*

70. M. Bercovier, M.J. Gander, R. Kornhuber, O. Widlund (eds.), *Domain Decomposition Methods in Science and Engineering XVIII.*

71. B. Koren, C. Vuik (eds.), *Advanced Computational Methods in Science and Engineering.*

72. M. Peters (ed.), *Computational Fluid Dynamics for Sport Simulation.*

73. H.-J. Bungartz, M. Mehl, M. Schäfer (eds.), *Fluid Structure Interaction II - Modelling, Simulation, Optimization.*

74. D. Tromeur-Dervout, G. Brenner, D.R. Emerson, J. Erhel (eds.), *Parallel Computational Fluid Dynamics 2008.*

75. A.N. Gorban, D. Roose (eds.), *Coping with Complexity: Model Reduction and Data Analysis.*

76. J.S. Hesthaven, E.M. Rønquist (eds.), *Spectral and High Order Methods for Partial Differential Equations.*

77. M. Holtz, *Sparse Grid Quadrature in High Dimensions with Applications in Finance and Insurance.*

78. Y. Huang, R. Kornhuber, O.Widlund, J. Xu (eds.), *Domain Decomposition Methods in Science and Engineering XIX.*

79. M. Griebel, M.A. Schweitzer (eds.), *Meshfree Methods for Partial Differential Equations V.*

80. P.H. Lauritzen, C. Jablonowski, M.A. Taylor, R.D. Nair (eds.), *Numerical Techniques for Global Atmospheric Models.*

81. C. Clavero, J.L. Gracia, F.J. Lisbona (eds.), *BAIL 2010 – Boundary and Interior Layers, Computational and Asymptotic Methods.*

82. B. Engquist, O. Runborg, Y.R. Tsai (eds.), *Numerical Analysis and Multiscale Computations.*

83. I.G. Graham, T.Y. Hou, O. Lakkis, R. Scheichl (eds.), *Numerical Analysis of Multiscale Problems.*

84. A. Logg, K.-A. Mardal, G. Wells (eds.), *Automated Solution of Differential Equations by the Finite Element Method.*

85. J. Blowey, M. Jensen (eds.), *Frontiers in Numerical Analysis - Durham 2010.*

86. O. Kolditz, U.-J. Gorke, H. Shao, W. Wang (eds.), *Thermo-Hydro-Mechanical-Chemical Processes in Fractured Porous Media - Benchmarks and Examples.*

87. S. Forth, P. Hovland, E. Phipps, J. Utke, A. Walther (eds.), *Recent Advances in Algorithmic Differentiation.*

88. J. Garcke, M. Griebel (eds.), *Sparse Grids and Applications.*

89. M. Griebel, M. A. Schweitzer (eds.), *Meshfree Methods for Partial Differential Equations VI.*

90. C. Pechstein, *Finite and Boundary Element Tearing and Interconnecting Solvers for Multiscale Problems.*

91. R. Bank, M. Holst, O. Widlund, J. Xu (eds.), *Domain Decomposition Methods in Science and Engineering XX.*

92. H. Bijl, D. Lucor, S. Mishra, C. Schwab (eds.), *Uncertainty Quantification in Computational Fluid Dynamics.*

93. M. Bader, H.-J. Bungartz, T. Weinzierl (eds.), *Advanced Computing.*

94. M. Ehrhardt, T. Koprucki (eds.), *Advanced Mathematical Models and Numerical Techniques for Multi-Band Effective Mass Approximations.*

95. M. Azaïez, H. El Fekih, J.S. Hesthaven (eds.), *Spectral and High Order Methods for Partial Differential Equations ICOSAHOM 2012.*

96. F. Graziani, M.P. Desjarlais, R. Redmer, S.B. Trickey (eds.), *Frontiers and Challenges in Warm Dense Matter.*

97. J. Garcke, D. Pflüger (eds.), *Sparse Grids and Applications - Munich 2012.*

98. J. Erhel, M. Gander, L. Halpern, G. Pichot, T. Sassi, O. Widlund (eds.), *Domain Decomposition Methods in Science and Engineering XXI.*

99. R. Abgrall, H. Beaugendre, P.M. Congedo, C. Dobrzynski, M. Ricchiuto, V. Perrier (eds.), *High Order Nonlinear Numerical Methods for Evolutionary PDEs - HONOM 2013.*

For further information on these books please have a look at our mathematics catalogue at the following URL: www.springer.com/series/3527

Monographs in Computational Science and Engineering

1. J. Sundnes, G.T. Lines, X. Cai, B.F. Nielsen, K.-A. Mardal, A. Tveito, *Computing the Electrical Activity in the Heart.*

For further information on this book, please have a look at our mathematics catalogue at the following URL: www.springer.com/series/7417

Texts in Computational Science and Engineering

1. H. P. Langtangen, *Computational Partial Differential Equations.* Numerical Methods and Diffpack Programming. 2nd Edition
2. A. Quarteroni, F. Saleri, P. Gervasio, *Scientific Computing with MATLAB and Octave.* 4th Edition
3. H. P. Langtangen, *Python Scripting for Computational Science.* 3rd Edition
4. H. Gardner, G. Manduchi, *Design Patterns for e-Science.*
5. M. Griebel, S. Knapek, G. Zumbusch, *Numerical Simulation in Molecular Dynamics.*
6. H. P. Langtangen, *A Primer on Scientific Programming with Python.* 4th Edition
7. A. Tveito, H. P. Langtangen, B. F. Nielsen, X. Cai, *Elements of Scientific Computing.*
8. B. Gustafsson, *Fundamentals of Scientific Computing.*
9. M. Bader, *Space-Filling Curves.*
10. M.G. Larson, F. Bengzon, *The Finite Element Method: Theory, Implementation, and Applications.*
11. W. Gander, M.J. Gander, F. Kwok, *Scientific Computing: An Introduction using Maple and MATLAB.*

For further information on these books please have a look at our mathematics catalogue at the following URL: www.springer.com/series/5151

MIX
Papier aus verantwortungsvollen Quellen
Paper from responsible sources
FSC® C105338

If you have any concerns about our products,
you can contact us on
ProductSafety@springernature.com

In case Publisher is established outside the EU,
the EU authorized representative is:
Springer Nature Customer Service Center GmbH
Europaplatz 3, 69115 Heidelberg, Germany

Printed by Libri Plureos GmbH
in Hamburg, Germany